Disclaimer

The publisher of this book is by no way associated with the National Institute of Standards and Technology (NIST). The NIST did not publish this book. It was published by 50 page publications under the public domain license.

50 Page Publications

Book Title: Backdraft Phenomena. Final Report. 1990-1992.

Book Author: Charles M. Fleischmann

Book Abstract: The purpose of this project was to develop a fundamental physical understanding of backdraft phenomena. The research was divided into three phases: exploratory simulations, gravity current modeling, and quantitative backdraft experiments. The primary goal of the first phase was to safely simulate a backdraft in the laboratory. A half-residential-scale compartment was built to conduct exploratory experiments. In the second phase, the gravity current speed and the extent of its mixed region was investigated in a series of scaled salt water experiments. In the final phase, 28 backdraft experiments were conducted in a 1.2m by 1.2m by 2.4m compartment. Histories recorded included: fuel flow rates, upper layer temperatures, lower layer temperatures, opening velocities, compartment pressures, upper layer species concentrations for O_2, CO_2, CO, and HC. Results

Citation: NISTGCR - 94-646

NIST-GCR-94-646

BACKDRAFT PHENOMENA

Charles M. Fleischmann
University of California
Berkeley, CA 94720

Issued June 1994
November 1993

Sponsored by:
U.S. Department of Commerce
Ronald H. Brown, *Secretary*
Technology Administration
Mary L. Good, *Under Secretary for Technology*
National Institute of Standards and Technology
Arati Prabhakar, *Director*

Notice

This report was prepared for the Building and Fire
Research Laboratory of the National Institute of
Standards and Technology under grant number 60NANB0D1042.
The statements and conclusions contained in this report
are those of the authors and do not necessarily reflect
the views of the National Institute of Standards and
Technology or the Building and Fire Research Laboratory.

Backdraft Phenomena

by

Charles Martin Fleischmann

B.S. (University of Maryland) 1985
B.S. (University of Maryland) 1985
M.S. (University of California at Berkeley) 1988

A dissertation submitted in partial satisfaction of the

requirement for the degree of

Doctor of Philosophy

in

Fire Protection Engineering Science

in the

GRADUATE DIVISION

OF THE

UNIVERSITY OF CALIFORNIA at BERKELEY

Committee in charge:

    Professor Patrick J. Pagni, Co-Chair
    Professor R. Brady Williamson Co-Chair
    Professor A. Carlos Fernandez-Pello
    Professor Robert W. Dibble
    Professor Gail E. Brager
    Dr. Nancy J. Brown

The Dissertation of Charles Martin Fleischmann is approved:

_____ 11/10/93
Co-Chair   Date

_____ 11/10/93
Co-Chair   Date

_____ 11/10/93
                                                                  Date

University of California at Berkeley

1993

Backdraft Phenomena

Copyright 1993

by

Charles Martin Fleischmann

ACKNOWLEDGMENTS

This work would never have been completed without the guidance and support of Professor Patrick J. Pagni. He has acted as mentor, advisor, editor and friend throughout this long and difficult process. I am grateful to Professor R. Brady Williamson who convinced me to come to Berkeley and has been mentor and friend to me ever since. He also provided me with the necessary laboratory space and gave free reign to pursue my research while always being available for advice.

I wish to thank the rest of my committee, Professor A. Carlos Fernandez-Pello, Professor Robert W. Dibble, Professor Gail W. Brager, and Dr. Nancy J. Brown, for their support of my interdisciplinary program and guidance through my research.

Bill MacCracken was extremely helpful throughout the entire process from conceptual design to video editing. Without his help the quality of this research would have been significantly reduced. My brother Jim Fleischmann assisted me greatly with all of the experiments from cleaning the salt water tank to wiring in the computer controls, all at the expense of his vacation. The assistance of Gus Revenaugh with the experiments are also noted. Nicholas Dembsey's help with the experiments as well as the many helpful discussions were appreciated. Fred Fisher has always been available for consultation and support.

I would like to specially thank Cecile Grant who has been editor, advisor, and friend throughout my graduate education.

I greatly appreciate the support my parents have given me throughout my academic career. Finally, I wish to thank my wife Carol for all the love and assistance she has given me and all the time she has spent editing this document.

It is impossible to address everyone to whom I wish to express my gratitude, and I hope that those not mentioned will accept my implied thanks.

This work was partially supported by the National Institute of Standards and Technology, Building and Fire Research Laboratory, Grant No. 60NANB1D1168.

TABLE OF CONTENTS

Abstract xix

Acknowledgments iii

Table of Contents iv

List of Figures and Tables vii

Nomenclature xiv

1. Introduction 1

   1.1 Impetus for this Research 1
   1.2 Backdraft Scenario 3
   1.3 Research Outline 9
   References - Chapter 1 11

2. Exploratory Backdraft Experiments 12

   2.1 Introduction 12
   2.2 Backdraft Scenario 13
   2.3 Description of Experimental Apparatus 14
   2.4 Results 17
   2.5 Backdraft Compartment Fire Modeling 25
   2.6 Discussion 29
   2.7 Conclusions 32
   References - Chapter 2 33

3. Salt Water Modeling of Fire Compartment Gravity Currents 35

   3.1 Introduction 35
   3.2 Gravity Current Scaling 36
   3.3 Apparatus and Procedure 39
   3.4 Compartment Gravity Current Structure 43
   3.5 Results of Salt Water Modeling 50
   3.6 Conclusions 55
   References - Chapter 3 56

4. Numerical and Experimental Gravity Currents Related to Backdrafts 57

    4.1 Introduction 57
    4.2 Numerical Modeling 58
    4.3 Salt Water Experimental Apparatus and Procedures 59
    4.4 Backdraft Experimental Apparatus 61
    4.5 Qualitative Results 64
    4.6 Quantitative Results 68
    4.7 Conclusions 75
    References - Chapter 4 77

5. Quantitative Backdraft Experiments 78

    5.1 Introduction 78
    5.2 Experimental Design & Procedures 79
        5.2.1 Apparatus 79
        5.2.2 Species Concentration 81
        5.2.3 Temperatures 82
        5.2.4 Compartment Pressure 83
        5.2.5 Hatch Flow 83
        5.2.6 Data Acquisition System 84
        5.2.7 Procedure 84
    5.3 Experimental Results 85
        5.3.1 70 kW Fuel Flow Rate 85
        5.3.2 200 kW Fuel Flow Rate 90
        5.3.3 Summary 93
    5.4 Conclusions 96
    References - Chapter 5 97

6. Backdraft Experiments Using a Simulated Window Opening 98

    6.1 Introduction 98
    6.2 Experimental Design & Procedures 99
        6.2.1 Apparatus 99
        6.2.2 Species Concentration 101
        6.2.3 Temperatures 102
        6.2.4 Compartment Pressure 103
        6.2.5 Opening Flow 103
        6.2.6 Data Acquisition System 104
        6.2.7 Procedure 104
    6.3 Experimental Results 105
        6.3.1 70 kW Fuel Flow Rate 105
        6.3.2 Summary 112
        6.3.3 Chemical Energy Accounting 114

	6.4 Future Research	122
	References - Chapter 6	123
7	Conclusions	124
	7.1 Summary	124
	7.2 Future Research	127
	Appendix A	A-1
	Appendix B	B-1
	Appendix C	C-1

LIST OF FIGURES AND TABLES

Figure 1.1a Photograph of the salt water model showing the entering gravity current with $\beta = 0.08$ and a $h_1/3$ horizontal slot opening at ~2.4 s after opening. 5

Figure 1.1b Photograph of the flame propagation along the top of the entering gravity current. The ignition spark is turned when the compartment is opened. Compartment is 1.2 m wide by 2.4 m long by 1.2 m high and the opening is 1.1 m wide by 0.4 m high. 5

Figure 1.2a Photograph of the salt water model showing the entering gravity current with $\beta = 0.08$ and a $h_1/3$ horizontal slot opening at ~6.0 s after opening. 7

Figure 1.2b Photograph of the hemispherical flame which is created when ignition is delayed until the gravity current has reflected off the wall opposite the opening. Compartment is 1.2 m wide by 2.4 m long by 1.2 m high and opening is 1.1 m wide by 0.4 m high. 7

Figure 1.3a Photograph of salt water model showing the compartment fluid trapped above the soffit. $\beta = 0.08$, opening was the $h_1/3$ horizontal slot, ~60 s after opening. 8

Figure 1-3b Photograph showing the flame propagating along the interface between the lower layer, made up primarily of air, and the fuel rich upper layer trapped above the soffit. Flame was ignited by a spark 300 s after opening. The spark ignitor was located 0.15 m from the wall opposite the opening and 0.8 m off the floor, i.e. the height of the soffit. Compartment is 1.2 m wide by 2.4 m long by 1.2 m high and the opening is 1.1 m wide by 0.4 m high. 8

Figure 2.1 Photograph of a gravity current in the salt water modeling experiments replicating the backdraft compartment as described in chapter 3. $\beta = 0.023$, $h_1/3$ horizontal slot opening, ~5 s after opening compartment. 14

Figure 2.2 A schematic diagram of the half-room-scale backdraft apparatus showing dimensions and component locations. 16

Figure 2.3 Photograph showing the observation window in the backdraft apparatus. 16

Figure 2.4   Figure 2.4 - Representative compartment temperature histories for
             Experiment 4 in Table 2.1. Locations are measured from the floor (— —) 1.02 m, (– – –) 0.72 m, (·······) 0.42 m, and (———) 0.12 m..................21

Figure 2.5a Photograph showing the dancing flame ~130 s after ignition of the
             burner in Experiment 4 in Table 2.1...22

Figure 2.5b Photograph showing the premixed flame in the mixed region at the
             interface between hot fuel rich and cold oxygen rich layers in
             Experiment 5 in Table 2.1..22

Figure 2.5c Photograph showing the large fireball bursting out of the
             compartment in Experiment 8 in Table 2.1.......................................23

Figure 2.6a Comparison of temperature data from Experiment 5 (x) (Table 2.1) at
             60 s with the idealized two layer approximation calculated from
             Equation (2-2) and (2-3) 27

Figure 2.6b Comparison of temperature data from Experiment 5 (x) (Table 2.1) at
             180 s with the idealized two layer approximation calculated from
             Equations (2-2) and (2-3)..27

Figure 2.7a Comparison of temperature data (x) from Experiment 5 (Table 2.1)
             calculated from Equations (2-2) and (2-3) with computer results from
             FIRST (———) 28

Figure 2.7b Figure 2-7b - Comparison of Experiment 5 (Table 2.1) data (x),
             calculated from Equations (2-2) with the thermal interface history
             computed from FIRST (———)..28

Figure 2.7c Figure 2.7c - Compartment upper layer hydrocarbon (– – –) and
             oxygen (———) mass fraction computed by FIRST. 29

Figure 2.8   A series of video images taken during Experiments 5 (Table 2.1)
             showing the flame propagation and the resulting backdraft
             deflagration. Time labels are from ignition of the backdraft...............31

Figure 3.1   Gravity current schematic. Velocities are indicated in a reference
             frame fixed   on the gravity current. Heights are indicated by h. 37

Figure 3.2a Sketch of the salt water compartment showing the elevation and plan
             views..41

Figure 3.2b Sketch of the four opening geometries for the salt water compartment. 42

Figure 3.3 Sketch of the gravity current with a no slip condition at the lower boundary. Both elevation and plan views are shown. Characteristic features of the gravity current are shown including the lobes, clefts, and billows. 45

Figure 3.4a Photograph of the gravity current approximately 3L/4 into the compartment for the fully open condition with $\beta = 0.018$. The grid shown on the model is 25 mm squares. 47

Figure 3.4b Photograph of the gravity current approximately 3L/4 into the compartment for a center $h_1/3$ horizontal slot opening with $\beta = 0.024$ 47

Figure 3.4c Photograph of the gravity current approximately 3L/4 into the compartment for a central, $h_1/3$ square, window opening with $\beta = 0.032$. 48

Figure 3.4d Photograph of the gravity current approximately 3L/4 into the compartment for a door opening of width $h_1/3$ and height $7h_1/9$ with $\beta = 0.026$. 48

Figure 3.5 Series of four photographs of the gravity current modeling showing the developing gravity current as it enters the compartment for the window opening $\beta = 0.032$ at times: a 1.4 s, b 2.2 s, c 3.3 s, d 4.6s 49

Figure 3.6 Nondimensional velocity versus density difference ratio for four opening conditions. The ◻, •, ○, ▵ represent the data for the full, slot, window and door opening. The (– – –), (·······),(– ··· –), (– ·– ·), represent the average values for the full (0.44), slot (0.32), door (0.35) and window (0.22), geometries 54

Figure 4.1 Sketch of salt water chamber showing the elevation and plan views as well as opening geometries. 61

Figure 4.2 - Sketch of the half scale backdraft compartment showing important features of the apparatus. 64

Figure 4-3a Density profile for the full opening case. Shown here after 4.0 nondimensional time units. Reynolds number is 20,000. 66

Figure 4.3b Photograph of the gravity current approximately 3L/4 into the compartment for the fully open condition. The grid shown on the model is 25 mm squares, β= 0.018. 66

Figure 4.4a Density profile for the $h_1/3$ centered slot opening case. Shown here after 6.8 nondimensional time units. Reynolds number is 20,000 67

Figure 4.4b Photograph of the gravity current approximately 3L/4 into the compartment for $h_1/3$ centered slot opening, β = 0.024. 67

Figure 4.5a Plot of the transit time versus density difference, β, for the full open condition comparing the numerical simulation (———) with salt water modeling results (□). 69

Figure 4.5b Plot of the transit time versus density difference, β, for the $h_1/3$ centered slot opening condition. Compares the numerical simulation (-----) with salt water modeling results (o). 69

Figure 4.6 - Plot of the time to ignition for the half scale backdraft experiments and numerical simulation transit time versus density difference, β. The (□) indicates ignition at the burner spark, (o) indicates ignition at the spark above the burner, and (———) indicates the computational results. 70

Figure 4.7a Plot showing the velocity history for the probe #1 (o) and #6 (□)in the opening of the half scale backdraft compartment compared with the numerical simulation results shown as, probe #1 (-----) and probe #6 (———). 72

Figure 4.7b Plot showing the velocity history for the probe 2 (o) and 5 (□) in the opening of the half scale backdraft compartment compared with the numerical simulation results shown as, probe 2 (-----) and probe 5 (———). 72

Figure 4.8 Plot of the velocity profiles in the opening of the backdraft compartment shown at 4 s and 8 s after opening. Numerical simulation results are shown as lines and the experimental data is shown as symbols. (4 s □ -----) (8 s o ———) 73

Figure 4.9a&b Plot of the quasisteady velocity in the opening of the backdraft compartment versus density difference ratio, β. The experimental data are indicated as symbols, the numerical simulation is shown as solid lines, and the potential flow result is shown as a dashed line. (probe 1 □)(probe 2 o)(probe 5 △)(probe 6 +) 74

Figure 5.1 - Sketch of the half scale backdraft compartment showing important features of the apparatus. 80

Figure 5.2a Species concentration histories from backdraft experiment for O_2 (———), CO(– – –), CO_2 (······), and HC (———) for the 70 kW (12th row in Table 1) fire source. 86

Figure 5.2b Idealized two zone upper layer (———), lower layer (– – –), and layer height (······) histories for the 70 kW (12th row in Table 1) backdraft experiment. 87

Figure 5.2c Figure 5.2d - Idealized two zone approximation compared with temperature data at 80 s (-----,□) and 780 s (———,○)...................87

Figure 5.2d Photograph showing flame structure for the same 70 kW fire source backdraft experiment. Photograph taken 1.5 s after ignition of the backdraft. 88

Figure 5.2e Compartment pressure (———) and total mass inflow (······) histories for the same 70 kW fire source backdraft. Arrow 1 indicates ignition of the backdraft and arrow 2 indicates flame out the opening 89

Figure 5.3a Species concentration histories from backdraft experiment for O_2 (———), CO (– – –), CO_2 (······), and HC (———) for the 200 kW (15th row in Table 1) fire source. 91

Figure 5.3b Idealized two zone upper layer (———), lower layer (– – –), and layer height (······) histories from the same 200 kW (15th row in Table 1) backdraft experiment. 91

Figure 5.3c Idealized two zone approximation compared with temperature data at 25 s (-----,□) and 180 s (———,○) for the same 200 kW (15th row in Table 1) backdraft experiment. 92

Figure 5.3d Photograph showing flame structure for the same 200 kW (15th row in Table 1) fire source backdraft experiment. Photograph taken 1.5 s after ignition of the backdraft. 92

Figure 5.3e Compartment pressure (———) and total mass inflow (······) histories for the 210 kW (15th row in Table 1) fire source backdraft. Arrow 1 indicates ignition of the backdraft and arrow 2 indicates flame out the opening. 93

Figure 6.1 Sketch of the half-scale backdraft compartment showing important features of the apparatus. 100

Figure 6.2a Upper layer species concentration histories from backdraft experiment for O_2 (———), CO (– – –), CO_2 (······), and HC (———). Run 9 in Table 6.1 and 6.2. 107

Figure 6.2b Lower layer species concentration histories from backdraft experiment for O_2 (———), CO (– – –), CO_2 (······), and HC (———). Run 9 in Table 6.1 and 6.2. 107

Figure 6.3a Idealized two zone upper layer (———), lower layer (– – –), and layer height (······) Run 9 in Table 6.1 and 6.2. 108

Figure 6.3b Idealized two zone approximation compared with temperature data at 60 s (-----, □) and 540 s (———, ○). Run 9 in Table 6.1 and 6.2. 108

Figure 6.4 Compartment pressure (———) and total mass inflow (······) and total mass outflow (– – –) histories. Arrow 1 indicates ignition of the backdraft and arrow 2 indicates flame out the opening. Run 9 in Table 6.1 and 6.2. 110

Figure 6.5 Series of video images showing flame structure of the backdraft. Times labels are from ignition of backdraft. Run 9 in Table 6.1 and 6.2. 111

Figure 6.6 Chemical energy histories for the energy within the compartment (———) and the chemical energy supplied by the burner (······). Run 9 in Table 6.1 and 6.2. 116

Table 2.1	Experimental parameters and results from exploratory backdraft experiments.	19
Table 2.2	Travel time determined from videotapes and gravity current velocities.	24
Table 3.1	Summary of the salt water modeling results for the entering and exiting current. Results are shown for all four openings: full, slot, door, and window	53
Table 3.2	Average values for v^* and h^* calculated from the data reported in Table 3.1	54
Table 5.1	Summary data from backdraft experiments showing burner characteristics, species concentrations at opening, idealized layer temperatures and height, ignition delay time, peak pressure, total mass flow into the compartment before ignition, and fire ball diameter.	95
Table 6.1	Summary of the 11 window experiments reporting the burner characteristics, layer temperatures and height, ignition delay, peak compartment pressure, and mass flow into and out of the compartment.	113
Table 6.2	Summary of chemical energy accounting reporting the species concentrations at compartment opening, the chemical energy stored in the compartment at opening, the chemical energy which exits the compartment prior to ignition, the chemical energy that exits the compartment after ignition and before flames exit the compartment, the chemical energy released in the compartment by the backdraft, the chemical energy that is available for the fireball, and an estimate of the fireball diameter.	121

NOMENCLATURE

b	width of the opening (m).
c	Salt water mass fraction.
C	flow coefficient for the opening, (0.68)
$C(Re)$	empirical calibration factor for the bidirectional probe
CO	carbon monoxide
CO_2	carbon dioxide
D	Diffusion coefficient (m²/s)
E	chemical energy (MJ)
g	Acceleration of gravity (m/s²)
h	height i.e. vertical dimension measured from the floor (m)
HC	hydrocarbons
H_2O	water
ΔH_C	heat of combustion (MJ/kg)
L	compartment length (m)
m	mass (kg)
$\dot{m}$	mass flow rate (kg/s)
M	molecular weight
N_2	nitrogen
O_2	oxygen
Δp	pressure difference across the bidirectional probe (Pa).
p	pressure (Pa)
t	time (s)
T	temperature (K)
$\mathbf{u}$	Velocity vector.
v	velocity (m/s)
w	compartment width (m)
w	specific humidity
x	horizontal direction
y	vertical direction
Y	species mass fraction
ν	kinematic viscosity (m²/s)
ρ	density of the fluid (kg/m³)
ϕ	relative humidity

Subscript

Air	air
b	sill
B	burner
c	characteristic
CO	carbon monoxide
CO_2	carbon dioxide
fo	flames out the opening

HC	hydrocarbons
H_2O	water
i	ignition
in	into the compartment
L	layer
n	neutral axis
N_2	nitrogen
out	out of the compartment
O	open
O_2	oxygen
sat	saturation
S	stored
t	soffitt
0	ambient
1	inside the compartment
2	see Figure 3.1

Superscripts

A	analyzer
C	combustion
D	dry
L	layer
LL	lower layer
mix	mixture
UL	upper layer
~	variations

Abstract

Backdraft Phenomena

by

Charles Martin Fleischmann

Doctor of Philosophy in Fire Protection Engineering Science

Professor P. J. Pagni, Co-Chair and Professor R. Brady Williamson Co-Chair

The purpose of this project was to develop a fundamental physical understanding of backdraft phenomena. The research was divided into three phases: exploratory simulations, gravity current modeling, and quantitative backdraft experiments. The primary goal of the first phase was to safely simulate a backdraft in the laboratory. A half-residential-scale compartment was built to conduct exploratory experiments. The initial experiments concluded with a scenario describing the fundamental physics of backdrafts. The importance of the gravity current which enters the compartment after opening was identified. In the second phase, the gravity current speed and the extent of its mixed region was investigated in a series of scaled salt water experiments. The scaled compartment (0.3m x 0.15m x 0.15m) was fitted with a variety of end openings: full, slot, door, and window. Video and photo data indicate that the mixing layer which rides on the gravity current in the full opening case, expands to occupy nearly the entire current in the partial opening cases. The Froude number and nondimensional head height are independent of β and are in good agreement with numerical simulations and special limits from the literature.

In the final phase, 28 backdraft experiments were conducted in a 1.2 m by 1.2 m by 2.4 m compartment. A methane burner was ignited inside a closed compartment and allowed to burn as long as oxygen was available. After the flame extinguished due to oxygen starvation, the burner was left on to allow the unburned fuel fraction to increase. Upon opening the hatch a gravity current enters the compartment and travels across the floor to the ignition source. After ignition a deflagration rips through the compartment and out the opening culminating in a large fireball. Histories recorded included: fuel flow rates, upper layer temperatures, lower layer temperatures, opening velocities, compartment pressures, upper layer species concentrations for O_2, CO_2, CO, and HC. Results indicate that unburned fuel mass fractions >15% are necessary for a backdraft to occur and that the backdraft severity strongly depends on the delay time and species concentrations.

Patrick J. Pagni
Co-Chair, Thesis Committee

R. Brady Williamson
Co-Chair, Thesis Committee

CHAPTER 1

INTRODUCTION

1.1 Impetus for this Research

A backdraft is defined as a rapid deflagration following the introduction of oxygen into a compartment filled with accumulated unburned fuel. Prior to 1991, the word backdraft was known only to firefighters and a few researchers. However, in 1991 Universal Studios released a major motion picture called "BACKDRAFT" and almost overnight backdraft became a household word. Prior to the movie there was a great deal of confusion about the concept of backdraft, and since the movie the situation has become even more confused. Articles on backdraft, sometimes called smoke explosions, first appeared in the literature in 1914[1]. This first article described backdraft as a dust explosion caused by the carbon particles in the smoke. Other more recently proposed "theories explaining backdrafts" require impossible conditions such as instantaneous transport of oxidizer to reactant when a compartment is opened, autoignition of gases at impossible compartment temperatures, and ignition of soot particles at temperatures less than as 500°K.[2]

A review of fire service literature reveals many narrative articles on backdrafts and the terrible consequences to firefighters caught in the backdraft[3,4,5]. Typically in the articles, firefighters are involved in initial search and rescue or suppression operations when the explosion occurs. Although the fire service training manuals specifically provide warning signs of backdrafts[6,7], in most reported occurrences, the warning signs were not observed by the firefighters. Current tactics for reducing the backdraft hazard are to vent the structure at the highest possible location prior to entry. However, the ventilation process is often a second priority to the rescue operation. Indeed, unless it is restricted to roof venting, ventilation may facilitate rather than prevent a backdraft. Lack of warning and poor ventilation practices are commonly reported in descriptions of backdraft.

As a result of a backdraft which killed two firefighters in Chatham, England in 1975[3], an extensive literature review on fires involving explosions was published by Croft[8]. This review covered the United Kingdom, the United States, and Canada during a seventy year period ending in 1976. A total of 127 fires involving explosions were reported in the literature with sufficient detail to be included in this study. The explosions were categorized as occurring in smoldering fires (52), developing fires (57), and developing fires with secondary explosions (18). The latter category is not relevant to backdrafts. Important conclusions from Croft's work are[8]:

> 1) More explosions were associated with developing fires than with smoldering fires.
>
> 2) More fire fighting personnel were injured in explosions associated with developing fires, than with smoldering fires.
>
> 3) A large number of fire fighting personnel were killed in explosions associated with smoldering fires because the firefighters had entered the premises believing that the fire had been quelled or that it had extinguished itself due to a lack of oxygen.
>
> 4) More fire fighting personnel were injured in explosions in shops and supermarkets than in any other type of occupancy.
>
> 5) More fire fighting personnel were killed in explosions in warehouses than in any other single occupancy.
>
> 6) Smoke explosions are more likely to occur in factories and warehouses than in any other occupancies.
>
> 7) More explosions resulted from fires involving cellulosic materials than any other material. This is most probably associated with the pattern of usage of different materials during the period studied.

8) The cellulosic materials which caused the largest number of explosions were varnished, painted, and polished woodwork and combustible fiberboard.

The results presented in Croft's study may be distorted. Although his study was extensive it was not comprehensive because it was limited to the available literature. Accurate statistical data were not available.

National statistics on the number of firefighters killed or injured as a result of backdrafts are not now available. The fire analysis and research division of the National Fire Protection Association does not identify backdraft as a separate cause of firefighters' death or injury; it simply combines backdraft casualties with others under the heading of "Rapid Fire Progress in Structures". The reason for this lack of distinction is that it is often difficult to identify the specific phenomena actually occurring during a fire[9]. Other phenomena in this category are flashover and ignition of fire gases. Other statistics on backdrafts such as the annual number of occurrences and number of civilian casualties are also not available.

In addition to the literature review, the Chatham fire also prompted some experiments. Smoldering foamed rubber was placed in a small enclosure, 1.4 m^3, and allowed to accumulate excess pyrolyzates[10]. When a diffusion flame was introduced near the bottom of the enclosure, an explosion occurred[11]. This was the only research on backdraft that this author could find within the commonly available literature.

1.2 Backdraft Scenario

As a result of the exploratory experiments presented in Chapter 2, a fundamentally sound backdraft scenario has been identified. Consider a fire in a closed compartment where only a minimal amount ventilation is provided by leakage. As the fire heats the compartment, leaks in the compartment bounding surfaces permit outflows that minimize any pressure differential[12]. A hot layer composed largely of combustion products descends around the fire causing some pyrolysis products to remain unburned. These products accumulate forming a deep layer which has insufficient oxygen to support combustion and becomes fuel rich as the fire continues to supply unburned fuel. We assume a small flame or glowing ember remains burning somewhere in the lower layer. Suddenly, a new ventilation source is provided by a window breaking[13] or door opening. The hot, vitiated atmosphere within the compartment flows out the upper portion of the vent. Simultaneously, cold, fresh air flows in the lower portion of the vent. This cold, density driven, flow is called a gravity current[14]. A mixed layer forms due to the instabilities at the shear interface between the outflow and the inflow and rides on the gravity current across the compartment[15]. For many opening geometries, large scale vortical structures with fuel rich and oxygen rich regions rolled up sequentially, fill the gravity current. Figure 1.1a shows a salt water model of a gravity current nearly completing its propagation across a compartment. A portion of this mixed current is within the flammable range and is ignited when it reaches a flame or glowing ember. After ignition, a new flame propagates back through the mixed region. Figure 1.1 b shows a flame approximately 1 s after ignition in an experimental compartment. The flame shape shows a precursor flame leading the primary burning zone. This preceding flame burns along the interface between the entering gravity current and the exiting compartment gases. The preceding flame and its wake are sufficiently unstable to generate a rapidly propagating turbulent flame. The resulting turbulent deflagration within the compartment drives some accumulated excess pyrolyzates out the vent and consumes that fuel outside the compartment in the dramatic fireball commonly associated with backdraft. This entire process: the accumulation of unburned gaseous fuel, the propagation of an oxygen rich gravity current creating a mixed region and carrying it to the ignition source, the ignition and propagation of an eventually turbulent deflagration and the external fireball, altogether constitutes a backdraft.

Figure 1.1a - Photograph of the salt water model showing the entering gravity current. $\beta=0.080$, opening was the $h_1/3$ horizontal slot, ~2.4 s after opening.

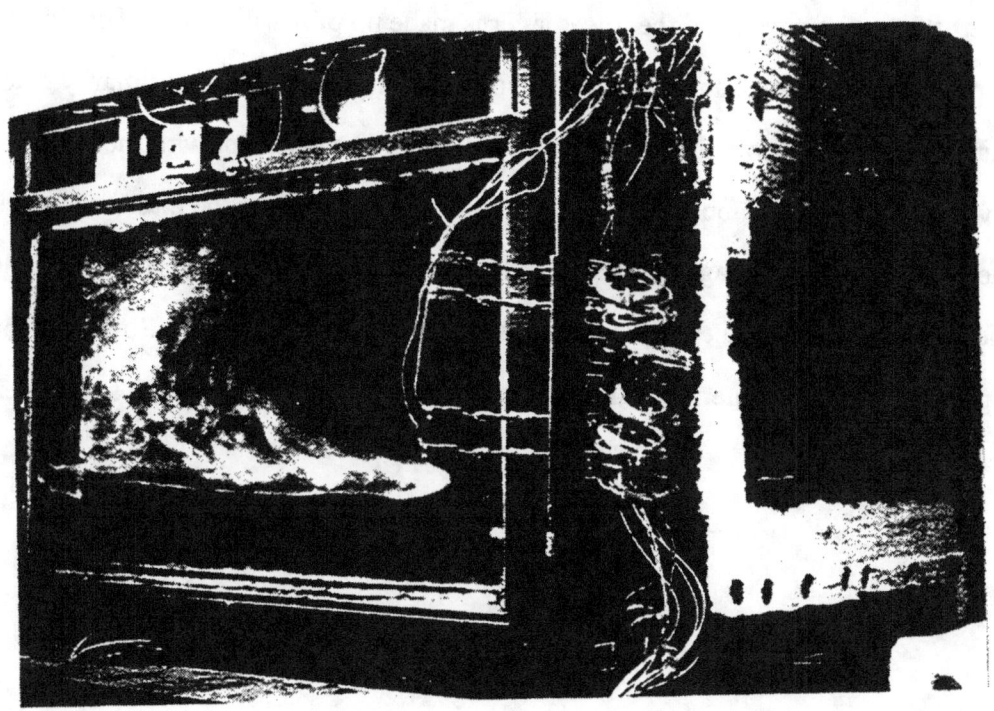

Figure 1.1b - Photograph of the flame propagation along the top of the entering gravity current. The ignition spark is turned when the compartment is opened. Compartment is 1.2 m wide by 2.4 m long by 1.2 m high and the opening is 1.1 m wide by 0.4 m high.

Two modifications to the above scenario were observed. Identical compartment conditions at opening are assumed but there is an increase in the time delay before ignition of the backdraft. This time delay results in a different deflagration flame structure and backdraft severity. In the first modification, the time delay is sufficiently long that the gravity current may have even returned to the front opening and the compartment will act like a reservoir in a reservoir filling problem. Figure 1.2a shows the current as it returns to the opening in the salt water model. The lower layer formed by the gravity current is made up of entrained fuel rich compartment gases along with the oxygen rich air entering the compartment. If ignition occurs during this time period, with the lower layer more uniformly mixed and within a flammable range, the flame structure is hemispherical in shape and the initial deflagration is more severe. Figure 1.2b shows the hemispherical flame identifying this scenario.

The second modification requires a long ignition delay where the gravity current has left the compartment and the lower layer is made up primarily of air. The flammable compartment gases are trapped above the soffit and the lower layer is primarily made up of air. Figure 1.3a shows the salt water model of the compartment with the gases trapped above the soffit. If ignition occurs now, a flame will travel along the interface as seen in a mine gallery[15]. The propagating flame and its wake are sufficiently unstable to generate a rapidly propagating turbulent, albeit less severe, deflagration. Figure 1.3b shows the flame propagating along the interface. In this experiment ignition occurred 300 s after vent opening.

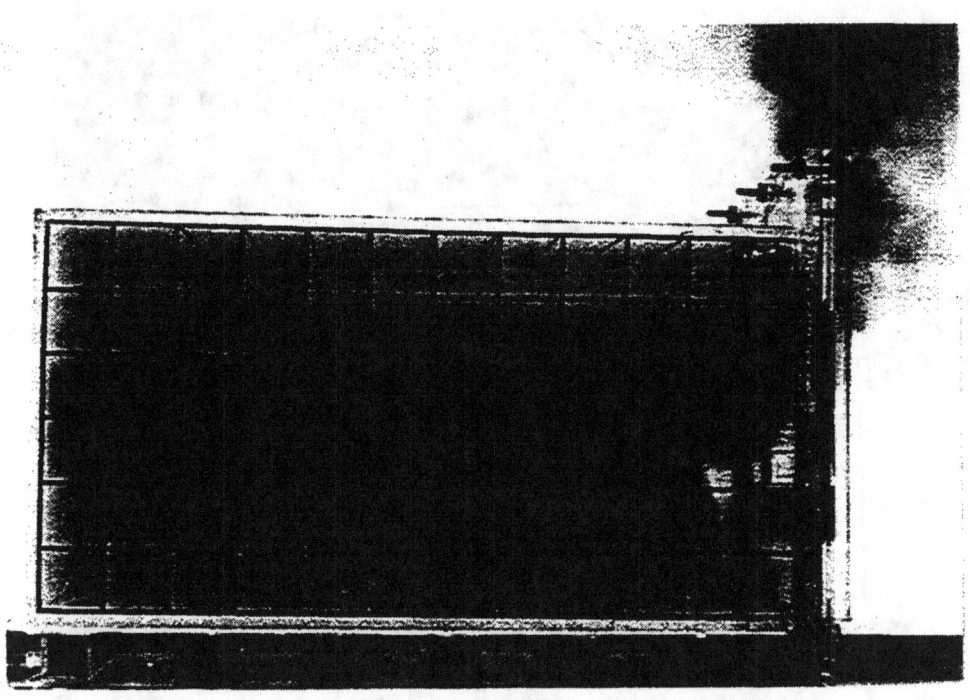

Figure 1.2a - Photograph of the salt water model showing the reflected gravity current. β=0.080, opening was the $h_1/3$ horizontal slot, ~6.0 s after opening.

Figure 1.2b - Photograph of the hemispherical flame which is created when ignition is delayed until the gravity current was reflected off the wall opposite the opening. Compartment is 1.2 m wide by 2.4 m long by 1.2 m high and opening is 1.1 m wide by 0.4 m high.

Figure 1.3a - Photograph of salt water model showing the compartment fluid trapped above the soffit. $\beta=0.080$, opening was the $h_1/3$ horizontal slot, ~60 s after opening.

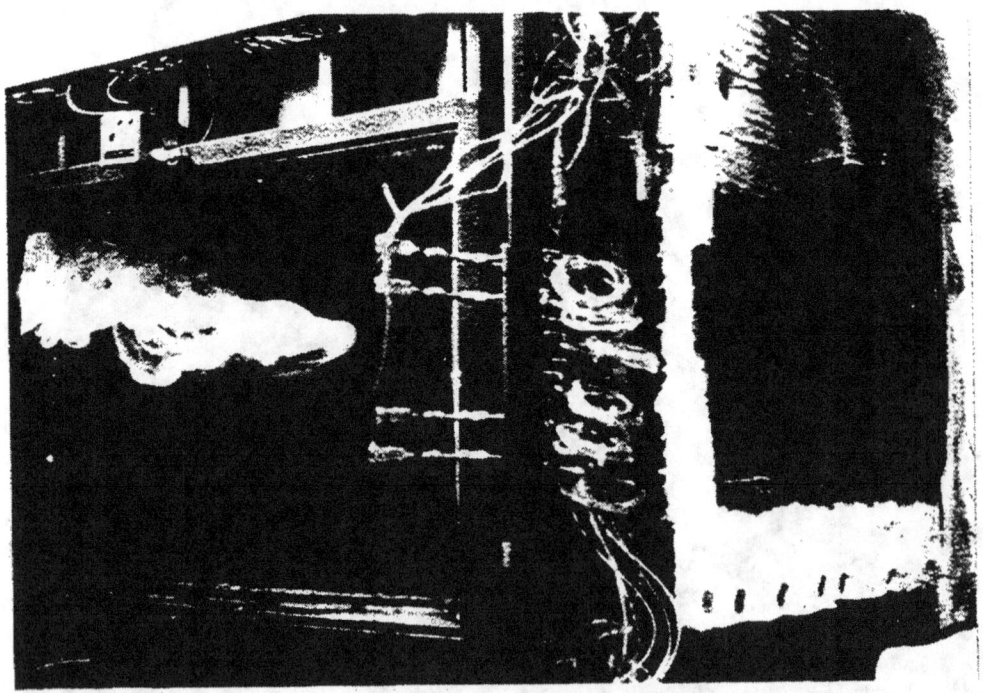

Figure 1.3b - Photograph showing the flame propagating along the interface between the lower layer, made up primarily of air, and the fuel rich upper layer trapped above the soffit. Flame was ignited by a spark 300 s after opening. The spark ignitor was located 0.15 m from the wall opposite the opening and at the height of the soffit. Compartment is 1.2 m wide by 2.4 m long by 1.2 m high and the opening is 1.1 m wide by 0.4 m high.

1.3 Research Outline

Since little research on backdraft is available, the first step was to formulate a working scenario as a physical explanation of the phenomena. A series of exploratory experiments were conducted in a special compartment designed to safely simulate backdraft experiments. A single horizontal slot opening was investigated. Description of the apparatus and results of these experiments are presented in Chapter 2 along with computer modeling results which were used to investigate the species concentrations within the compartment prior to backdraft.

Once the feasibility of the scenario was demonstrated, a series of salt water experiments were conducted to study the geometry, mixing, and velocity of the gravity current which enters the compartment prior to backdraft. The results of the salt water experiments are given in Chapter 3. The velocity and height of the entering gravity current are presented. In addition, a two dimensional numerical simulations of the gravity current experiments were performed at the National Institute for Standards and Technology. These numerical predictions are compared with the salt water modeling and backdraft experimental results in Chapter 4.

Chapter 5 discusses results from a second series of backdrafts which were more fully instrumented than the exploratory experiments. Data collected include burner flow rates, compartment temperatures, species concentrations, layer height, vent flow rates, and compartment pressure. Exemplar data are presented along with a summary table for the experiments using a horizontal slot opening. A series of experiments were also conducted using a simulated window opening. The results of these experiments are given in Chapter 6. The fractions of the chemical energy stored as unburned hydrocarbons within the compartment at the opening time which were released inside and outside the compartment are also discussed in this chapter. Chapter 7 gives the conclusions of this research and discusses possible directions for future work. Appendix A gives the details of the species balance performed on the gas samples to determine the concentrations of water and nitrogen and as a check on the hydrocarbon concentration. Details of the experimental apparatus are covered in appendix B including design drawings and construction material information sheets. Appendix C is a summary of the data collected in all of the backdraft experiments. Although all of the experiments performed are not discussed in this work, all of experiments are included here for completeness and future reference.

References - Chapter 1

[1] Steward, P.D.C., "Dust and Smoke Explosions", National Fire Protection Association Quarterly, 7, 424-428, 1914.

[2] Roblee, C. L., "Backdraft", Fire Chief, 33-35, December 1977.

[3] "Fatal Mattress Store Fire At Chatham Dockyard", Fire, 67, 388, 1975.

[4] Russel, D., "Seven Fire Fighters Caught In Explosion", Fire Engineering, 22-23, April, 1983.

[5] "Backdraft: A Horrible Reality that Kills or Maims in Seconds", Fire Fighting in Canada, 4-5, April-May, 1980.

[6] Fire Ventilation Practices, International Fire Service Training Association, Oklahoma, 32-33, 1970.

[7] Dunn, V., "Beating the Backdraft", Fire Engineering, 44-48, 1988.

[8] Croft, W.M., "Fires Involving Explosions - A Literature Review", Fire Safety Journal, 3, 3-24, 1980/81.

[9] Washburn, A. E., Leblanc, P. R., and Fahy, R. F., "1989 Fire Fighter Fatality Report", Fire Command, 30-49, June 1990.

[10] Pagni, P. J., and Shih, T. M., "Excess Pyrolyzates," 16th Symposium (Int'l) on Combustion, 1329-1343, The Combustion Institute, Pittsburgh, PA, 1976.

[11] Wooley, W.D., and Ames, S.A., "The Explosive Risk of Stored Foamed Rubber", Building Research Establishment, Current Paper 36/75, Borehamwood, UK, 1975.

[12] Emmons, H.W., "The Calculation of a Fire in a Large Building," ASME Paper No. 81-HT-2, American Society for Mechanical Engineers, New York, 1981.

[13] Pagni, P.J., and Joshi, A.A., "Glass Breaking in Fires," in Fire Safety Science - Proceedings of the Third International Symposium, 791-802, Elsevier Science Pub., London, 1991.

[14] Simpson, J.E., "Gravity Currents in the Laboratory, Atmosphere, and Ocean," Annual Review of Fluid Mechanics, 14, 213-234 1982.

[15] Britter, R. E., and Simpson, J.E., "Experiments on the Dynamics of a Gravity Current Head," Journal. of Fluid Mechanics, 88, pt. 2, 223-240 (1978).

[16] Phillips, H., "Flame in a Buoyant Methane Layer", 10th Symposium (Int'l) on Combustion, 1277-1283, The Combustion Institute, Pittsburgh, PA, 1965.

CHAPTER 2

EXPLORATORY BACKDRAFT EXPERIMENTS

2.1 Introduction

Fires can produce more fuel than the locally available oxygen can consume. This surplus fuel is called excess pyrolyzates[1]. If the compartment containing the fire is well-ventilated, the excess pyrolyzates fuel long flames that extend out openings in the compartment, rapidly spreading the fire[2]. If the compartment is closed, the excess pyrolyzates accumulate, ready to burn when a vent is suddenly opened, e.g., by a window breaking due to the fire-induced thermal stress[3,4] or by a firefighter entering the compartment[5,6,7,8]. Upon venting, a gravity current carries fresh air into the compartment. This air mixes with the excess pyrolyzates producing a flammable pre-mixed gas which can be ignited in many ways. The rapid deflagration moving through the compartment after ignition, consuming the accumulated excess pyrolyzates, is called a backdraft.

The fire service community has long recognized the hazards associated with backdrafts[9,10,11]. The literature provides a definition of backdraft[12]: "Backdraft is the burning of heated gaseous products of combustion when oxygen is introduced into an environment that has a depleted supply of oxygen due to fire. This burning often occurs with explosive force." This definition would be correct if the word pyrolysis were substituted for combustion since it is primarily the unburned pyrolyzates which provide the fuel for backdrafts.

None of the quantitative models of compartment fires currently available[13,14,15] incorporate backdraft phenomena because the underlying fundamentals are not well understood. This research is aimed at advancing our understanding of the physics and chemistry of backdrafts. A backdraft scenario is presented. Half-scale room fire experiments attempting to confirm this scenario and results of computer modeling of pre-backdraft compartment conditions are described.

2.2 Backdraft Scenario

Consider a fire in a closed compartment where the only ventilation provided is by leakage. As the fire heats the compartment, leaks in the compartment bounding surfaces permit outflows that minimize any pressure differential[16]. A hot layer composed largely of combustion products descends around the fire causing some fuel pyrolysis products to remain unburned. These products accumulate forming a deep, fuel-rich layer. We assume a small flame or glowing ember remains burning. Suddenly a new vent is opened. The hot, vitiated atmosphere within the compartment flows out of the upper portion of the vent. Simultaneously, cold, fresh air flows into the lower portion of the vent. The propagation of the leading edge of this cold, density driven, flow is called a gravity current[17]. Figure 2-1 shows a gravity current flowing into a compartment in a salt water model. A mixed layer forms due to instabilities at the shear interfaces between the outflow and the inflow, and is carried across the compartment by the gravity current[18,19]. A portion of this mixed current is within the flammable range and is ignited when it reaches a flame or glowing ember. After ignition, the new flame propagates back through the mixed layer. That flame and its wake are sufficiently unstable to generate a rapidly propagating turbulent flame. The resulting turbulent deflagration within the compartment drives any remaining unburned fuel and pyrolysis products out the opening to burn outside the compartment in a dramatic fireball.

Figure 2.1 - Photograph of a gravity current in the salt water modeling experiments replicating the backdraft compartment as described in chapter 3. $\beta = 0.023$, $h_1/3$ horizontal slot opening, ~5 s after opening compartment.

2.3 Description of Experimental Apparatus

To test the hypothesized physical explanation of backdraft an experimental program was undertaken. The primary goal of this program was to safely simulate backdrafts in the laboratory.

Because of the explosive nature of backdraft, a special chamber was constructed to replicate a small room at approximately half scale. Figure 2-2 shows a schematic of the apparatus giving the internal dimensions of the compartment. Figure 2-3 is a photo of the apparatus. In order to control the overpressure hazard, one long wall was designed as a pressure relief panel. The entire wall was hinged along the bottom and closed with a single nylon fastener at the top. Failure of the fastener was designed to relieve any overpressure greater than 1 kPa, as recommended for venting[20,21]. Tests with a large, pressurized plastic bag showed that the blow-out panel released at 0.6 ± 0.1 kPa. It is also recommended that the pressure relief panel weight be limited to <15 kg/m² to reduce inertia and opening time. The pressure relief wall weighed approximately 13 kg/m². It was constructed of 1.2 mm (18 gauge) steel studs, 50 mm wide, 1.6 m long, 0.6 m on center. The sheathing over the studs was 1.2 mm galvanized sheet steel. The panel interior was covered with a 25 mm thick layer of refractory fiber blanket.

All the stationary walls, ceiling, and floor were designed to withstand 5 kPa, five times the expected maximum overpressure. The three stationary walls used 1.2 mm steel studs, 0.10 m wide, 1.4 m long, 0.4 m on center. The ceiling was constructed of 1.2 mm steel studs, 0.2 m wide, 1.4 m long, 0.4 m on center. Two layers of 16 mm Type X, fire rated, gypsum wallboard were mounted as interior sheathing to the stationary walls and ceiling to provide structural strength and secondary thermal fire resistance. Gypsum wallboard was also installed on the floor to protect the plywood platform. A 50 mm thick refractory fiber blanket was installed over the gypsum wallboard on the walls and ceiling to provide the primary thermal resistance for the structure. This insulation allows for repeated experiments without the need to rebuild the chamber.

A 0.9 m high and 1.5 m wide observation window was installed in the wall opposite the pressure relief panels, as shown in Fig. 2-3. The window glass was Neoceram[22], a ceramic with a negative coefficient of expansion below 900K and is capable of withstanding continuous exposure to temperatures of 1000K. The glass was mounted in a standard steel frame protected from the hot compartment gases by a refractory insulation blanket.

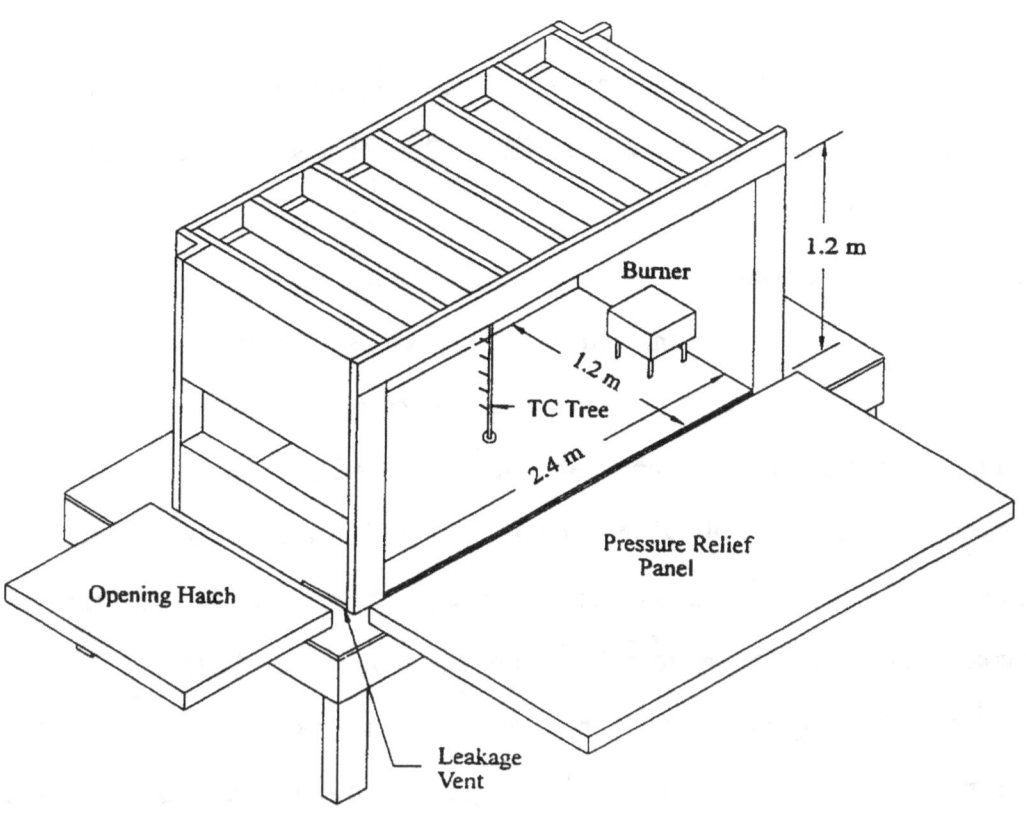

Figure 2.2 - A schematic diagram of the half-room-scale backdraft apparatus showing dimensions and component locations.

Figure 2.3 - Photograph showing the observation window in the backdraft apparatus.

To simulate a window or door, a 0.4 m high, 1.2 m wide opening was centered in the short wall opposite the burner, see Figs. 2-2 and 2-3. This vent was covered with a Both the pressure relief panel and the opening hatch are in the open position. manually operated hatch which was opened after the fire had been burning for several minutes. The hatch was hinged at the bottom and held closed by a single throw latch at the top. Figures 2-2 and 2-3 show the hatch in the open position.

A gas burner, 0.3 m square and 0.3 m high, was used in all these experiments. A spark ignitor mounted 50 mm above the burner, centered on the edge toward the compartment center, was the ignition source for both the burner and the backdraft. A 10,000 volt transformer was used to produce the arc between two electrodes 5 mm apart. The burner was placed against the wall opposite the opening, as seen in Fig. 2-2. Every effort was made to seal all construction holes to control leakage. Two small rectangular vents, 25 mm high, 0.3 m wide, were placed at the floor and ceiling in the wall with the hatch to allow for controlled leakage. In most experiments only the floor vent was open. A vertical thermocouple tree was placed in the geometric center of the compartment, as shown in Fig. 2-2. The thermocouples were 0.5 mm type K thermocouple wire with a stainless steel overbraid. The average bead diameter was 1.1 mm. Seven thermocouples were located at 0.15 m intervals, measured from the ceiling. An additional thermocouple was placed 50 mm from the ceiling to measure the ceiling jet temperature.

2.4 Results

A total of 23 backdraft experiments were conducted. The experimental parameters are summarized in columns 2 through 6 of Table 2-1 including: fuel, fuel flow rate, burner time, opening time, and number of leakage vents. The experiment numbers given in column 1 are used throughout this paper. The last column in Table 2-1 indicates if a large fire ball was observed outside the compartment. Density relative to air was the criterion for selecting the two gaseous fuels, natural gas and propane. Different fuel flow rates, burn times, and ventilation configurations were used in order to vary the excess pyrolyzates (unburned fuel) stored within the upper layer. Two different ventilation configurations were used. In experiments 1 through 11 and 20 through 23 a single vent at the floor provided leakage. In experiments 12 through 19, an additional vent was placed at the ceiling. None of the two vent experiments resulted in a backdraft.

Only 8 of the 23 experiments resulted in backdrafts, i.e., 1 - 8. The experiments were not considered to result in a backdraft unless a large fire ball was observed outside the compartment. The 8 experiments which resulted in a backdraft used natural gas as fuel. In experiments 4, 5 and 7, the experimental parameters were constant as indicated in Table 2-1: a burner time of 175 s, a single floor vent and a 5 s delay between burner shut off and the opening of the hatch to allow the fluid mechanics caused by the burner to subside. In experiment 1 the burn time was shorter at 170 s and a 2 s delay. The burn times in experiments 6 and 8 were 180 s and 185 s, respectively.

In experiment 9, a similar flame structure was seen but it did not result in a large fire ball. When opening the compartment, the latch was released and the hatch opened slightly but did not fall open. The hatch was then pulled open by the operator. This manner of opening allowed flow into the compartment ahead of the gravity current. Ignition had occurred before the incoming gravity current reached the ignitor, causing the compartment pressure to increase and thus reduce the incoming air flow and lowering the energy release.

In experiments 10 and 11 the burner was on at the time of opening which caused additional mixing in the area of the ignitor and interfered with the gravity current ignition. When the ceiling vent was opened as in experiment 12, insufficient hydrocarbons were available to fuel the backdraft.

Table 2.1 - Experimental parameters and results from exploratory backdraft experiments.

Experiment Number	Fuel	Fuel Flow rate (kW)	Burner Time (s)	Opening Time (s)	Number of vents	Fire Ball
1	Natural Gas	150	170	172	1	yes
2*	Natural Gas	150	175	180	1	yes
3*	Natural Gas	150	175	180	1	yes
4	Natural Gas	150	175	180	1	yes
5	Natural Gas	150	175	180	1	yes
6	Natural Gas	150	180	185	1	yes
7	Natural Gas	150	175	180	1	yes
8	Natural Gas	150	185	190	1	yes
9	Natural Gas	150	200	205	1	no
10	Natural Gas	150	180	150	1	no
11	Natural Gas	150	210	180	1	no
12	Natural Gas	150	360	240	2	no
13	Propane	80	240	122	2	no
14	Propane	80	300	248	2	no
15	Propane	80	420	360	2	no
16	Propane	120	300	240	2	no
17	Propane	120	420	360	2	no
18	Propane	150	330	210	2	no
19	Propane	150	420	300	2	no
20	Propane	150	175	180	1	no
21	Propane	150	150	155	1	no
22	Propane	150	120	125	1	no
23	Propane	150	90	95	1	no

* Glare on the observation window washed video recording

In the propane experiments, experiments 13 through 23, the density of the propane relative to the other compartment gases resulted in the accumulation of the propane low in the compartment. With the fuel near the floor, a small dancing flame stabilized near the small open floor vent. When the hatch was opened, the fuel was quickly ignited by the dancing flame and burned as dark orange and yellow flames which slowly propagated throughout.

The data collected in these experiments has been limited to the temperatures measured on the thermocouple tree and data recorded through the window using 35 mm cameras and video camcorders. Typical temperature histories at different heights within the compartment for experiment 4 are shown in Fig. 2-4. The burner was ignited at time zero. The temperature at the top of the compartment rose quickly to a maximum of 820 K at 25 seconds after ignition. After this peak the temperature dropped as the compartment lost energy through its boundary surfaces and the burning rate was limited by the lack of oxygen. After 120 seconds when the fire was nearly out the flames detached from the burner and began to dance across the floor, as seen in Fig. 2-5a. The dancing lasted ~ 30 seconds and was responsible for the temperature rise shown between 120 and 150 seconds in Fig. 2-4. These dancing flames occurred in some but not all of the experiments. Similar behavior has been described by Sugawa et al.[23] in their work on poorly ventilated pool fires within a compartment. Figure 2-5b, taken ~ 3.5 s after the vent is opened, just after the gravity current reaches the spark, shows the propagation of a mostly premixed flame through the mixed layer formed between the hot, fuel-rich, upper layer gases and the cold, oxygen-rich, fresh air entering the compartment through the lower portion of the open vent. Similar premixed flames have been reported by Phillips[24] on a buoyant natural gas layer interface within a model mine gallery, open at the bottom to allow free expansion of the combustion products. Phillips identified three flames: a premixed U-shaped flame burning where flammable natural gas concentrations occurred, a diffusion flame at the natural gas/air interface behind the premixed flame, and an unstable flame formed in the hot product layer sandwiched between the cold natural gas and air layers. In the backdraft experiments described in this paper, the burning occurs within a closed chamber which restrains the hot products. As the burnt gases expand, they force the unburned fuel and air ahead of the advancing flame front, out the vent. This behavior is demonstrated by the large fire ball which burns outside the compartment, shown in Fig. 2-5c. The spike in the temperature, seen in Fig. 2-4 after 180 s, is the deflagration wave flame front as it moves past the thermocouple tree on its way out of the compartment.

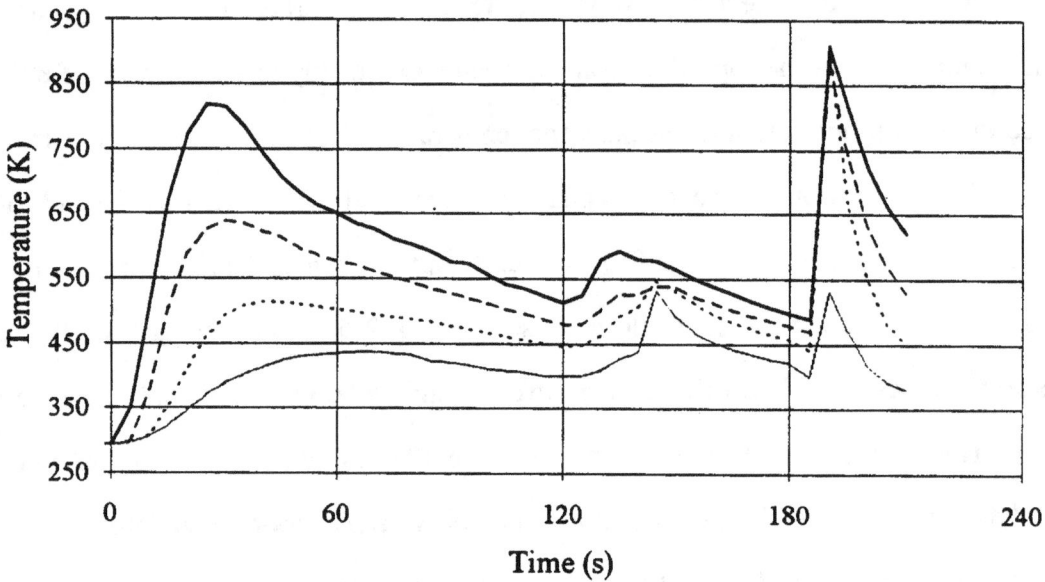

Figure 2.4 - Representative compartment temperature histories for Experiment 4 in Table 2.1. Locations are measured from the floor (———) 1.02 m, (– – –) 0.72 m, (······) 0.42 m, and (———) 0.12 m.

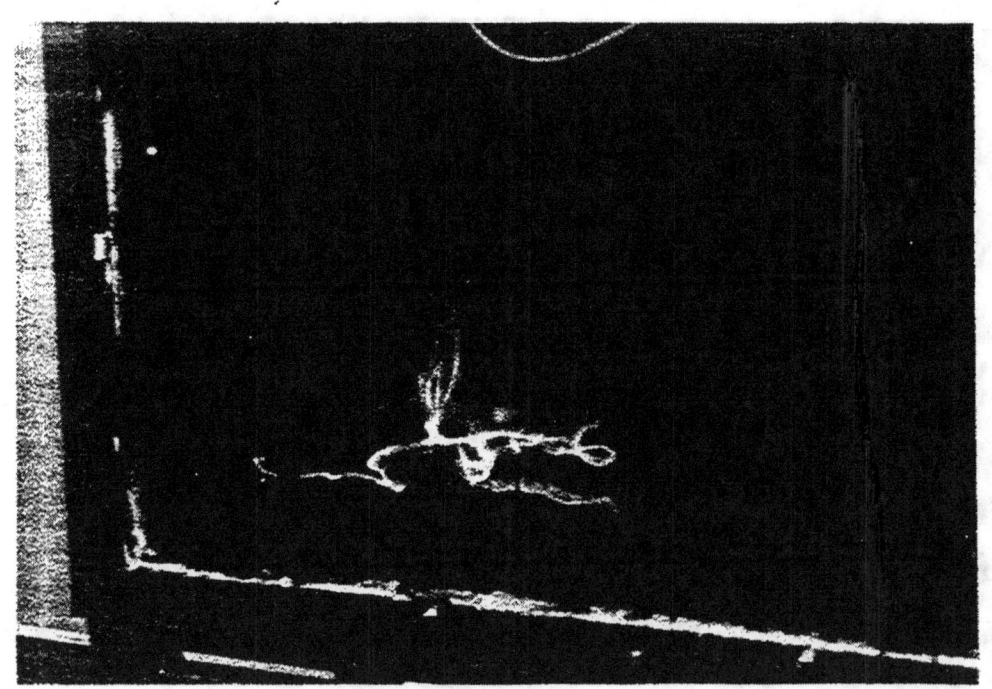

Figure 2.5a - Photograph showing the dancing flame ~130 s after ignition of the burner in Experiment 4 in Table 2.1.

Figure 2.5b - Photograph showing the premixed flame in the mixed region at the interface between hot fuel rich and cold oxygen rich layers in Experiment 5 in Table 2.1.

Figure 2.5c - Photograph showing the large fireball bursting out of the compartment in Experiment 8 in Table 2.1.

In Table 2-2, the ambient temperature, average compartment temperature, calculated gravity current travel time, ignition delay time, deflagration wave travel time, and total travel time are shown in columns 2 through 7, respectively. The experiment numbers given in column 1 correspond to those in Table 2-1. In experiments 2 and 3 the sun's glare on the window washed out the video camera image. The ignition delay time and deflagration travel time are determined from the video tapes by counting the individual frames at 30 frames per second. Taking into account all errors, accuracy is conservatively estimated at ± 0.2 s. The time of opening was taken as the time when the hatch was at 45° with the horizontal. This was done to avoid including slight variations in the hatch movement upon opening. The time from the opening of the compartment to ignition, shown in column 5, ranges from 2.1 s to 5.2 s. The gravity current travel time, shown in column 4, is calculated from gravity current velocity,

$$v = \frac{1}{3}\sqrt{\beta g h}, \qquad (2\text{-}1)$$

where v is the speed of the leading edge of the gravity current, $\beta = \Delta\rho/\rho_1$ with $\Delta\rho$ is the initial density difference across the opening, ρ is the compartment density, h is the compartment height, and g is 9.8 m/s². This expression is developed in chapter 3. Densities were calculated using the average compartment temperature at the opening time and the ambient temperature.

The deflagration wave travel time, shown in column 6, is the time from ignition to the time the leading edge of the flame leaves the compartment; it ranges from 1.2 s to 2.6 s. In experiment 8, the ignition delay time was significantly longer, and the premixed flame was not observed. When ignition finally occurred, the flames immediately filled a large turbulent hemisphere. It should also be noted that the deflagration travel time was considerably faster than for previous experiments. The reason for the delayed ignition may be that the initial gravity current was still too rich and additional flow into the compartment was required to produce a mixed layer within the flammable range. The last column in Table 2-2 is the total time required for the gravity current to move across the compartment, ignition to occur, and the deflagration to leave the compartment.

Table 2.2 - Travel time determined from videotapes and gravity current velocities.

Experiment Number	Ambient Temperature (K)	Average Compartment Temperature (K)	Gravity Current Time (s)	Ignition Delay Time (s)	Deflagration Travel Time (s)	Total Travel Time (s)
1	293	465	2.8	2.1	2.0	4.1
4	295	467	2.8	3.2	1.9	5.1
5	295	422	3.2	2.4	2.6	5.0
6	291	462	2.8	3.0	2.2	5.2
7	289	470	2.7	4.3	2.2	6.5
8	289	485	2.6	5.2	1.2	6.4

The variation of the ignition times, shown in Table 2-2, for similar experimental conditions may be caused by nonuniformities within the gravity current head. Calculations discussed in chapter 4 show highly non-uniform large scale vortical structures within the gravity current. While the mixed region may be flammable on average, it may not ignite because the fuel and oxygen have not yet mixed.

2.5 Backdraft Compartment Fire Modeling

The computer program FIRe Simulation Technique, FIRST[25], was used to analyze the compartment conditions by predicting the histories of the upper layer temperature, depth, and species concentrations. The FIRST model is a direct descendant of the Harvard compartment fire code[26] and is currently available from the National Institute of Standards and Technology in Gaithersburg, Maryland, USA. It provides a time dependent solution of simplified species and energy equations governing a compartment fire. The model assumes that the compartment can be broken into four large zones, the upper layer, the lower layer, the fire plume and the burning objects. Detailed explanations of compartment fire zone modeling are available in the literature[27].

The input data were well defined since all difficult parameters such as leakage rate and rate of heat release were specified in the experiments. The following assumptions were made: 1) The fire source was a gas burner, 0.34 m in diameter, flowing natural gas at 150 kW, with an initial 100% efficiency, placed in a corner. FIRST requires a circular burner, therefore an equivalent diameter was used. The corner location was chosen since the plume entrainment model used in FIRST overestimates entrainment close to burner[28]. 2) The only leakage into and out of the compartment was through the 2.5 cm high, 30 cm wide vent at the floor. 3) The thermal response of the compartment was governed by the refractory fiber blanket. 4) A list of all the other assumptions inherent to FIRST is available[29].

Figure 2-6 shows a comparison of the FIRST upper layer temperature history with data from experiment 5. The experimental temperature histories from the thermocouple tree, e.g. see Fig. 2-4, were converted into unsteady average upper and lower layer temperatures using the method Quintiere et. al.[30] applied to steady state temperature profiles. Quintiere assumed an upper layer temperature from the data and then used the following equations to solve for the lower layer temperature and the thermal layer interface height:

$$\int_0^{h_1}\left(\frac{1}{T}\right)dx = [h_1 - h_L]/T^{UL} + h_L/T^{LL}, \qquad (2.2)$$

and $$\int_0^{h_1} T dx = [h_1 - h_L]T^{UL} + h_L T^{LL}, \qquad (2.3)$$

where T^{UL} and T^{LL} are the upper and lower layer temperatures, and h_1 and h_L are the heights of the compartment and the layer interface. Equation (2-2) is a mass balance and Eq. (2-3) retains the same mean temperature as in the data. Here, the lower layer temperature was specified as the arithmetic average of the two lower thermocouples and the upper layer temperature and thermal interface location were calculated from Eqs. (2-2 and 2-3). Comparisons for experiment 5 are shown in Figs. 2-6a and b at 60 s and 180 s respectively. The upper layer temperature and thermal interface height data are indicated by the x's in Fig. 2-7a & b, respectively. Both the temperature and thermal interface compare well with the FIRST results shown as solid lines. The oxygen and unburned hydrocarbon mass fraction histories calculated by FIRST are shown in Fig. 2-7c. Future experiments will obtain species concentrations in the upper layer. The hatched vent is opened at 180 s in the modeling, causing the rise in oxygen and decrease in hydrocarbons shown in Fig 2-7c. The vitiated layer also rises and cools, as shown in Figs. 2-7b and 2-7a,

calculations are only applicable to the point of ignition of the backdraft. The experimental temperature spikes shown in Figs. 2-4 and 2-7a are not depicted by the computer model.

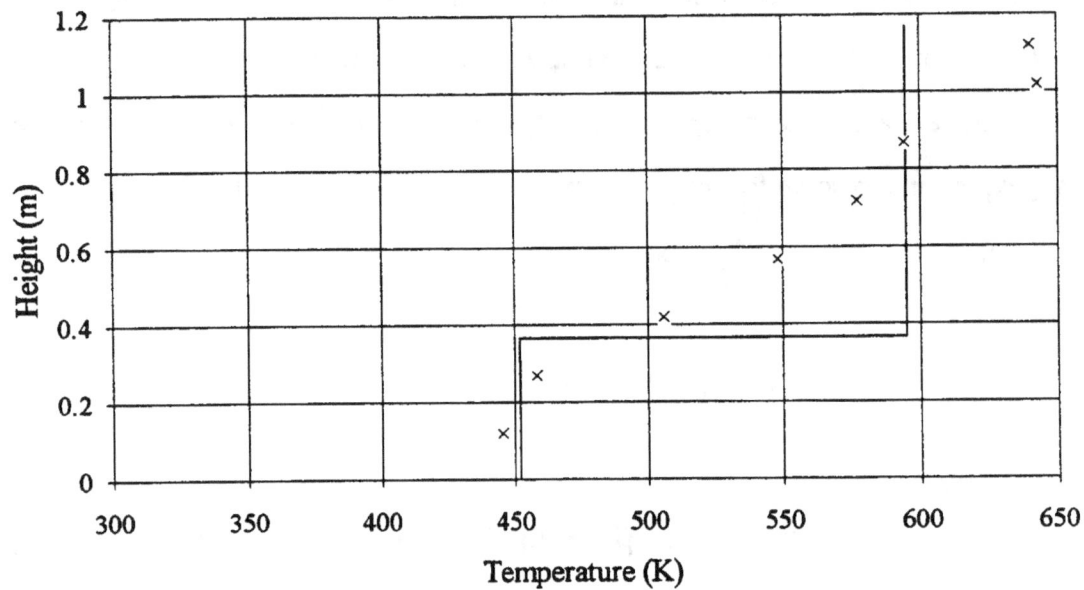

Figure 2.6a - Comparison of temperature data from Experiment 5 (x) (Table 2.1) at 60 s with the idealized two layer approximation calculated from Equation (2-2) and (2-3).

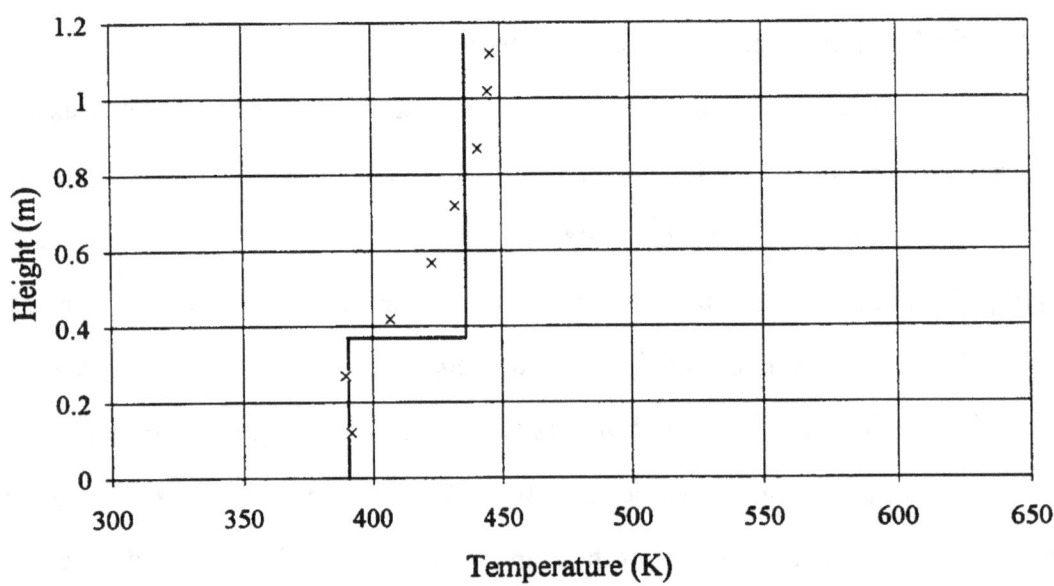

Figure 2.6b Comparison of temperature data from Experiment 5 (x) (Table 2.1) at 180 s with the idealized two layer approximation calculated from Equations (2-2) and (2-3).

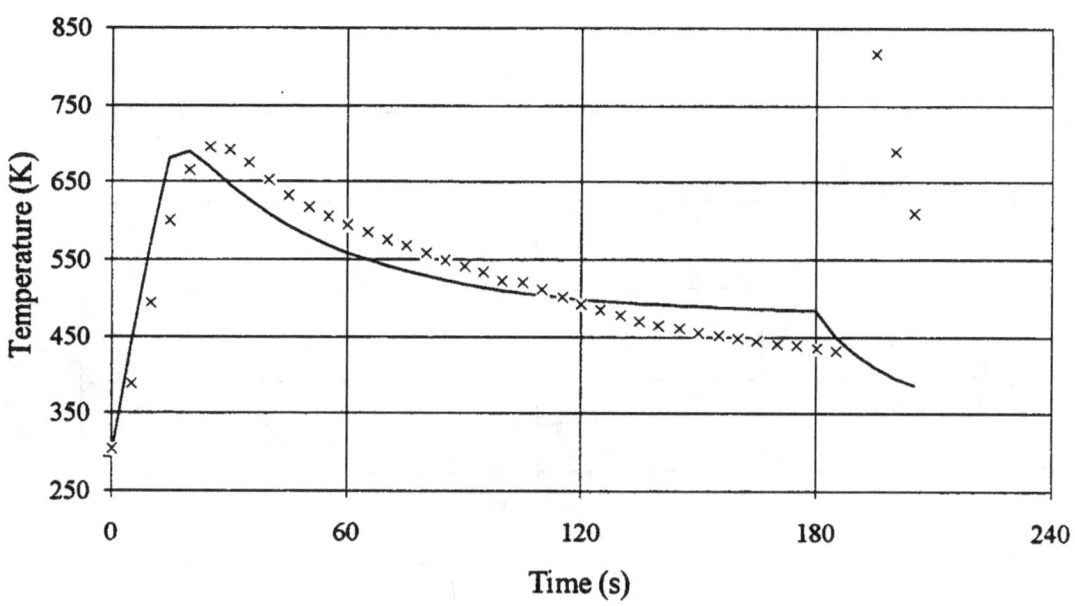

Figure 2.7a - Comparison of temperature data (x) from Experiment 5 (Table 2.1) calculated from Equations (2-2) and (2-3) with computer results from FIRST (———).

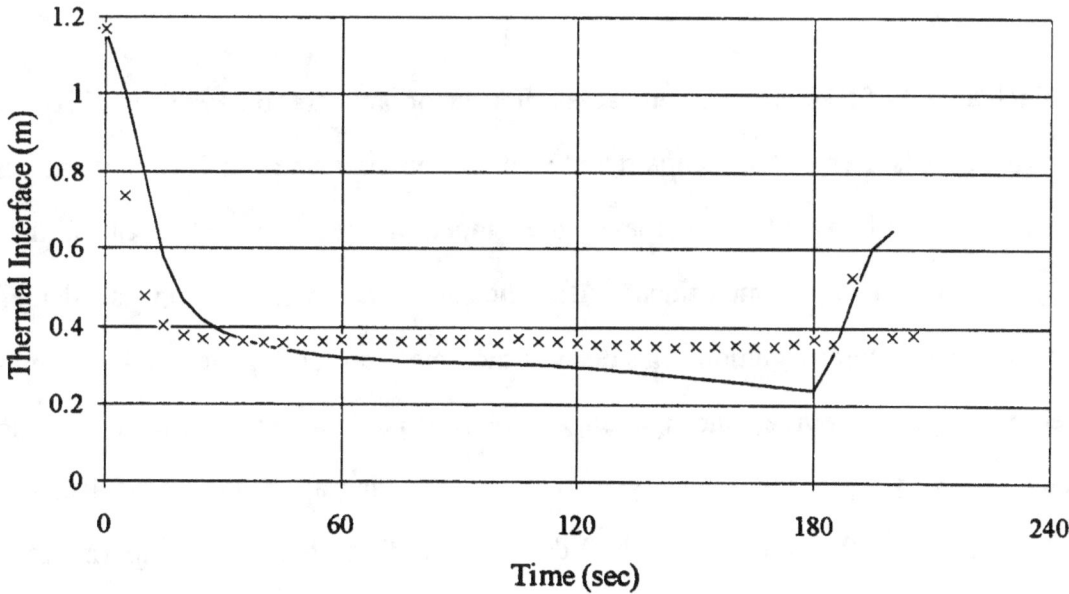

Figure 2-7b - Comparison of Experiment 5 (Table 2.1) data (x), calculated from Equations (2-2) with the thermal interface history computed from FIRST (———).

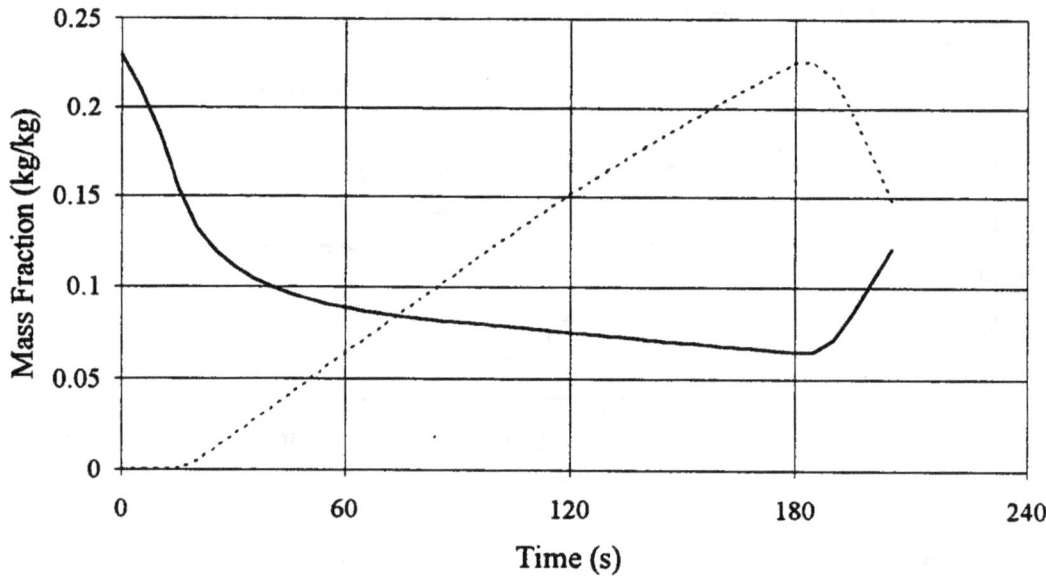

Figure 2.7c - Compartment upper layer hydrocarbon (– – –) and oxygen (———) mass fraction computed by FIRST.

2.6 Discussion

Each aspect of the hypothesized backdraft scenario appears to have been observed in these experiments. Before the compartment was opened there were high concentrations of unburned hydrocarbons in the upper layer, the temperature was relatively cool and the upper layer nearly filled the compartment. After the hatch was opened there was a delay of at least 2 seconds before ignition. This delay is the time required for the gravity current to travel to the ignition source. Once ignited, a small premixed flame front travels into the flammable mixed layer preceding the large non-premixed deflagration. Figure 2-8 is a series of video images showing the backdraft development from ignition for experiment 5 in Tables 2-1 & 2-2. The images are taken at 0.4 s intervals. In the first three images, the flame in the mixed region can be clearly seen. Behind this flame is a turbulent combustion region which develops from the buoyant instability of the hot combustion products, as shown in the next three images. The hot products rise and displace the fuel-rich upper layer downward into the oxygen-rich air. Combustion then further enhances the mixing. As the quasi-premixed flame approaches the center of the compartment, Fig. 2-8, the expanding turbulent flame has accelerated sufficiently to engulf the laminar flame and produce a deflagration which advances rapidly through the compartment and bursts out of the opening in a dramatic ball of flame, see in image six.

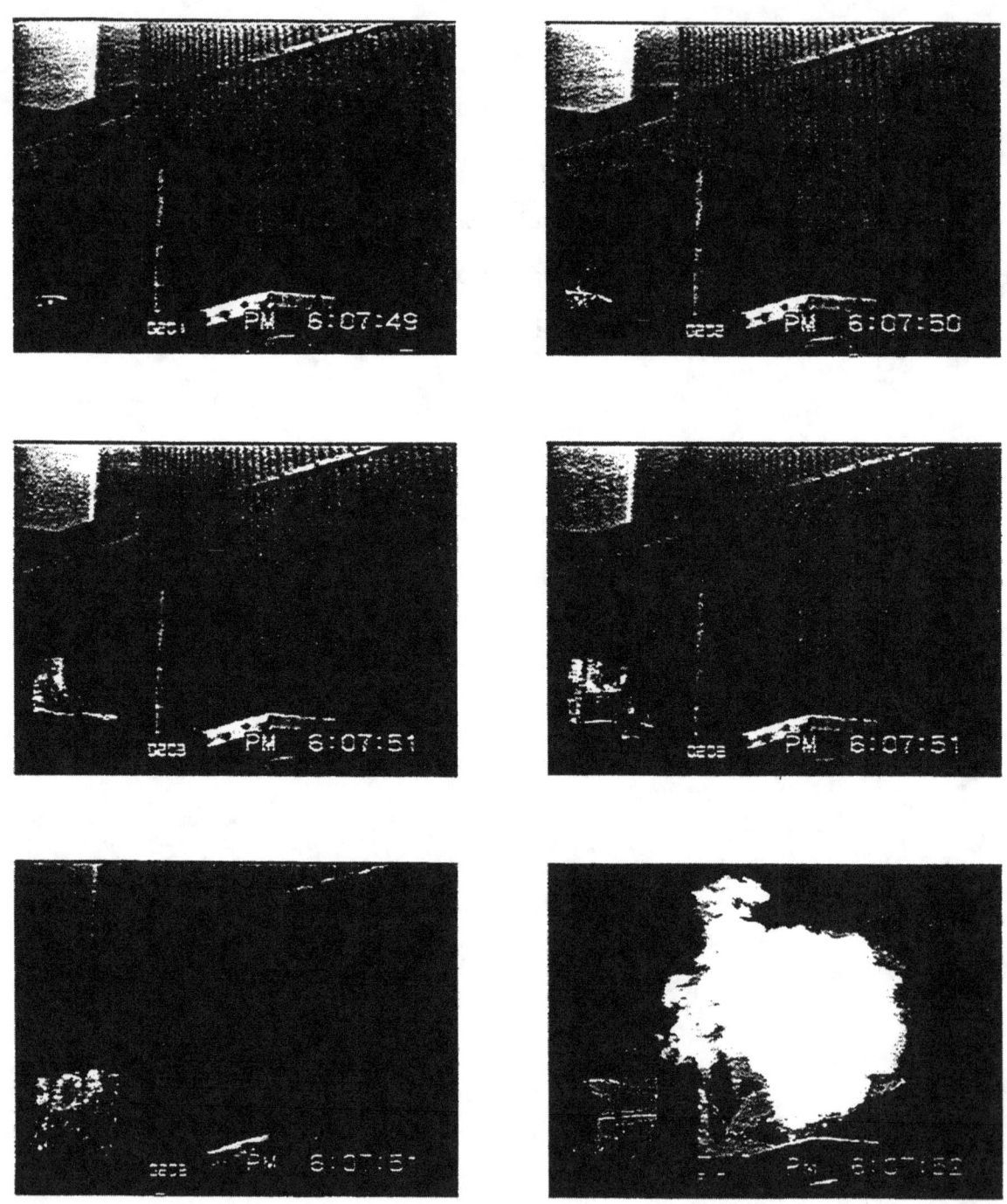

Figure 2.8 - A series of videoimages taken during Experiments 5 (Table 2.1) showing the flame propagation and the resulting fireball.

2.7 Conclusions

The physical model postulated here appears to accurately describe the backdraft phenomena observed in these half-scale experiments. Ignition does not occur immediately upon opening a vent to a fuel rich compartment. For a fuel rich backdraft to occur, a gravity current, on which a mixed layer rides to an ignition source, is required. It is this time delay caused by the gravity current propagation which creates a hazard to firefighters who may enter a compartment and become trapped in the backdraft process.

Future work will focus on improved instrumentation in the compartment. A variety of opening geometries will be examined. Salt water modeling has been performed to determine the size and location of the mixing layer as a function of the density difference and the opening configuration. Future compartment experiments will also be more fully instrumented with additional thermocouples, pressure transducers, analysis of the upper layer gas concentrations including HC, O_2, CO_2, and CO, bi-directional velocity probes in the hatch and vent openings, and improved video recording for better flow visualization. Hopefully, a foundation will be laid for more sophisticated compartment fire models which can incorporate backdraft phenomena.

References - Chapter 2

[1] Pagni, P. J., and Shih, T. M., "Excess Pyrolyzates," 16th Symposium (Int'l) on Combustion, 1329-1343, The Combustion Institute, Pittsburgh, PA, 1976.

[2] Pagni, P.J., "Diffusion Flame Analyses," Fire Safety Journal, 3, 273-286, 1980/81.

[3] Pagni, P.J., and Joshi, A.A., "Glass Breaking in Fires," in Fire Safety Science - Proceedings of the Third International Symposium, 791-802, Elsevier Science Pub., London, 1991.

[4] Keski-Rahkonen, O., "Breaking of Window Glass Close to Fire," Fire and Materials, 12, 61-69, 1988.

[5] Russell, D., "Seven Fire Fighters Caught in Explosion," Fire Engineering, 136, 20-23, 1983.

[6] Bowen, J.E., "Flashover/Backdraft Explosive Situation," Western Fire Journal., 34, 33-38, 1982.

[7] "Fatal Mattress Store Fire at Chatham Dockyard," Fire, 67, 388, 1975.

[8] Croft, W.M., "Fires Involving Explosions - A Literature Review," Fire Safety Journal, 3, 3-24, 1980/81.

[9] Steward, P.D.C., "Dust and Smoke Explosions," National Fire Protection Association Quarterly, 7, 424-428, 1914.

[10] Dunn, V., "Beating the Backdraft," Fire Engineering, 141, 44-48, 1988.

[11] Woolley, W.D., and Ames, S.A., "The Explosion Risk of Stored Foamed Rubber," Fire International, 50, 45-54, 1975.

[12] Burklin, R.W., and Purington, R.G., Fire Terms: A Guide to their Meaning and Use, National Fire Protection Association, Boston, MA, 1980.

[13] Jones, W.W., "A Review of Compartment Fire Models," NBSIR 83-2684, National Bureau of Standards, Gaithersburg, MD, 1983.

[14] Mitler, H.E., "The Harvard Fire Model," Fire Safety Journal, 9, 7-16, 1985.

[15] Friedman, R., "Survey of Computer Models for Fire and Smoke", Factory Mutual Research Corporation, Norwood, MA, 1991.

[16] Emmons, H.W., "The Calculation of a Fire in a Large Building," ASME Paper No. 81-HT-2, American Society for Mechanical Engineers, New York, 1981.

[17] Simpson, J.E., "Gravity Currents in the Laboratory, Atmosphere, and Ocean," Annual Review of Fluid Mechanics, 14, 213-234, 1982.

[18] Britter, R. E., and Simpson, J.E., "Experiments on the Dynamics of a Gravity Current Head," Journal of Fluid Mechanics, 88, 223-240, 1978.

[19] Simpson, J.E. and Britter, R.E., "The Dynamics of the Head of a Gravity Current Advancing over a Horizontal Surface," Journal of Fluid Mechanics, 94, 477-495, 1979.

[20] Zalosh, R.G., "Explosion Protection," in SFPE Handbook of Fire Protection Engineering, (P.J. DiNenno, ed.) pp. 2-88 to 2-105, National Fire Protection Association, Quincy, MA, 1988.

[21] Factory Mutual Engineering Corporation, "Damage-Limiting Construction," Loss Prevention Data Sheet 1-44, Norwood, MA, 1991.

[22] Neoceram Super Heat-Resistant Glass for Industrial Use, Nippon Electric Glass Co. Ltd.

[23] Sugawa, O., Kawagoe, K., Oka Y., and Ogahara, I., "Burning Behavior in a Poorly-Ventilated Compartment Fire -- Ghosting Fire," Fire Science and Technology, 9, 5-14, 1989.

[24] Phillips, H., "Flame in a Buoyant Methane Layer", 10th Symposium (Int'l) on Combustion, pp. 1277-1283, The Combustion Institute, Pittsburgh, PA, 1965.

[25] Mitler, H.E., and Rockett, J.A., "Users' Guide to FIRST, a Comprehensive Single-Room Fire Model," NBSIR 87-3595, National Bureau of Standards, Gaithersburg, MD, 1987.

[26] Mitler, H.E., and Emmons, H.W., "Documentation for CFC V, the Fifth Harvard Computer Fire Code," Home Fire Project Technical Report No. 45, Harvard University, Division of Applied Sciences, Cambridge, MA, 1981.

[27] Rockett, J.A., Morita, M., and Cooper, L.Y., "Comparison of NBS/Harvard VI Simulations and Full-Scale, Multi-Room Fire Test Data, NBSIR 87-3567, National Bureau of Standards, Gaithersburg, MD, 1987.

[28] Zukoski, E.E., Kubota, T., and Cetegen, B., "Entrainment of Fire Plumes," Fire Safety Journal, 3, 107 1980/81.

[29] Mitler, H.E., "The Physical Basis for the Harvard Computer Fire Code," Home Fire Project Technical Report No. 45, Harvard University, Division of Applied Sciences, Cambridge, MA, 1981.

[30] Quintiere, J.G., Steckler, K., and Corley, D., "An Assessment of Fire Induced Flows in Compartments" Fire Science and Technology, 4, 1-14, 1984.

CHAPTER 3

SALT WATER MODELING OF FIRE COMPARTMENT GRAVITY CURRENTS

3.1 Introduction

A gravity current is the flow of one fluid into another caused by a difference in density. This density difference may be due to a dissolved chemical or a difference in the temperature between the two fluids. There are many common examples of gravity currents including sea-breeze fronts, avalanches, lock exchanges, and flows following volcanic eruptions. A large body of research is available on the subject of gravity currents.[1,2] In many cases, the flow field in gravity currents is sufficiently complex that the problem is difficult to solve from first principles. For this reason, physical models, typically salt water models, are used to analyze these problems. Salt water models have been applied to many fire problems including corridor smoke flow, ship board fires, and compartment fires.[3,4,5]

When a fire occurs in a closed compartment where the only ventilation is due to leakage, the fire initially burns independent of the surroundings and a hot upper layer develops within the compartment. If the leakage rate is small, the hot layer descends over the fire and the burning becomes limited by the available oxygen thus producing large amounts of unburned fuel. Left undisturbed, the heat release rate will decrease and the fire may enter a smoldering stage. When the compartment is opened, a gravity current of dense ambient air enters the compartment mixing with the lighter, fuel rich, compartment gases. If the fuel concentrations are high enough the mixed region carried with the gravity current may ignite, resulting in a backdraft. The important role of gravity currents in backdrafts has been demonstrated in a series of half-scale experiments discussed in chapter 2.

This study attempts to quantify the gravity current which enters a compartment prior to a backdraft. A salt water scale model with two different density fluids is used to visualize the flow into the compartment. Because the fire is assumed to be small or smoldering, plume effects are ignored and the compartment is filled with a uniform density fluid lighter than the fluid outside the compartment. Data collected from these experiments include gravity current propagation velocity and geometry. In addition to the entering current, the current which is reflected off the wall opposite to the opening wall is examined.

3.2 Gravity Current Scaling

As a simplified limit of a compartment fire gravity current, consider the steady flow of a perfect fluid in a semi-infinite horizontal box of arbitrary width as shown in Fig. 3.1. At time zero, far to the right, the entire end of the box is instantaneously removed. High density ambient fluid, state 0, flows into the box, as low density compartment fluid, states 1 and 2, flows out, due to buoyancy. The parameter indexing that buoyancy is the normalized positive density difference,

$$\beta = \frac{(\rho_0 - \rho_1)}{\rho_1}, \qquad (3.1)$$

where ρ_0 is the higher density, ambient, fluid within the gravity current and ρ_1 is the lower density fluid, (at opening) within the hot compartment ahead of the gravity current.

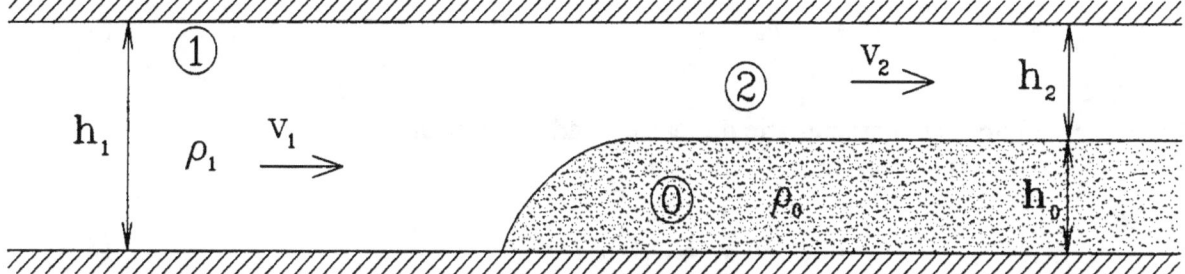

Figure 3.1 - Gravity current schematic. Velocities are indicated in a reference frame fixed on the gravity current. Heights are indicated by h.

Benjamin[1] has shown that, in this perfect fluid limit with no mixing or dissipation, conservation of mass, momentum and energy can be written, respectively, as

$$v_1 h_1 = v_2 h_2, \tag{3.2}$$

$$v_1^2 h_1 + \beta g h_1^2 = 2 v_2^2 h_2 + \beta g h_2^2, \tag{3.3}$$

$$v_2^2 = 2\beta g (h_1 - h_2). \tag{3.4}$$

Eliminating v_1 from Eqs. 2 and 3 and v_2 from Eqs. 3 and 4, the height of the exiting compartment fluid is

$$h_2 = \frac{h_1}{2} = h_0. \tag{3.5}$$

This is also the height of the gravity current since $h_0 = h_1 - h_2$. Benjamin[1] shows that for flows with energy losses, $h_0 < h_1/2$. The nondimensional velocity or Froude number of the fluid exiting the compartment, from Eqs. 4 and 5 is

$$\frac{v_2}{\sqrt{(\beta g h_2)}} = \sqrt{2}, \tag{3.6}$$

since this is >1 a dissipative hydraulic jump is possible. The velocity of the compartment fluid approaching the gravity current, or in the laboratory reference frame, the gravity current velocity, from Eqs. 2, 5 and 6, is,

$$v^* = \frac{v_1}{\sqrt{\beta g h_1}} = \frac{1}{2}. \qquad (3.7)$$

Thus for scaling, the characteristic dimension and velocity are

$$x_c = h_1 \text{ and } v_c = \sqrt{\beta g h_1}, \qquad (3.8)$$

from Eqs. 5 and 7 respectively. The characteristic time is then

$$t_c = \frac{x_c}{v_c} = \sqrt{\frac{h_1}{\beta g}}. \qquad (3.9)$$

Typical dwellings have room heights of 2.4 m (8 ft), so the compartment used for backdraft experiments, height 1.2 m (4 ft), is 1/2 scale. The salt water compartment described here is 0.15 m (0.5 ft) high, so it is 1/8 scale to the experimental backdraft compartment and 1/16 scale to a dwelling.

The salt water experiments are necessary to quantify the effect of transients, mixing, energy dissipation, opening geometry, and aspect ratio on the simple gravity current size and nondimensional velocity expressions given by Eqs. 5 and 7. They will provide confirmation and corrected formulas useful for modeling backdrafts in fire compartments.

The salt water experiments are limited to $0.003 \leq \beta \leq 0.101$, while the backdraft compartment and full scale fires produce higher β, up to 1.2. However, the literature suggests,[6] as confirmed by experimental Froude numbers developed here, that v^* is independent of β and depends only on the opening geometry. Therefore, these scaled velocity and geometry results are expected to apply directly to actual and modeled backdrafts. For example, a velocity of 0.09 m/s at a $\beta = 0.05$ for a salt water current in a slot opening geometry gives $v^* = 0.32$, which would correspond to 0.8 m/s at a $\beta = 0.5$ in the 1.2 m high model compartment and 1.1 m/.s at the same β in a 2.4 m high dwelling.

Heat transfer effects at the boundaries and between fluids are not included[4]. Separate analyses and experiments will be required for gravity currents submerged at great depths [1,7] as would occur upon opening a small door in a large, high warehouse with a ventilation limited fire.

3.3 Apparatus and Procedure

Salt water experiments were conducted by placing an acrylic compartment within a larger glass tank. The large tank, 0.3 m wide, 0.6 m long, and 0.45 m deep, contained a dense water and salt solution ranging in density from 1.003 kg/m^3 to 1.101 kg/m^3. The solution temperature was 18°C. Standard "Rock Salt" crystals were dissolved in tap water to raise the density to the desired level. Densities less than 1.003 kg/m³ were too difficult to measure accurately and with densities above 1.10 kg/m^3 the solution became opaque making visual observation unreliable.

The compartment was constructed from 6 mm thick acrylic with interior dimensions of 0.15 m wide, 0.30 m long (L) and 0.15 m high. Figure 3.2a shows the plan and elevation views of the compartment. A flange was built at one end of the compartment so that the opening geometry could be easily modified by replacing a face plate bolted to the flange. Four opening geometries were used, as seen in Fig. 3.2b, the cross hatched area indicates the opening. Opening #1 was a fully open wall, 0.15 m by 0.15 m, used to demonstrate the similarity between the transient results presented in Section 5 and the steady state inviscid theory of Benjamin[1]. Opening #2 is a horizontal slot, 0.15 m wide by 0.05 m high, centered vertically in the end wall and corresponds to the opening used in the backdraft experiments. Opening #3 was a 0.05 m square, centered vertically and horizontally in the wall designed to simulate a window opening. Opening #4 was 0.12 m high by 0.05 m wide centered horizontally with the bottom of the opening at floor level, to simulate a door. Openings 1 & 2 can be considered two-dimensional in the large scale whereas openings 3 & 4 are clearly three-dimensional. The end opening was covered with a vertical sliding partition that was removed to start the experiment. A neoprene gasket was placed between the face plate and sliding partition to prevent leakage before the experiment began. The compartment was made negatively buoyant by adding 1.6 kg of lead shot in ballast channels beneath the compartment as indicated in Fig. 3.2a. The solution in the compartment was regular tap water with: pH 6.8, density 1.000 kg/m^3, and temperature 18°C.

In the early two dimensional experiments, blue vegetable dye was added to the compartment. A small amount phenolphthalein (4×10^{-5}M concentration) was also added to the compartment fluid to visualize the gravity current mixing. When phenolphthalein mixes with a base, in this case sodium hydroxide, the product of the reaction is red. This reaction is believed to be diffusion limited. The red product is strongly visible even in dilute concentrations. Turbulence is unaffected by the reaction since it produces little surface tension, buoyancy, or heat release. Although reversible, the disappearance of the red product can be minimized by keeping the pH in the large tank high, in this case 11.7. The pH in the large tank was raised using sodium hydroxide crystals. Unlike the passive scalar techniques, such as dye, the chemical reaction of the phenolphthalein gives a much better indication of the mixing that occurs in the gravity current. A thorough discussion of this technique is given by Breidenthal[8].

In the three dimensional experiments, a mirror was placed above the compartment at a 45° angle to show the plan view of the gravity current in the same plane as the elevation view for video recordings. When the mirror was used, the blue dye was eliminated and the phenolphthalein concentration was increased by a factor of 4 to 1.6×10^{-4}M to produce a more visible gravity current.

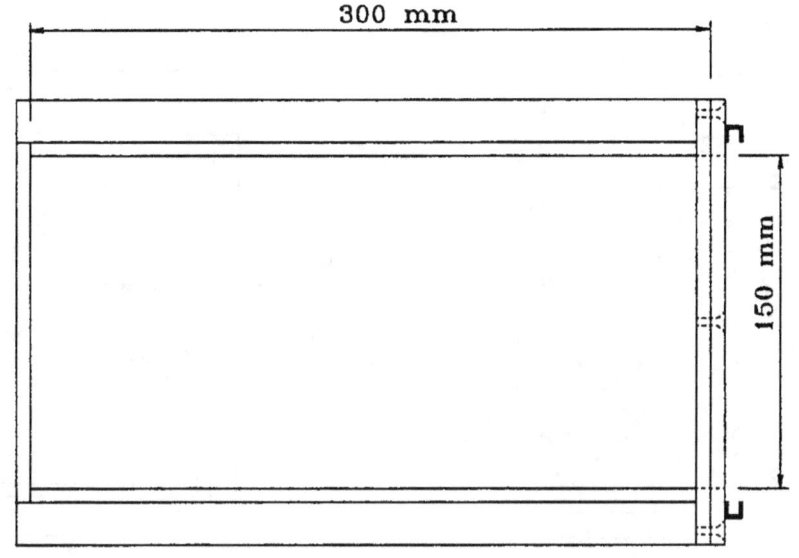

Plan View

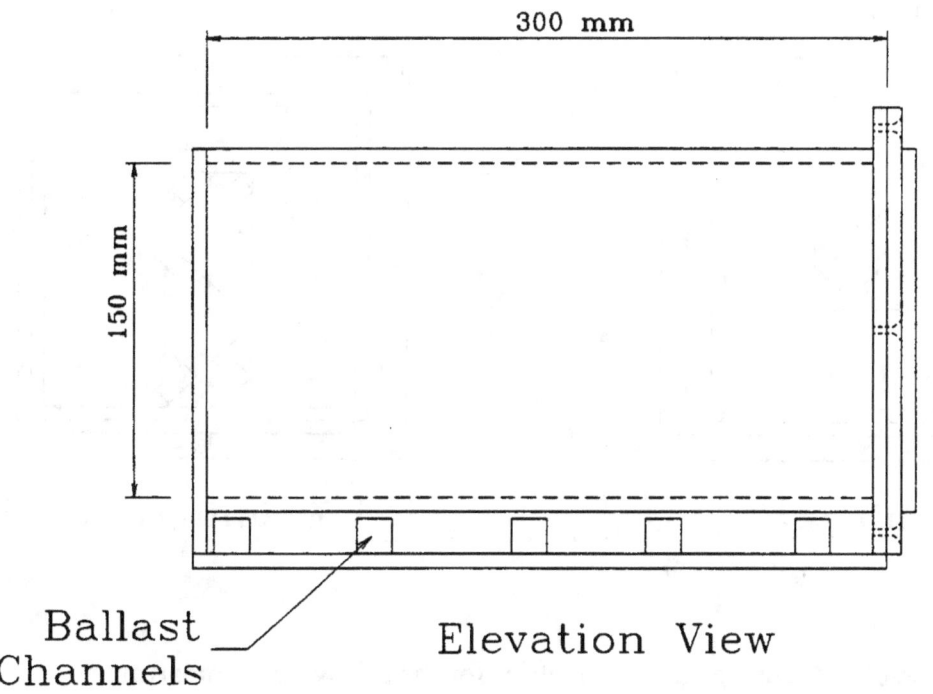

Ballast Channels

Elevation View

Figure 3.2a - Sketch of the salt water compartment showing the elevation and plan views.

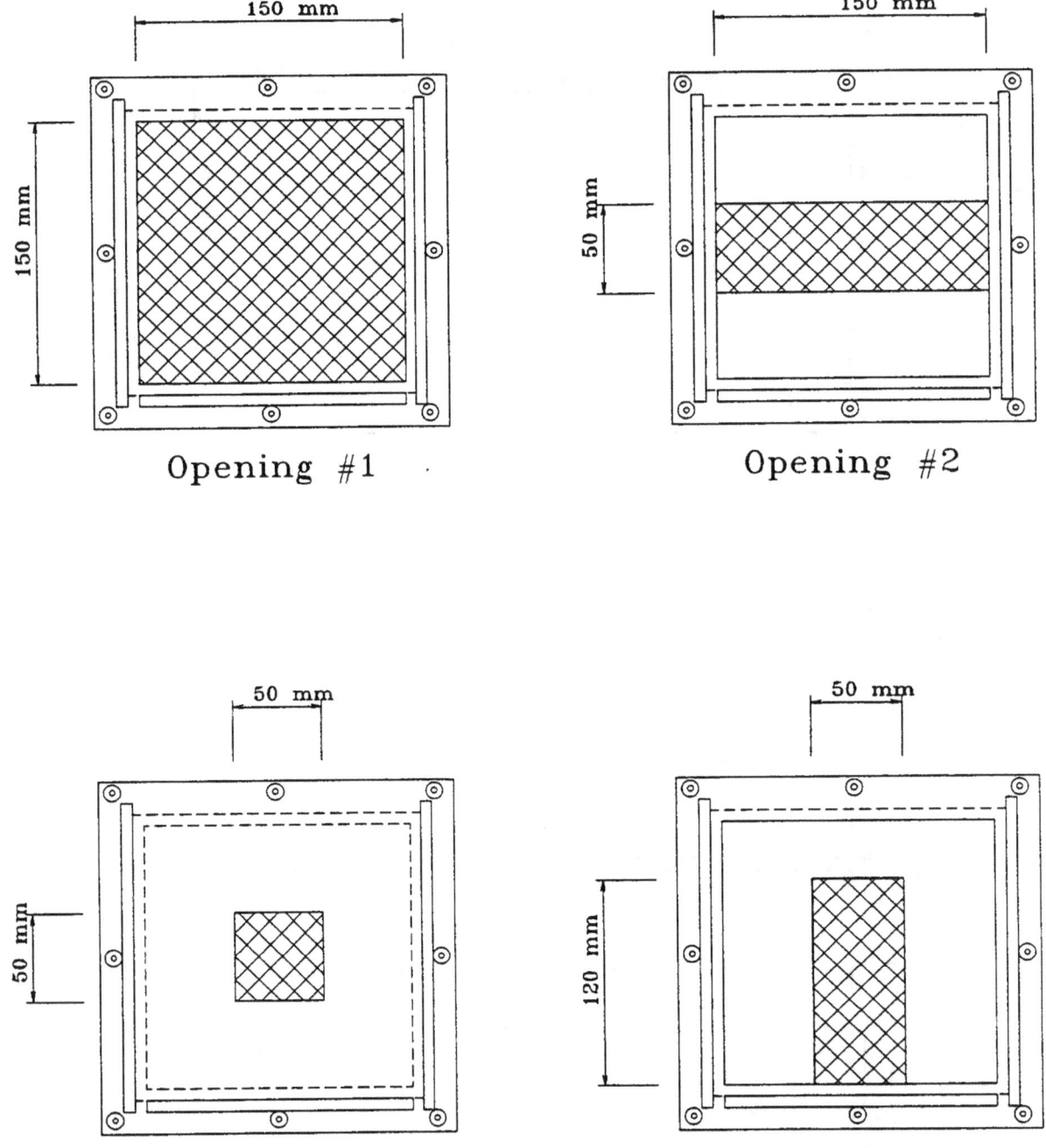

Figure 3.2b - Sketch of the four opening geometries for the salt water compartment.

Once the two solutions were prepared, specific gravity, temperature, and pH were recorded. The compartment was then lowered into the tank and the partition on the compartment was removed within 120 s to avoid leakage effects. Within 0.1s the partition was completely clear of the opening. The gravity current was recorded using a high 8 mm video camcorder at 30 frames per second. Typically these experiments lasted less than five minutes.

3.4 Compartment Gravity Current Structure

Some characteristic features of a steady state gravity current include: a head at the front of the current, mixing at the shear interface between the two fluids, and a series of advancing lobes and clefts at the leading edge. Figure 3.3 is a simple sketch showing plan and elevation views displaying some of these features on a steady state gravity current. The foremost point of the current is slightly raised above the bottom surface to a height of h_n. This lifting of the head is a result of the faster moving heavier fluid overrunning the slower light fluid. The lighter fluid is forced under the gravity current as a result of the no slip condition at the lower bounding surface.

The overrun fluid causes a gravitational instability which is largely responsible for the three-dimensional effects which are seen in natural gravity currents. The instability is manifested as the lobes and clefts which make up the leading edge of the current and the billows which form above and behind the head of the current[9,10]. In Fig. 3.3, the plan view of the leading edge shows the lobe and cleft structure. The width of the lobe is $b \sim O(h_0)$ as reported by Simpson[9]. As a lobe widens it will split and a portion of the dense fluid, mixed with the lighter fluid overrun by the current, is swept up and over the head forming a new billow behind the head of the advancing current. As a result of the split two smaller lobes are formed and a cleft develops between them. The billows which form behind the head of the gravity current are both qualitatively and quantitatively similar to the Kelvin-

Helmholtz instability of a shear layer separating two flowing fluids of different densities[6,11].

Figure 3.4a is a photograph showing the gravity current approximately 3L/4 into the compartment for the full open condition. The lower half of the photograph shows the profile of the gravity current. The top half of the photograph shows the plan view reflected in the inclined mirror. The mixed region, confined to a shallow layer between the two fluids along the shear interface, is red due to the chemical reaction of the phenolphthalein but appears gray in the black and white photograph. In profile the gravity current head is raised above the lower boundary as described above. The fluid that is overrun by the current is visible in the photograph as the gray area under the head. Along the interface between the two fluids the large billows can be seen developing behind the head. In plan view the lobes and clefts which make up the leading edge can be clearly seen. The lobes and clefts make the leading edge of the current difficult to determine and contribute significantly to the overall error analysis described in Section 3.5.

The photograph in Figure 3.4b of the gravity current approximately 3L/4 into the compartment for the slot opening shows an important conclusion from this salt water modeling. The current is mixed throughout, as indicated by the increased size of the gray (red) region compared to the full open case. The increased mixing is a result of the rearward facing step caused by the opening being placed above the floor. Otherwise, the profile of the gravity current shows a similar structure to the traditional gravity currents discussed above i.e., the slightly raised head and the billows formed behind the head. The plan view of the current clearly shows the presence of lobes and clefts at the leading edge. The gravity current head is not as high as in the full open case and the lobes are also smaller. The retarding effect of the no slip boundary condition along the walls is also apparent in the plan view.

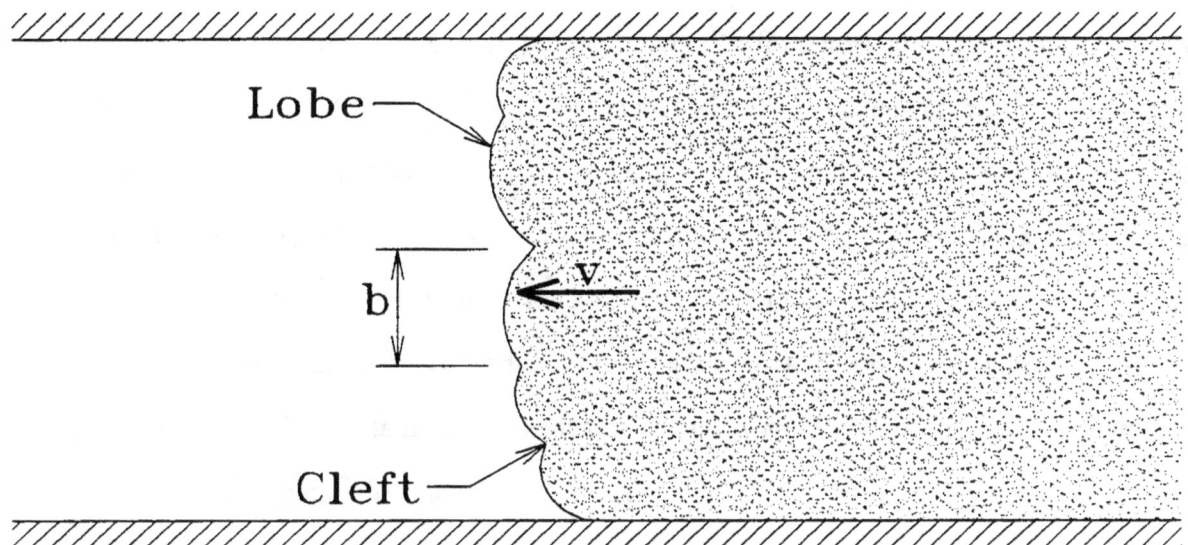

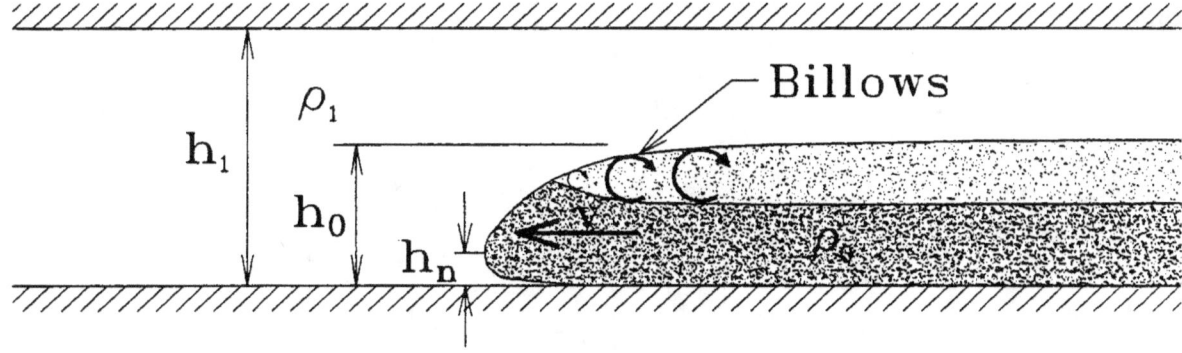

Figure 3.3 - Sketch of the gravity current with a no slip condition at the lower boundary. Both elevation and plan views are shown. Characteristic features of the gravity current are shown including the lobes, clefts, and billows.

Figures 3.4c & d are photographs showing the gravity currents approximately 3L/4 of the way into the compartment for the window and door opening conditions respectively. By the time the current has reached the 3L/4 point, the effects of the three-dimensional opening are reduced and the gravity current is qualitatively similar to the two-dimensional slot opening. The effect of the three-dimensional opening can be seen in the series of photographs shown in Fig. 3.5. These photos show how the gravity current enters the compartment for the window geometry. The current initially spreads radially from the opening but by the time the current reaches approximately L/2, the leading edge is moving into the compartment in a similar manner to the slot opening. The three dimensional opening will increase the amount of entertainment that occurs as the fluid cascades over the edges forming the opening. This entrainment is caused by relatively large coherent structures somewhat similar to those occurring in the plume as it exits the compartment.

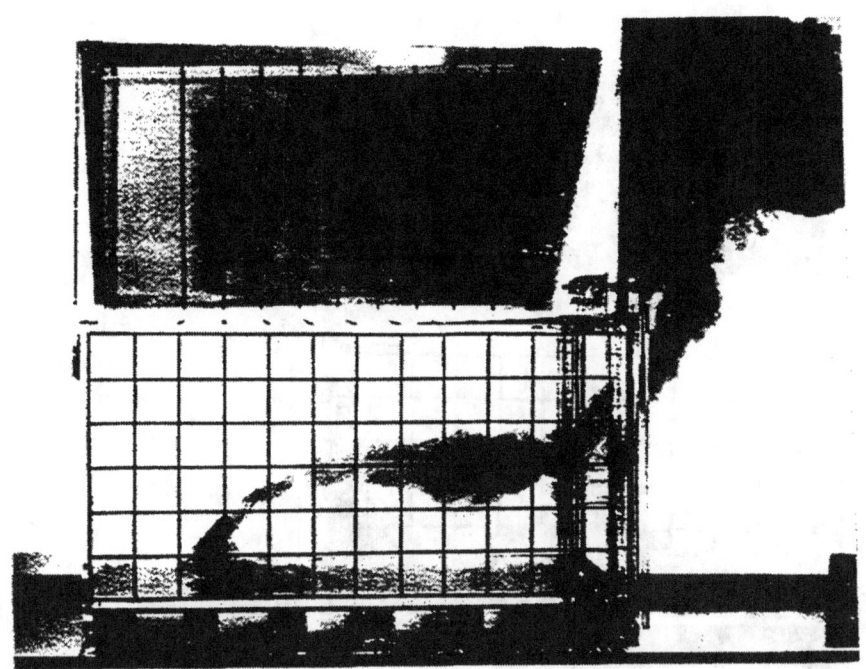

Figure 3.4a - Photograph of the gravity current approximately 3L/4 into the compartment for the fully open condition with $\beta = 0.018$. The grid shown on the model is 25 mm squares.

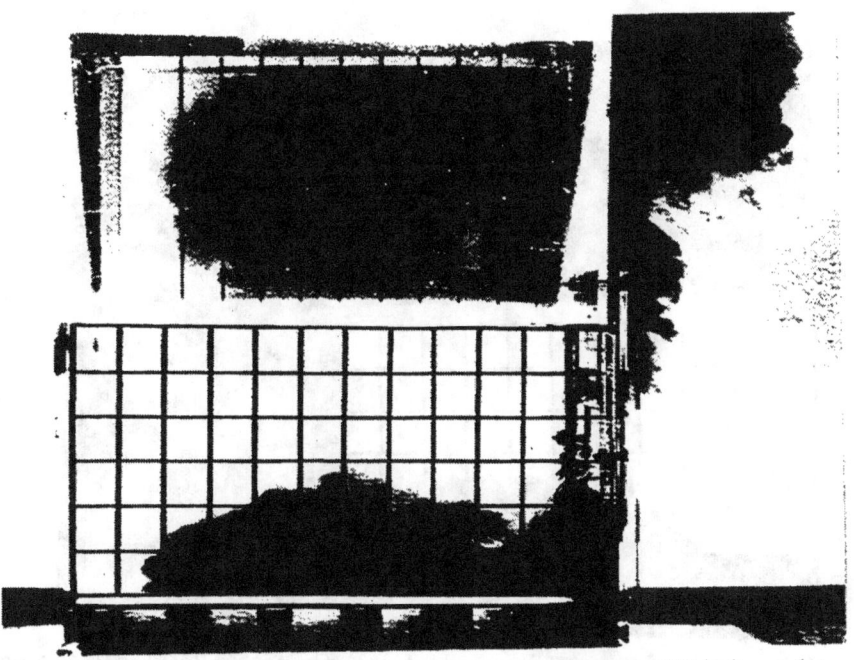

Figure 3.4b - Photograph of the gravity current approximately 3L/4 into the compartment for a center $h_1/3$ horizontal slot opening with $\beta = 0.024$.

Figure 3.4c - Photograph of the gravity current approximately 3L/4 into the compartment for a central, $h_1/3$ square, window opening with $\beta = 0.032$.

Figure 3.4d - Photograph of the gravity current approximately 3L/4 into the compartment for a door opening of width $h_1/3$ and height $7h_1/9$ with $\beta = 0.026$.

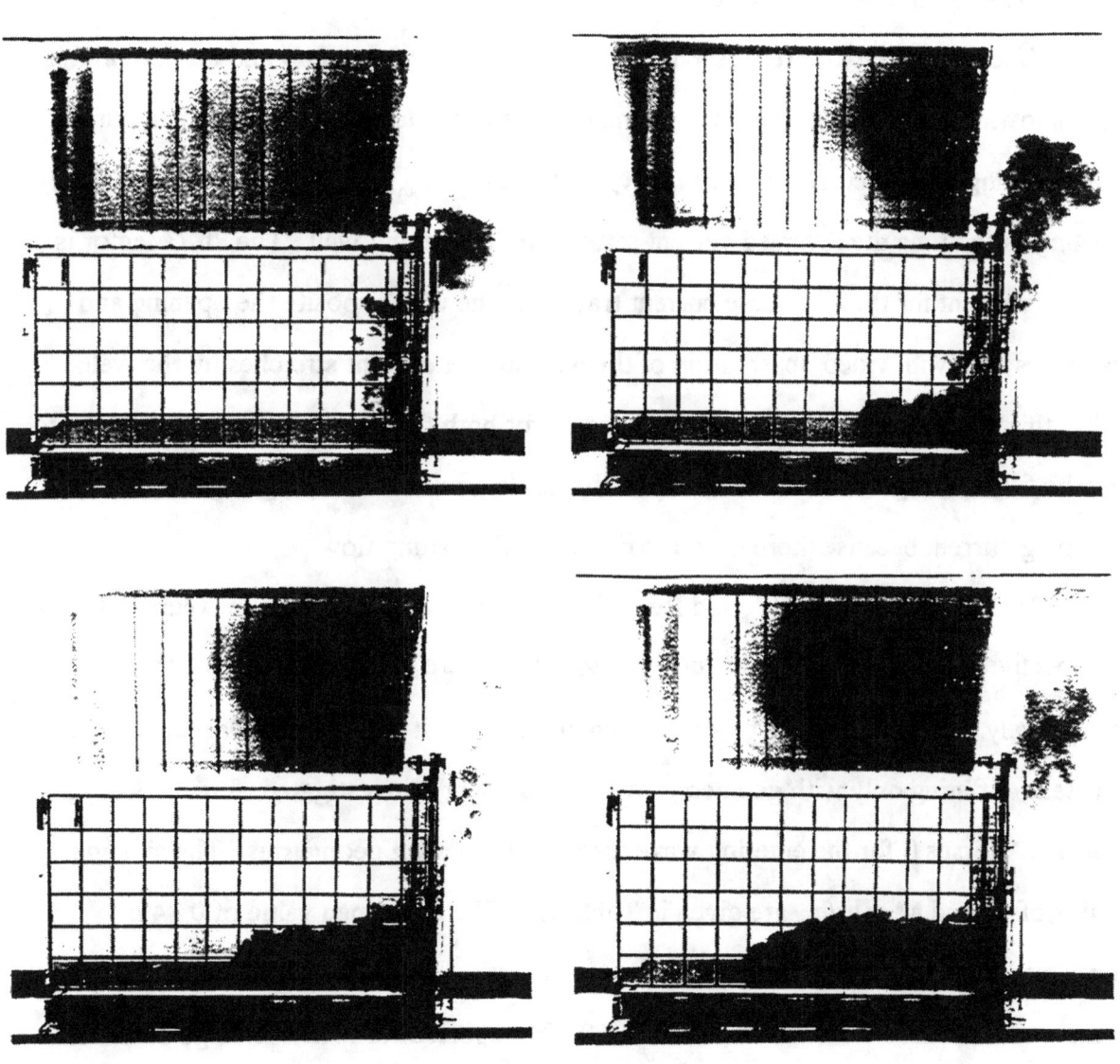

Figure 3.5 Series of four photographs of the gravity current modeling showing the developing gravity current as it enters the compartment for the window opening β = 0.032 at times: a 1.4 s, b 2.2 s, c 3.3 s, d 4.6s

3.5 Results of Salt Water Modeling

The experimental nondimensional velocity Froude Number, as given in Eq. 7, is defined in terms of the gravity current velocity, $v = L/t_{in}$, where t_{in} is the time required for the leading edge of the gravity current to reach the wall opposite the opening. Times reported for these experiments were obtained from frame by frame analysis of the video recordings at 30 frames per second.

Once the gravity current reaches the rear wall it is reflected up and around until it travels toward the opening. A Froude number is also calculated for the returning current. The returning gravity current velocity is, $v_e = (2L+2h_1/3)/t_{out}$, where t_{out} is the time from opening to the time the reversed current returns to the opening wall. The $2h_1/3$ factor is used to account for the length the current travels up the wall opposite the opening and compares well with video observation of the distance the current stretches up the wall. Using this factor results in v^* being nearly equal for both the entering and exiting currents. For the full opening condition it is not possible to determine the leading edge of the returning current because there is no restriction on the exiting flow.

In Table 3.1 β, t_{in}, v, v^*, and Re for the entering wave are given in columns 1 thru 5 respectively. For the exiting current, t_{out}, v_e, and v_e^*, are given in columns 6 thru 8 respectively. Looking at the v^* values given in Table 3.1 it can be seen that the value is constant, differing only with geometry, over the range of β investigated. Figure 3.6 is a plot of v^* versus β for the entering wave for all four opening geometries. The average values of v^* and $h^* = h_0/h_1$, are given in Table 3.2. The fully open value of 0.44 is confirmed by independent two dimensional computations presented in chapter 4. This value also compares well to the $0.47 \leq v^* \leq 0.50$[2] reported for lock exchange problems and $v^* = 0.5$ derived in section 2. The experimental $v^* = 0.44$ is lower than the $v^* = 0.5$ derived in section 3.2, due to the mixing and the transient flow in this compartment. The average v^* values for the slot, door and window decrease as the mixing increases. As the slot, door, or window size relative to h_1 decreases $v^* \to 0$ from the Table 3.2 values. Similarly, increases in relative size cause $v^* \to 0.44$. As the aspect ratio (L/h_1) increases this limit may approach 0.50.

The error bars shown in Fig. 3.6 were calculated by compounding the errors for each parameter in Eqs. 1, 7 and 10. The large relative error bounds for the window opening are a result of the reduction in v^*. The absolute error remains unchanged. The nondimensional height of the entering gravity current head, h^* in Table 3.2, is based on visual observation from the video recordings of the experiments. The grid on the compartment seen in Figs. 3.4 and 3.5 is used to determine the heights. The grid lines are 25 mm apart and can be visually divided into four equal parts giving an accuracy of $\pm h_1/24$.

The average head height of the gravity current is viewed over the distance $3L/4$ to L to reduce any effects caused by the opening. The average head height did not change over the length of $3L/4$ to L, within the accuracy of the measurement. The fully open head height, $h^* = 0.5$ is consistent with the Eq. 5. The decreasing nondimensional head height with increasing mixing is also consistent with the $h^*<0.5$ suggested by Benjamin[1] for dissipative flows.

The Reynolds number shown in column 5 of Table 3.1 ranged from $939<Re<13407$. The Reynolds number is defined as:

$$Re = \frac{vh_0}{\nu}. \tag{3.10}$$

In order to compare with other gravity current results, h_0 is used in this definition. Over this range of Reynolds number the nondimensional velocities were found to be constant for each opening geometry. This Reynolds number independence is consistent with the results of Keulegan[12] and Barr[13] who found that the nondimensional velocity was strongly dependent on the Reynolds number for $Re < O(10^3)$ and independent for large Re.

Abraham and Vreugdenhil[14] indicate a slight increase in the nondimensional velocity for large Re. Simpson and Britter[15] indicate that the nondimensional velocity is either independent or only slightly dependent on the Reynolds number for $Re > O(10^3)$. For a 3m compartment fire, which is a candidate for a backdraft, the expected Reynolds number range would be $5 \times 10^3 < Re < 5 \times 10^4$. The independence suggested in the literature, and shown in Fig 3.6 and Table 3.1 for $10^3 < Re < 10^4$ indicates that the salt water results are directly applicable to typical fire compartments.

Table 3.1 Summary of the salt water modeling results for the entering and exiting current. Results are shown for all four opening: full, slot, door, and window.

$\beta=\dfrac{(\rho_o-\rho_i)}{\rho_i}$	Entering Current					Exiting Current		
	t_{in} (s)	v (m/s)	$v^*=\dfrac{v}{\sqrt{\beta gh_1}}$	$Re=\dfrac{vh_o}{v}$		t_{out} (s)	v (m/s)	$v^*=\dfrac{v_E}{\sqrt{\beta gh_1}}$
Full Opening								
0.010	5.80	0.053	0.43	3999				
0.018	4.20	0.073	0.44	5523		No Clearly defined exiting current was observed		
0.040	2.80	0.109	0.45	8284				
0.070	2.23	0.137	0.42	10401				
0.101	1.73	0.176	0.45	13407				
Slot Opening								
0.003	14.23	0.021	0.32	1239		32.53	0.02	0.33
0.005	10.43	0.029	0.34	1690		24.63	0.03	0.33
0.009	8.67	0.035	0.30	2033		20.47	0.03	0.30
0.010	8.57	0.036	0.29	2057		19.50	0.04	0.30
0.012	7.50	0.041	0.30	2350		17.87	0.04	0.30
0.022	5.70	0.053	0.30	3093		12.83	0.06	0.31
0.024	5.17	0.059	0.31	3410		12.13	0.06	0.31
0.030	4.43	0.069	0.32	3977		10.33	0.07	0.33
0.043	3.73	0.082	0.32	4726		9.27	0.08	0.30
0.043	3.67	0.083	0.33	4803		8.80	0.08	0.32
0.045	3.53	0.086	0.33	4990		8.60	0.08	0.32
0.050	3.33	0.092	0.33	5294		8.03	0.09	0.32
0.070	3.03	0.101	0.31	5818		6.93	0.10	0.32
0.075	3.00	0.102	0.30	5876		6.77	0.11	0.31
0.080	2.67	0.114	0.33	6602		6.40	0.11	0.32
0.090	2.50	0.122	0.33	7051		5.93	0.12	0.33
0.100	2.47	0.123	0.32	7137		5.73	0.12	0.32
Door Opening								
0.012	6.60	0.046	0.35	2319		16.37	0.04	0.32
0.026	4.50	0.068	0.34	3402		10.67	0.07	0.34
0.040	3.50	0.087	0.36	4374		8.73	0.08	0.33
0.055	2.97	0.103	0.36	5154		7.50	0.09	0.33
0.070	2.80	0.109	0.34	5467		6.57	0.11	0.33
0.085	2.43	0.125	0.35	6300		5.97	0.12	0.33
0.100	2.27	0.134	0.35	6744		5.50	0.13	0.33
Window Opening								
0.005	14.33	0.021	0.25	939		33.23	0.02	0.25
0.010	11.93	0.026	0.21	1128		27.63	0.03	0.21
0.025	7.00	0.044	0.23	1922		16.37	0.04	0.22
0.032	6.37	0.048	0.22	2112		14.80	0.05	0.22
0.040	5.47	0.056	0.23	2459		13.13	0.05	0.22
0.055	5.10	0.060	0.21	2638		11.53	0.06	0.22
0.070	4.23	0.072	0.22	3180		10.23	0.07	0.21
0.084	3.77	0.081	0.23	3568		9.23	0.08	0.22
0.102	3.33	0.092	0.23	4040		7.97	0.09	0.23

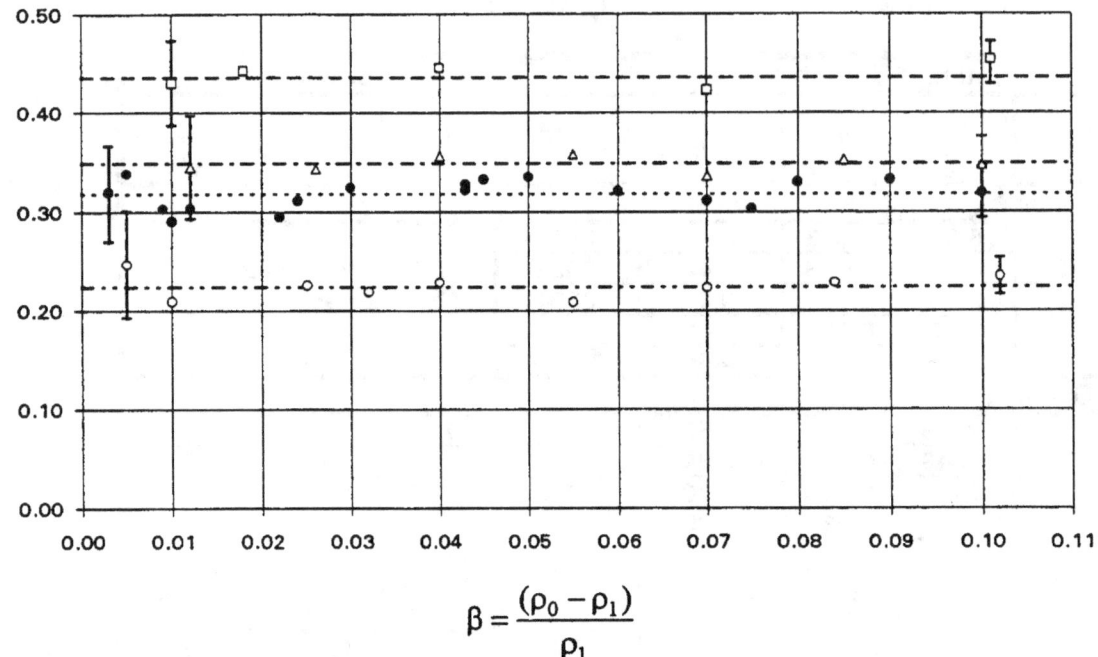

$$\beta = \frac{(\rho_0 - \rho_1)}{\rho_1}$$

Figure 3.6 - Nondimensional velocity versus density difference ratio for four opening conditions. The □, ●, o, △ represent the data for the full, slot, window and door opening. The (– – –), (········), (– ·· –), (– · – -), represent the average values for the full (0.44), slot (0.32), door (0.35) and window (0.22), geometries.

Table 3.2 - Average values for $\bar{v}^*$ and $\bar{h}^*$ calculated from the data reported in Table 3.1.

Opening	Full	Slot	Door	Window
$\bar{v}^*$	0.44	0.32	0.35	0.22
$\bar{h}^*$	0.50	0.38	0.33	0.29

3.6 Conclusions

This experimental work shows that the gravity current entering a compartment is both qualitatively and quantitatively similar to other naturally occurring gravity currents. The nondimensional velocity for the full opening compares well with the perfect fluid theory presented in section 2. The structure of the entering gravity current head for the full open condition shows a shallow mixed region riding on the current at the interface. It also has the detailed features of the steady state gravity currents, i.e. billows, lobes, and clefts. Entering currents for the slot, window, and door openings show a different gravity current structure with the mixed region occupying nearly the entire current due to the enhanced mixing near the opening. Similar detailed features appear on the these currents.

The values of v^* and h^* obtained here for a variety of opening geometries, are independent of the density difference ratio, β. The exiting gravity current is also independent of β, and has a v^* approximately equal to the entering current. These results can be applied to predict the time of ignition for a backdraft with compartment and opening geometries similar to the conditions reported here.

Additional research is necessary to investigate other compartment and opening geometries. Future work should focus on using larger aspect ratios (L/h_1) and a variety of opening ratios (side/h_1) as well as openings offset from the wall center line. More sophisticated instrumentation may also be used to measure concentrations within the current.

References - Chapter 3

[1] Benjamin, T.B., "Gravity Currents and Related Phenomenon", Journal of Fluid Mechanics, 31, 209-248, 1968.

[2] Simpson, J.E., "Gravity Currents in the Laboratory, Atmosphere, and Ocean: Annual Review of Fluid Mechanics, 14, 213-234, 1972.

[3] Steckler, K.D., "Fire Induced Flows in Corridors -- A Review of Efforts to Model Key Features" NISTIR-89-4050, National Institute of Standards and Technology, Gaithersburg, MD, 1987.

[4] Steckler, K.D., Baum, H.R., and Quintere, J.G., "Salt Water Modeling of Fire Induced Flows in Multicompartment Enclosures", 21st Symposium (Int'l) on Combustion, 143-149, The Combustion Institute, Pittsburgh, PA, 1986.

[5] Chobotov, M. V., Zukoski, E. E., and Kubota, T., "Gravity Currents With Heat Transfer" NBS-GCR-87-522, National Bureau of Standards, Gaithersburg, MD, 1987.

[6] Britter, R.E. and Simpson, J.E., "Experiments on the Dynamics of a Gravity Current Head", Journal of Fluid Mechanics, 88, 223-240, 1978.

[7] Kármán, T. von, "The Engineer Grapples with Nonlinear Problems", Bulletin of American Mathematics Society, 46, 615, 1940.

[8] Breidenthal, R. "Structure in Turbulent Mixing Layers and Wakes Using a Chemical Reaction", Journal of Fluid Mechanics, 109, 1-24, 1981.

[9] Simpson, J.E., "Effects of the Lower Boundary on the Head of a Gravity Current", Journal of Fluid Mechanics, 53, 759-768, 1972.

[10] Simpson, J. E., "A Comparison Between Laboratory and Atmospheric Density Currents", Quarterly Journal of the Royal Meteorological Society, 95, 758-765, 1969.

[11] Thorpe, S. A., "Experiments on Instability and Turbulence in a Stratified shear Flow", Journal of Fluid Mechanics, 61, 731-751, 1973.

[12] Keukeganm G.H., "An Experimental Study of the Motion of Saline Water from Locks into Fresh Water Channels", National Bureau of Standards Report Number 5168, 1957.

[13] Barr, D.I.H., "Densimetric Exchange Flow in Rectangular Channels. III. Large Scale Experiments", Houille Blanche, 22, 619-631, 1967.

[14] Abraham, G. and Vreugdenhill, C.B., "Discontinuities in Stratified Flow", Journal of Hydraulic Research, 9, 292-308, 1971.

[15] Simpson, J.E., Briter, R.E., "The Dynamics of the Gravity Current Advancing Over a Horizontal Surface", Journal of Fluid Mechanics, 94, 477-495, 1979.

CHAPTER 4

NUMERICAL AND EXPERIMENTAL GRAVITY CURRENTS RELATED TO BACKDRAFTS

4.1 Introduction

When a fire occurs in a closed compartment where the only ventilation is due to leakage the fire can become limited by the available oxygen and produce large amounts of unburned fuel. If the leakage rate is low enough the fire may enter a smoldering stage. Temperatures within the compartment will be low compared with flashover temperatures but significantly higher than the ambient. The higher temperatures decrease the average compartment density below the ambient density. When the compartment is opened a density driven flow referred to as a gravity current enters the compartment. The dense ambient air pours through the opening mixing with the hot compartment gas as the current travels across the floor. If the fuel concentrations are high enough and the gravity current comes in contact with and ignition source, a backdraft will rip through the compartment injuring any unsuspecting firefighters trapped in the wake of the explosion. A better understanding of gravity currents which enter the compartment is necessary to improve our knowledge of the backdraft phenomena.

A large body of research is available on the subject of gravity currents.[1,2] Much of this research involves laboratory scale salt water models. Typically, salt water models of gravity currents do not provide complete answers to gravity current questions. Some important detail is lost when a salt water model is used including the large vortical structure typically seen in shear flow problems.

Here, a two-dimensional numerical simulation is used to analyze the gravity current as it enters a fire compartment. The results of the two dimensional computations provide valuable insight into the detailed structure of the gravity current. The results of the numerical simulation are validated by comparison with the results from two different experiments. In the first set of experiments, a salt water model with two fluids of different density is used to visualize the flow into the compartment as described in chapter 3. The second set of experiments uses a large scale, 1.2m x 1.2m x 2.4m, compartment to produce backdrafts in the laboratory, see chapter 2. Data collected in the highly instrumented backdraft experiments on the entering gravity current are compared with both the salt water and numerical results.

4.2 Numerical Modeling

Consider the buoyancy-driven flow induced by the interaction of salt water and fresh water initially separated by a vertical interface. The equations of motion for this incompressible, isothermal mixture are:

$$\mathbf{div}(\mathbf{u}) = 0, \tag{4.1}$$

$$\partial(\rho c)/\partial t + \mathbf{div}(\rho c \mathbf{u}) = \mathbf{div}(\rho D \nabla c), \tag{4.2}$$

$$\rho(\partial \mathbf{u}/\partial t + \mathbf{u}\nabla\mathbf{u}) + \nabla p - \rho \mathbf{g} = \rho \nu \nabla^2 \mathbf{u}, \tag{4.3}$$

where c is the salt water mass fraction (defined as the ratio of the mass of salt water to the total mass of fluid in a given volume element) $\mathbf{u}$, is the velocity vector, p is the pressure, $\mathbf{g}$ is the acceleration of gravity, ν is the kinematic viscosity, and D is the diffusion coefficient. The latter two quantities will be assumed to be constant. The density of the mixture may be expressed as $\rho = \rho_1(1+\beta\tilde{\rho})$, where $\beta = (\rho_0 - \rho_1)/\rho_1$; and ρ_1 and ρ_0 are the densities of the fresh and salt water, respectively. In terms of β and $\tilde{\rho}$, the mass fraction c may be written as:

$$c = \tilde{\rho}(1+\beta)/(1+\beta\tilde{\rho}). \qquad (4.4)$$

We are interested here in the motion of the fluid mixture in a two dimensional polygonal configuration, consisting of a small chamber initially filled with a fresh water, separated from the salt water outside by a vertical interface. Equations (4.1)-(3) are solved numerically in nondimensional form using finite differences. The Reynolds and Schmidt numbers that result from the scaling are given by $Re = v_c h_1/\nu$ and $Sc = \nu/D$ where v_c is a characteristic velocity related to the Froude scaling $v_c = \sqrt{\beta g h_1}$, and h_1 is the height of the enclosure. An alternating direction implicit numerical scheme was used to solve the above equations. Details of the numerical method may be found in Ref. 3. For the comparisons with the salt water and backdraft experiments shown here, computations were performed on the IBM RISC System/6000 Model 550 of the Mathematics Laboratory and the Building and Fire Research Laboratory at NIST. A typical computation required between 20 and 80 megabytes of memory and 5 to 25 hours of CPU time. The resolution of the computational grid determines the maximum Reynolds number for a given run. Roughly, this maximum value scales as K^2, where K is the number of grid cells in the direction of the length scale h_1. The largest Reynolds number reported here is 50,000, and this simulation required a grid of dimension 1024×256.

4.3 Salt Water Experimental Apparatus and Procedures

Salt water experiments were conducted by placing an acrylic chamber within a larger glass tank. The chamber was constructed from 6 mm thick acrylic with interior dimensions of 0.15 m wide, 0.30 m long and 0.15 m high. Figure 4.1 shows the plan and elevation views along with the opening geometries for the chamber. Two two-dimensional opening geometries were used, as seen in Fig. 4.1, the cross hatched area indicates the opening. The end opening was covered with a vertical sliding partition that was removed to start the experiment. The large tank, 0.3 m wide, 0.6 m long, and 0.45 m deep, contained a dense water and salt solution ranging in density from 1.003 kg/m^3 to 1.101 kg/m^3.

The solution in the chamber was regular tap water with a pH of 6.8 and a density of 1.000 kg/m^3. A small amount of phenolphthalein ($>2 \times 10^{-4}$M) was added to the chamber. Phenolphthalein, a common pH indicator, was used to visualize the gravity current. When phenolphthalein mixes with a base, in this case sodium hydroxide crystals were added to the large tank to raise the pH to 11.7, the product of the reaction is red. This reaction is believed to be diffusion limited. The red product is strongly visible even in dilute concentrations. Turbulence is unaffected by the reaction since there is little surface tension, buoyancy, or heat release produced by the reaction. Unlike passive scalar techniques, such as dye, this chemical reaction is a much better indicator of the mixing within the gravity current. A formal discussion of this technique is given by Breidenthal[4].

Once the two solutions were prepared, specific gravity, temperature, and pH were recorded. The chamber was then lowered into the tank and the partition on the chamber was removed within 120 s to avoid leakage effects. Within 0.1s the partition was completely clear of the opening. The gravity current was recorded using a high 8 mm video camcorder at 30 frames per second. A more complete discussion of the apparatus and procedures used in the salt water modeling can be found in chapter 3.

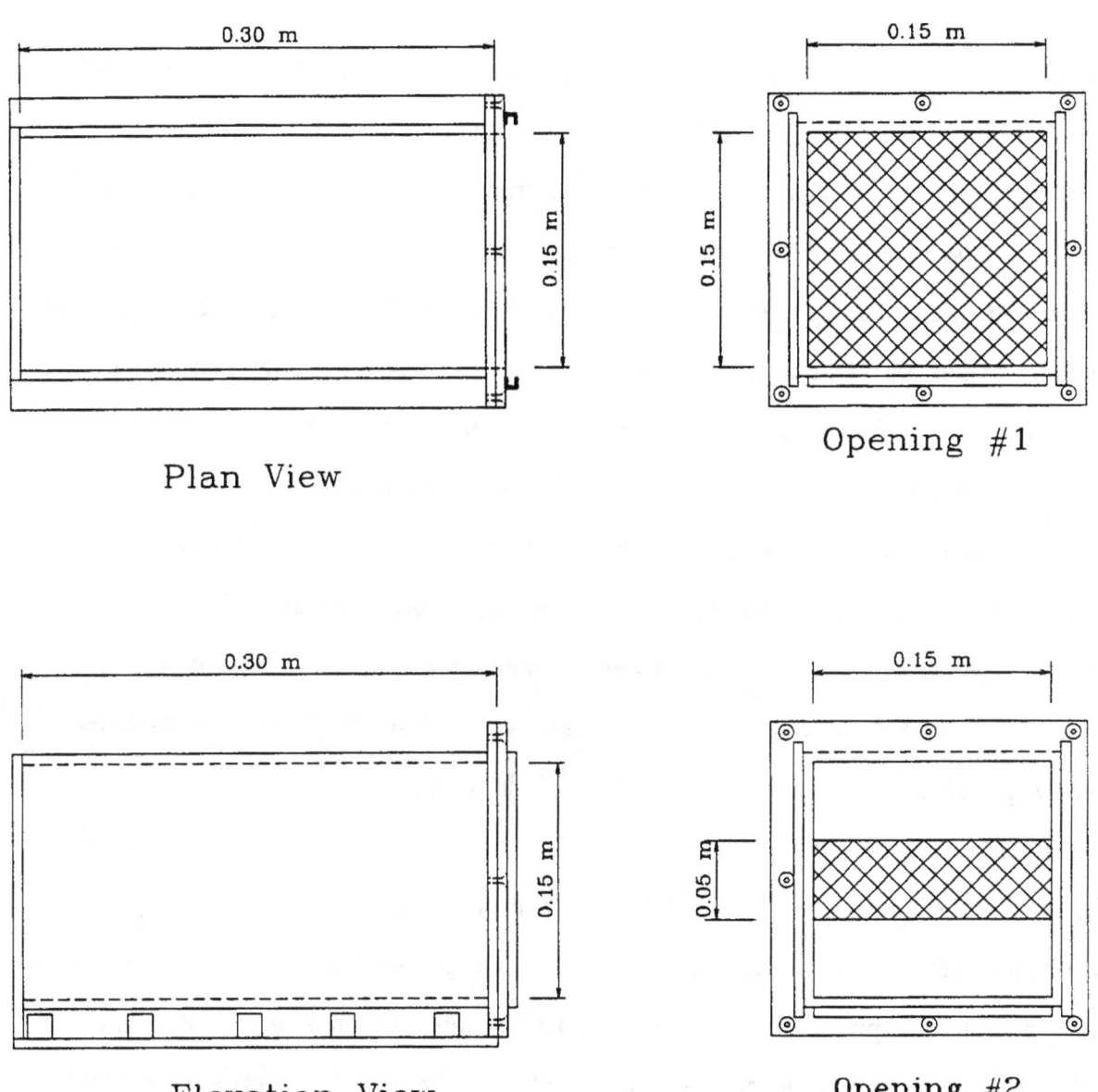

Figure 4.1 - Sketch of salt water chamber showing the elevation and plan views as well as opening geometries.

4.4 Backdraft Experimental Apparatus

A second series of experiments were conducted using a half scale compartment filled with hot gases from a methane fueled gas burner. Figure 4.2 is a sketch of the compartment. In one of the short walls, a 0.4 m high, 1.1 m wide opening was centered vertically. A 0.3 m square gas burner, with a horizontal surface 0.3 m off the floor and centered horizontally along the wall opposite the opening, provided the initial fire within the compartment. Gas flow to the burner varied from 70 to 200 kW of technical (98% pure) methane. The burner was ignited with an electric arc located 50 mm above the burner and centered along the burner edge facing the room (SPK2). The backdrafts were ignited using either SPK2 or another spark generator placed 0.15 m above and centered over the burner surface (SPK1). The sparks were generated using a 10,000 VDC transformer and two 3 mm diameter 308 stainless steel electrodes 5 mm apart.

A computer controlled hatch, hinged at the bottom, covered the slot until a predetermined opening time was reached. In the slot, six bidirectional probes were installed in a vertical rake. The probes were 17 mm in diameter and designed in accordance with the guidelines given by McCaffrey and Heskestad.[5] The probes were evenly spaced, 65 mm apart centered horizontally in the opening and numbered sequentially from the top down. The top (#1) and bottom (#6) probes were 43 mm from the soffit and sill, respectively. The bidirectional probes were oriented horizontally to avoid incorrect readings due to buoyancy effects. The pressure differential was measured using a differential pressure transducer which had a calibrated range of ± 25 Pa. The response time of the transducer was 30µs. Readings were recorded approximately 50 times a second and a 11 point smoothing routine was applied to the data. Bare bead thermocouples made from 0.5 mm type K wire, with an average bead diameter of 1.1 mm were placed at each probe to measure the temperature as required for density calculations. No correction was applied to the thermocouple data. Aspirated thermocouples placed in close proximity to the probes indicated that a radiation correction was unnecessary. The response time of these thermocouples was adequate for the gravity current, but much too slow to characterize backdraft velocities. Additional details can be found in the references 1 and 8.

In each experiment, the gas burner was ignited in the closed room. The fire burned until the burner could no longer support combustion due to a lack of oxygen. The burner was then left on to allow unburned fuel to accumulate. At a predetermined time, the burner was turned off and 5 s later the hatch was opened to allow the gravity current to enter. In some of the experiments, the spark ignitor at the burner edge (SPK2) was activated continuously and when the gravity current reached the spark a backdraft occurred. In other experiments, SPK2 was not used and the spark above the burner (SPK1) was activated at opening. When the gravity current reached SPK1, ignition of the backdraft occurred. A video recording of each experiment was used to determine event timing. Video tape and computer data times are synchronized using a computer controlled light, visible in the video frame, which turned on when the hatch was activated.

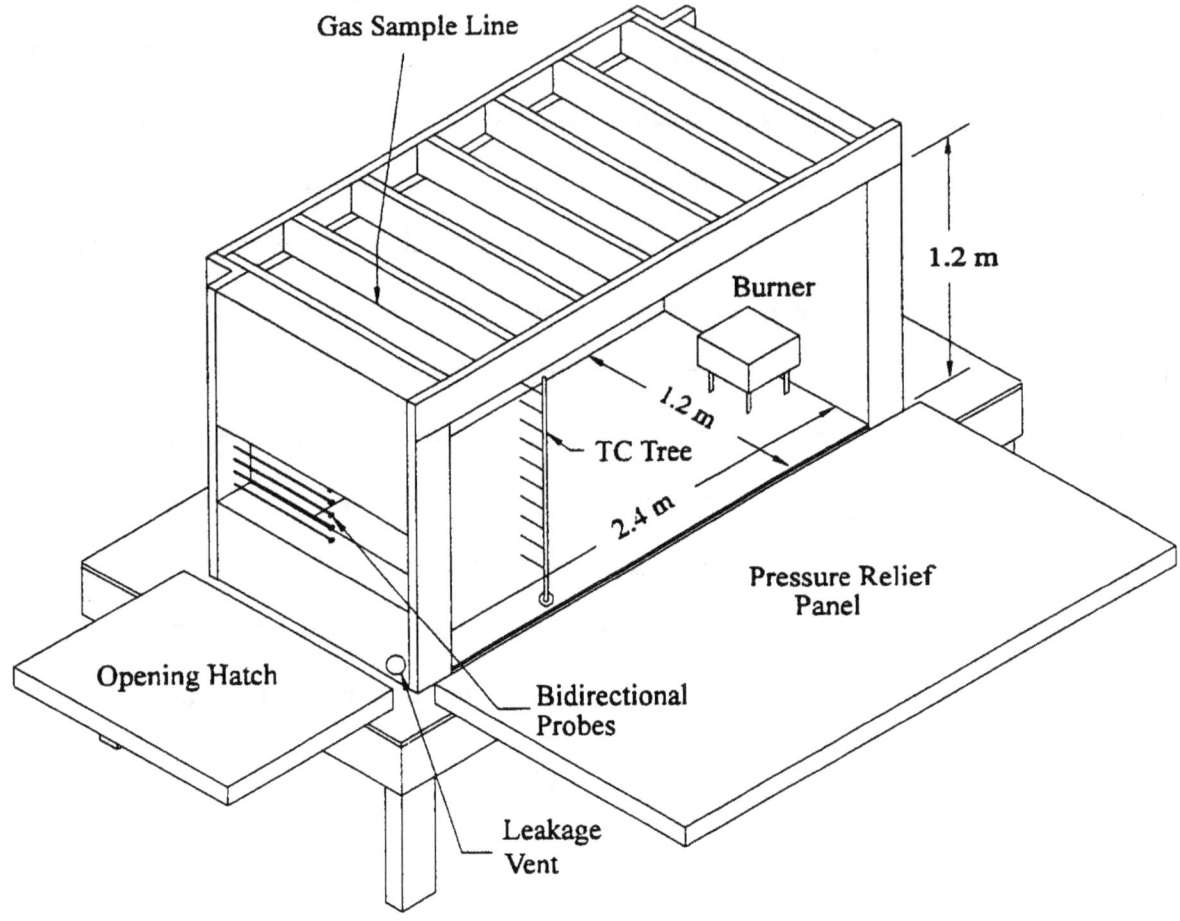

Figure 4.2 - Sketch of the half scale backdraft compartment showing important features of the apparatus.

4.5 Qualitative Results

Figure 4.3a shows the density field from the numerical simulation for the fully open condition. The black color represents the compartment fluid and the lightest gray color is the ambient fluid. Looking at Fig. 4.3a the gravity current can be divided into two regions: region 1 is purely ambient, cold, fluid which is moving toward the head of the gravity current and region 2 is the mixed layer along the shear interface which consists of hot, fuel rich, compartment fluid and rolled up within long coherent structure of cold, oxygen rich ambient fluid.

Figure 4.3b is a photograph from the salt water experiments for the fully open case at approximately the same location as Fig. 4.3a. The photograph closely resembles the numerical simulation results. Region 1 is the clear salt water which indicates that no phenolphthalein has reached that portion of the gravity current. The mixed layer, region 2, is the gray (red) area along the shear interface. The large vortical structure seen in the numerical simulation is not visible here because the photograph of the salt water experiment is an integral along a line of sight across the entire width of the chamber and the details of the structure are lost. Experiments by Simpson[6] show similar large vortices when vertical slit lighting and fluorescent dye are used to illuminate a more two dimensional image of a steady state gravity current.

Figure 4.4a shows the computed density profile for the $h_1/3$ centered slot opening condition. The structure of the gravity current is significantly different from the full opening case. The current can not be divided up into two distinct regions as in fully open case. Large vortices make up entire gravity current for the slot opening. The coherent structure making up the gravity current indicates that large scale mixing is occurring. The increased mixing is caused by the $h_1/3$ centered slot opening acting as a rearward facing step which is a well known source of large vortical structures.

Figure 4.4b is a photograph of the salt water experiment with the $h_1/3$ centered slot opening showing the gravity current in approximately the same location as Fig. 4.4a. The large
scale mixing predicted by the computations is seen as the dark gray color throughout the gravity current. From these results, it can seen that the simple two region model used for steady state gravity currents works for the full opening but cannot be applied to the slot opening condition.

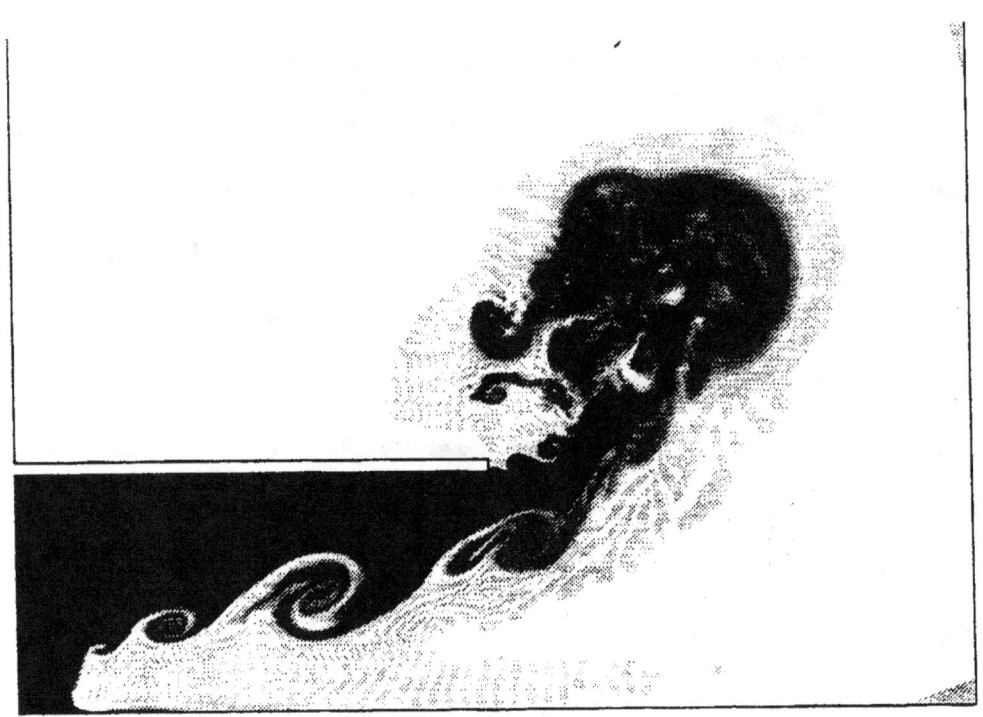

Figure 4-3a - Density profile for the full opening case. Shown here after 4.0 nondimensional time units. Reynolds number is 20,000.

Figure 4.3b - Photograph of the gravity current approximately 3L/4 into the compartment for the fully open condition. The grid shown on the model is 25 mm squares, $\beta = 0.018$.

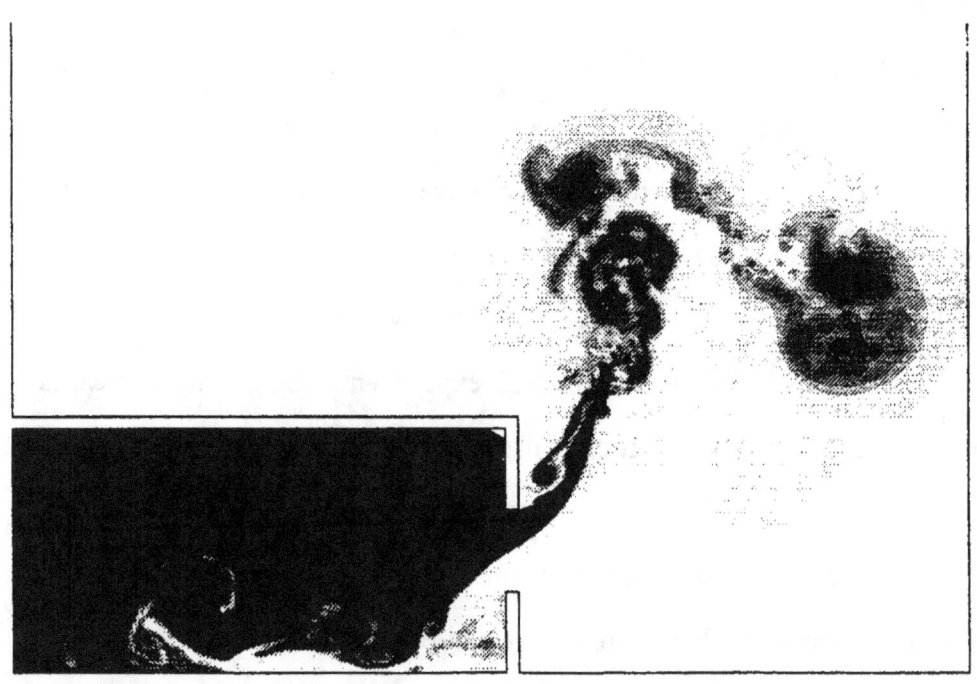

Figure 4.4a - Density profile for the $h_1/3$ centered slot opening case. Shown here after 6.8 nondimensional time units. Reynolds number is 20,000

Figure 4.4b - Photograph of the gravity current approximately 3L/4 into the compartment for $h_1/3$ centered slot opening, $\beta = 0.024$.

4.6 Quantitative Results

To quantitatively compare the numerical simulation with the experiments, the gravity current transit time is used. The transit time is the time required for the leading edge of the gravity current to reach the wall opposite the opening. For the numerical simulation the transit time was determined when the density at the rear wall changed by 10%. In the salt water experiments, the transit time was taken from video recordings of the gravity current. In the half scale backdraft experiments, the gravity current is not directly measured. The time to ignition is used to approximate the transit time assuming that the gravity current is ignited as soon as the current reaches an ignition source.

Figures 4.5a and b show the transit times versus relative density difference, β, for the full and slot opening conditions, respectively. Figure 4.5a shows the excellent agreement between the salt water model and the numerical simulation for the fully open condition. For the $h_1/3$ centered slot opening, Fig. 4.5b, the agreement is also good but computed values are consistently longer than the values measured in the salt water experiments.

Figure 4.6 shows the comparison between the time to ignition from the half scale experiments and the numerical results versus β. The □ indicates that ignition occurred at the SPK2 and the o indicates that ignition occurred at the spark above the burner, SPK1. Although the data are somewhat scattered, the comparison indicates the expected trend. For the burner spark ignition the results are excellent. The longer times seen for the spark above the burner maybe due to the burner's effect on the flow field. The two data points which occurred earlier and the three which occurred later than expected may demonstrate the dependence of the ignition phenomenon on the steep concentration gradients shown in Fig. 4.4a.

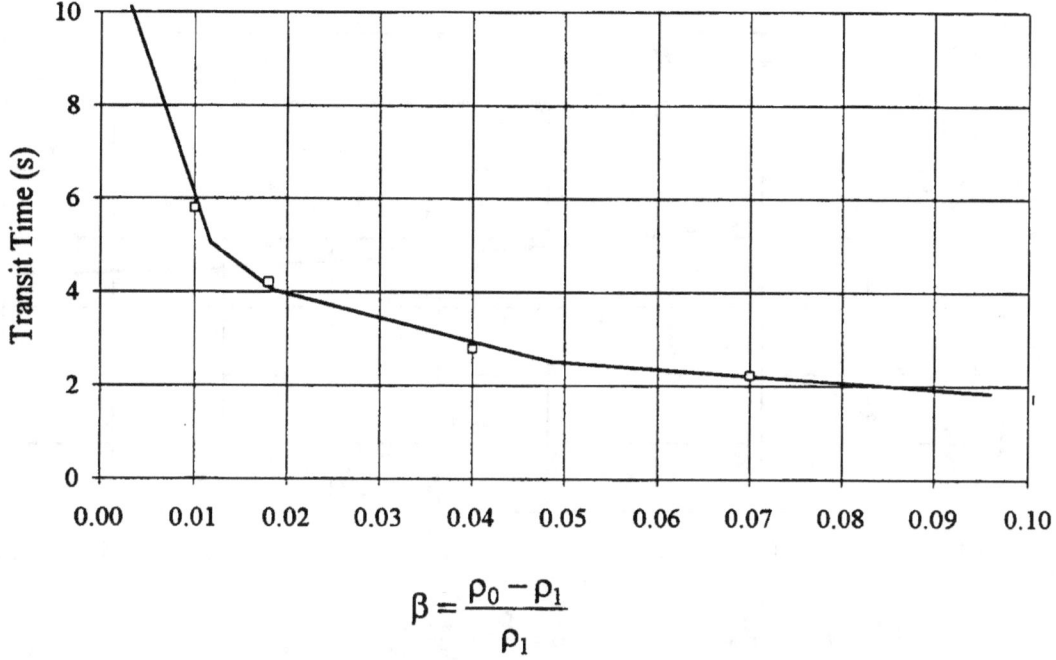

Figure 4.5a - Plot of the transit time versus density difference, β, for the full open condition comparing the numerical simulation (———) with salt water modeling results (□).

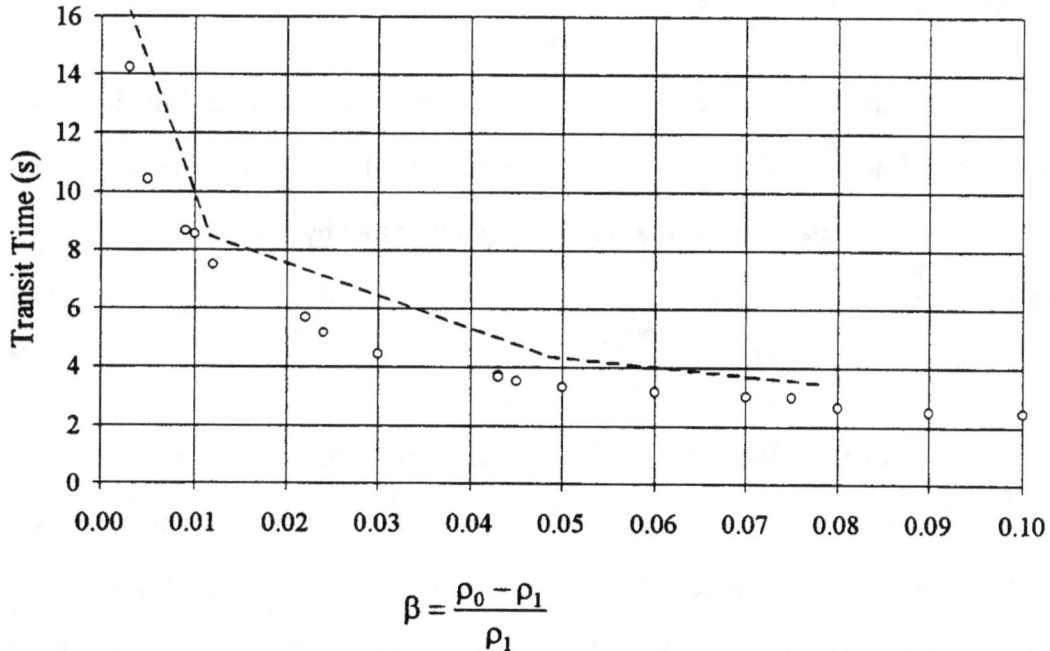

Figure 4.5b - Plot of the transit time versus density difference, β, for the $h_1/3$ centered slot opening condition. Compares the numerical simulation (-----) with salt water modeling results (o).

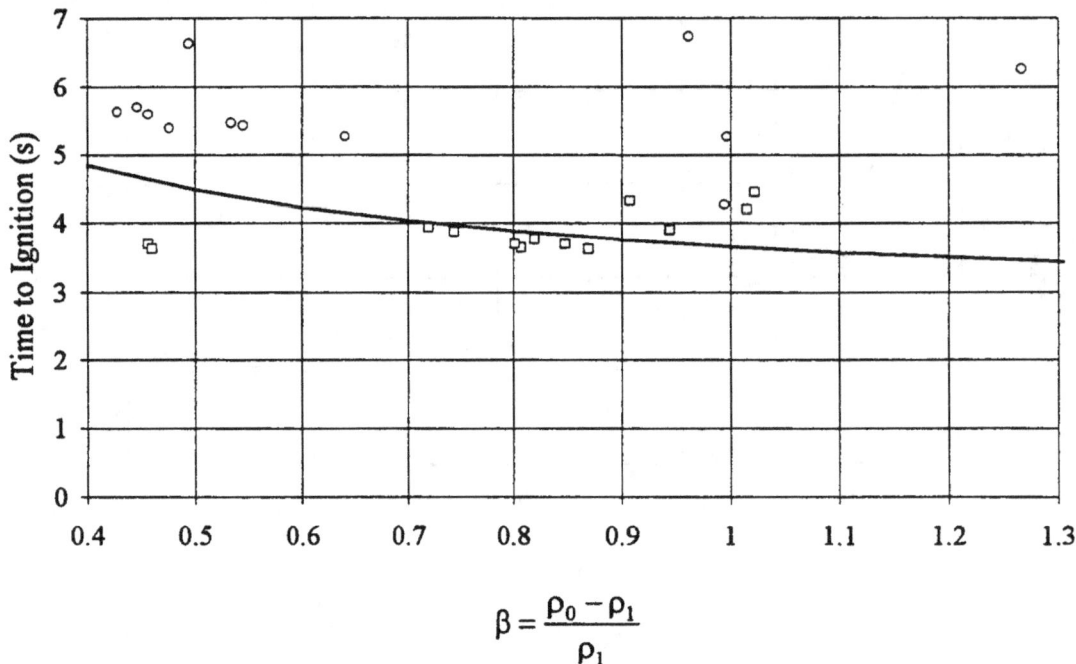

Figure 4.6 - Plot of the time to ignition for the half scale backdraft experiments and numerical simulation transit time versus density difference, β. The ($\square$) indicates ignition at the burner spark, (o) indicates ignition at the spark above the burner, and (———) indicates the computational results.

A more direct comparison between the numerical simulation and the half scale backdraft experiments can be made using the bidirectional probe measurements. The probe velocity is calculated using the relationship suggested by McCaffrey and Heskestad:[9]

$$v = C(Re)\sqrt{\frac{2\Delta p}{\rho}} \qquad (4.5)$$

where v is the velocity at the probe, Δp is the pressure difference measured across the probe, ρ is the local density, and C(Re) is an empirical calibration constant which is a function of the Reynolds number. Typically, C(Re) is taken as a constant of 0.926 which gives a maximum error of 7% for Re > 520[9]. In this case, the Re was smaller, Re ~ 300, so the following calibration constant was required:

$$\frac{1}{C(Re)} = 1.533 - 1.366 \times 10^{-3} Re + 1.688 \times 10^{-6} Re^2 - 9.706 \times 10^{-10} Re^3 + 2.55 \times 10^{-13} Re^4 - 2.484 \times 10^{-17} Re^5 \quad (4.6)$$

Equations (4.5 and 4.6) are applicable for $40 < Re < 3800$. For $Re < 40$, $v < 0.07$ m/s which is considered negligible.

Figures 4.7a and b compare the velocity history for probes 1, 2, 5, & 6 with the numerical simulation for $\beta=0.52$. Negative velocities indicate flows into the compartment. The hatch was released at 0 s. At approximately 1.5 s the opening hatch strikes the table and causes excessive noise in the data. From 0 to 4 s the flow in the opening is developing. After approximately 4 s the flow can be assumed to be quasisteady. At 4 s the total mass flow into the chamber is 0.28 kg and the gravity current is approximately 3L/4 into the compartment as shown in Fig. 4.3. If ignition does not occur, the flow will slowly diminish to zero as the compartment cools. The flow into and out of the compartment which continues long after the initial gravity current has subsided is the result of the thermal energy stored in the chamber heating the incoming air and driving the flow. In one experiment where the ignition was delayed for 600s, the velocity dropped to ~0.2 m/s.

Figure 4.8 shows the vertical velocity profile in the opening at different times. The lines show the numerical simulation and the symbols are the experimental data. At 4 s the data indicate that the flow has developed into the expected in/out profile and show excellent agreement with the numerical simulation. The velocity profile shows little change at 8 s when the spark is activated and ignition occurs.

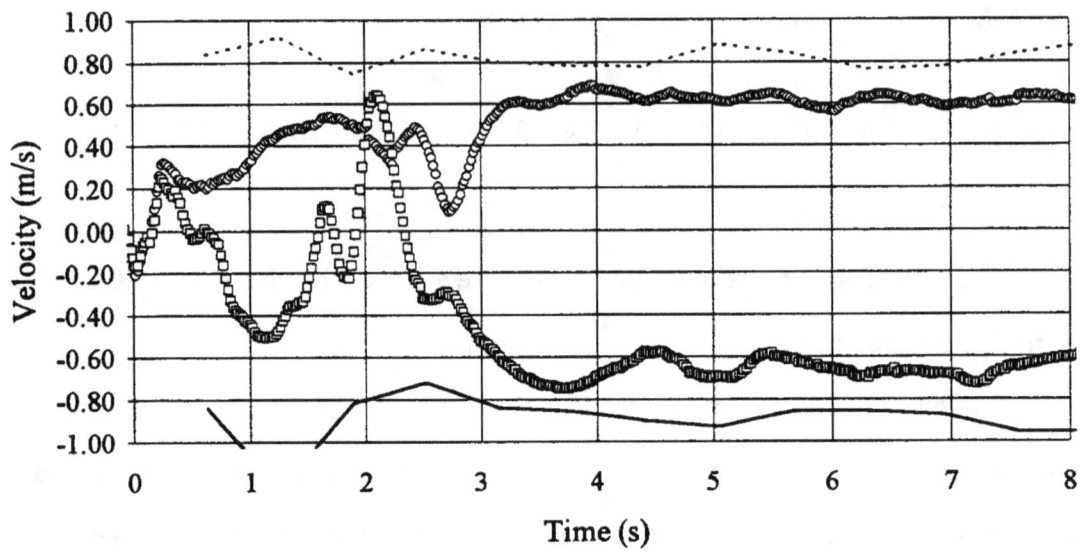

Figure 4.7a - Plot showing the velocity history for the probe #1 (o) and #6 (□)in the opening of the half scale backdraft compartment compared with the numerical simulation results shown as, probe #1 (-----) and probe #6 (———).

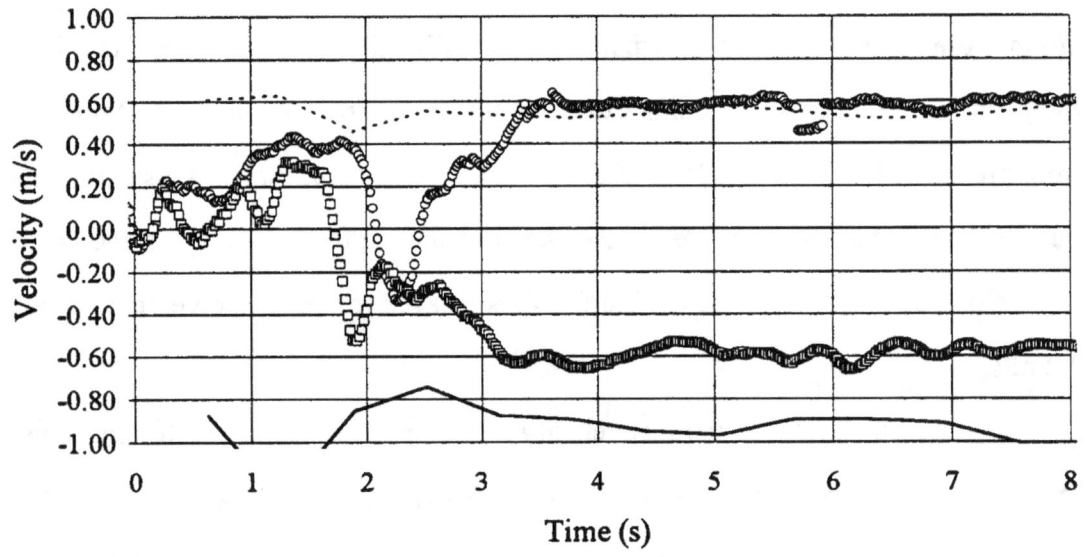

Figure 4.7b - Plot showing the velocity history for the probe 2 (o) and 5 (□) in the opening of the half scale backdraft compartment compared with the numerical simulation results shown as, probe 2 (-----) and probe 5 (———).

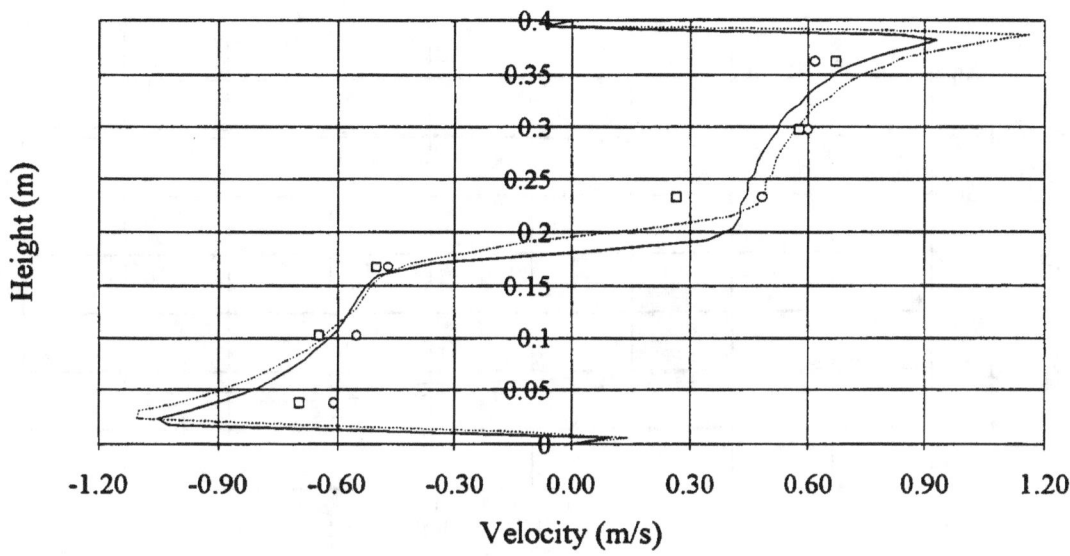

Figure 4.8 - Plot of the velocity profiles in the opening of the backdraft compartment shown at 4 s and 8 s after opening. Numerical simulation results are shown as lines and the experimental data is shown as symbols. (4 s □ -----) (8 s o ———)

Figures 4.9a and b are plots of quasisteady velocity in the opening versus β. Data are shown as discrete points, numerical simulation results are the solid lines, and potential flow results are the dashed lines[7]. There is good agreement between the experimental data and the numerical results, with the numerical result being consistently higher than the data. The potential flow results is almost a factor of 2 greater than the measured result which indicates that assumption, as expected, oversimplifies the problem. The numerical simulation is much closer to the measured values. The differences between experiments and computation are likely to be due to turbulence effects not included in the simulation and the large errors which can be expected in the measurements at this low velocity. For velocities of O(1 m/s), the experimental error is as high as ±40%.

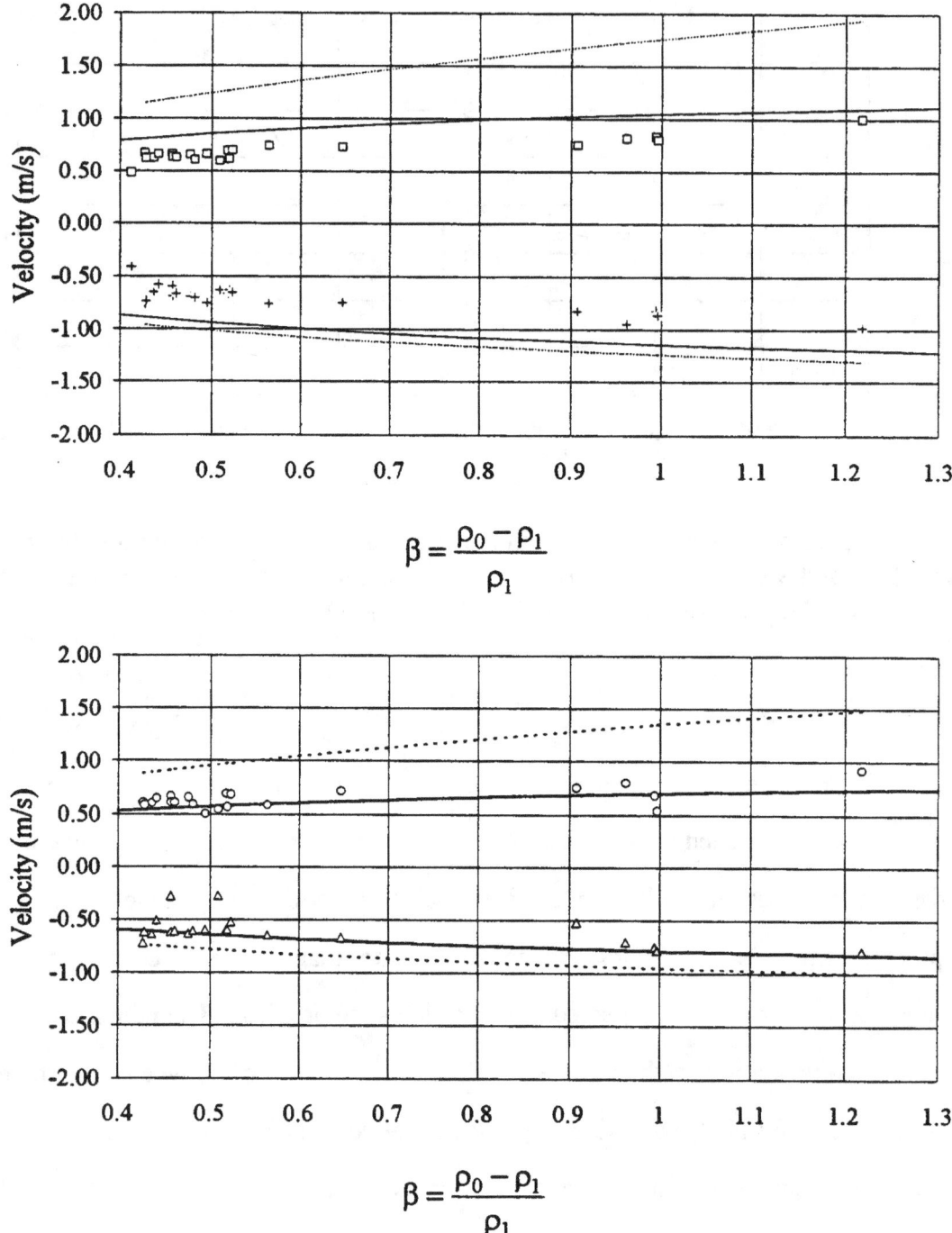

Figure 4.9a & b - Plot of the quasisteady velocity in the opening of the backdraft compartment versus density difference ratio, β. The experimental data are indicated as symbols, the numerical simulation is shown as solid lines, and the potential flow result is shown as a dashed line. (probe 1 □)(probe 2 ○)(probe 5 △)(probe 6 +)

4.7 Conclusions

The two dimensional numerical simulation presented here accurately predicts the compartment gravity current for the fully open condition. The two dimensional computational density profile shows that the main body of the current is made up ambient fluid while a mixed layer made up of large vortical structures exists along the shear interface between the two fluids. Such structure is also observed in the salt water experiments although the individual vortices are not resolved due to the limitations of the flow visualization techniques. Results of other researchers[10] indicate that large vortices do occur behind the head of a steadily propagating gravity current which are similar in appearance to the transient compartment gravity current results shown here.

For the slot opening, the results of the two dimensional simulation still compare very favorably with the experimental results. The computed density profile shows that the structure of the gravity current for the $h_1/3$ centered slot opening is significantly different from the fully open condition. For the slot opening, the entire gravity current is filled with a complex structure of large scale vortices which translates into large scale mixing. This large scale mixing for the $h_1/3$ centered slot opening results from the rearward facing step, formed by the lower edge of the slot.

Transit times predicted by the numerical simulation compare well with the salt water experiments although the computed times are slightly longer for the slot opening. Comparing the computational transit time with the time to ignition for the backdraft experiments gravity current also shows good agreement. Numerical velocity profile predictions in the opening compared well with backdraft experiment results, falling well within the experimental error bounds. Future work should focus on different compartment aspect ratios, smaller openings, and openings at the floor level. Improved flow visualization techniques including slit lighting to obtain a two dimensional image of the gravity current are recommended. Time dependent concentrations should be measured within the gravity current for comparison with computational results.

References - Chapter 4

[1] Simpson, J.E., "Gravity Currents in the Laboratory, Atmosphere, and Ocean: <u>Annual Review of Fluid Mechanics</u>, **14**, 213-234, 1972.

[2] Steckler, K.D. "Fire Induced Flows in Corridors -- A Review of Efforts to Model Key Features" <u>NISTIR-89-4050</u>, National Institute of Standards and Technology, Gaithersburg, MD, 1987.

[3] McGrattan, K.B., Rehm, R.G., Tang, H.C., and Baum, H.R., "A Boussinesq Algorithm for Buoyant Convection in Polygonal Domains" <u>NISTIR 4831</u>, National Institute of Standards and Technology, Gaithersburg, MD, 1992.

[4] Breidenthal, R. "Structure in Turbulent Mixing Layers and Wakes Using a Chemical Reaction", <u>Journal of Fluid Mechanics</u>, **109**, 1-24, 1981.

[5] McCaffrey, B. J. and Heskestad, G., "A Robust Bidirectional Low-Velocity Probe for Flame and Fire Application", <u>Combustion and Flame</u>, **26**, 125-127, 1976.

[6] Simpson, J. E. "Effects of the Lower Boundary on the Head of a Gravity Current", <u>Journal of Fluid Mechanics</u>, **53**, 759-768, 1972.

[7] Babrauskas, V. and Williamson, R.B., "Post Flashover Compartment Fires: Basis of a Theoretical Model", <u>Fire and Materials</u>, **3**, 39-53.

CHAPTER 5

QUANTITATIVE BACKDRAFT EXPERIMENTS

5.1 Introduction

The dangerous consequences of a backdraft are documented in numerous fire service publications and training manuals.[1,2,3,4] However, little research has been done in the area of backdrafts, and only recently has a scenario been presented describing the fundamental physics underlying backdraft phenomena. A backdraft is defined as a rapid deflagration following the introduction of oxygen into a compartment filled with accumulated unburned fuel.

The scenario presented here assumes a fire in a closed room. The fire heats up the room, and leakage in the bounding surfaces minimize the pressure differential. The hot layer descends over the fire as the oxygen concentration is reduced and the combustion efficiency decreases. Excess pyrolyzates accumulate in the upper layer forming a fuel rich mixture of low oxygen content. A small flame or glowing ember exists as a source of ignition. Suddenly, a new ventilation opening is provided and cold, oxygen rich, air enters the compartment and propagates across the floor as a gravity current. Large scale mixing in the gravity current provides areas within the flammable range which can ignite when they contact a source of ignition. Once ignited, a flame propagates through the compartment and drives the remaining unburned fuel out through the opening to burn outside the compartment in a spectacular fireball.

In this paper, experimental results are presented from a series of half scale experiments attempting to quantify backdraft. Experimental variables included fuel flow rate, burn time, ignition location, ignition delay time, burner height, species sample location, and opening size. This paper focuses on 17 experiments in which the opening geometry was a vertically centered, horizontal slot, in one wall; the fuel source was a 0.30 m square burner 0.30 m above the floor; and the ignition source was a spark located opposite the opening. Two different burner flow rates were used, 70 kW and 200 kW. Data collected in these experiments had two goals: 1) to characterize the conditions in the compartment prior to backdraft and 2) to quantify the severity of the deflagration.

5.2 Experimental Design & Procedures

5.2.1 Apparatus: Experiments were conducted in a special compartment designed to safely control the dangerous overpressures expected in backdrafts. The experimental apparatus dimensions were limited to half a small residential room to minimize the expected hazard and to allow the experiments to be conducted inside a 900 m³ facility. Figure 5.1 shows a schematic of the apparatus giving the internal dimensions of the compartment and the locations of the instrumentation. In order to control the overpressure hazard, one long wall was a pressure relief panel designed according to refs. 5 and 6. The interior surfaces of the compartment were lined with a 50 mm thick refractory fiber blanket installed over the gypsum wallboard on the walls and ceiling to provide the primary thermal resistance for the structure. This insulation allowed for repeated experiments without the need to rebuild the compartment. A 0.9 m high by 1.5 m wide observation window of Neoceram[7] was installed in the wall opposite the pressure relief panel.

To simulate a window or door, a 0.4 m high by 1.1 m wide opening was centered in the short wall opposite the burner, see Fig. 5.1. This opening was covered with a computer activated hatch which was opened after the fire had been burning for several minutes. A methane burner, 0.3 m square and 0.3 m high, was used in all of these experiments. The burner was placed against the wall opposite the opening, as seen in Fig.

5.1. A pilot flame was used to ignite the burner and was turned off 10 s after the start of the experiment. The primary ignition source for the backdraft was a spark ignitor located 0.45 m above the floor and centered over the top of the burner. For 3 of the experiments, the primary spark malfunctioned and the backup spark ignitor used to ignite the pilot light had to be used to ignite the backdraft. The backup spark was located 0.35 m above the floor and centered on the side of the burner facing the opening. A 10,000 volt transformer was used for each spark ignitor to produce an arc between two 3 mm diameter 308 stainless steel electrodes 5 mm apart.

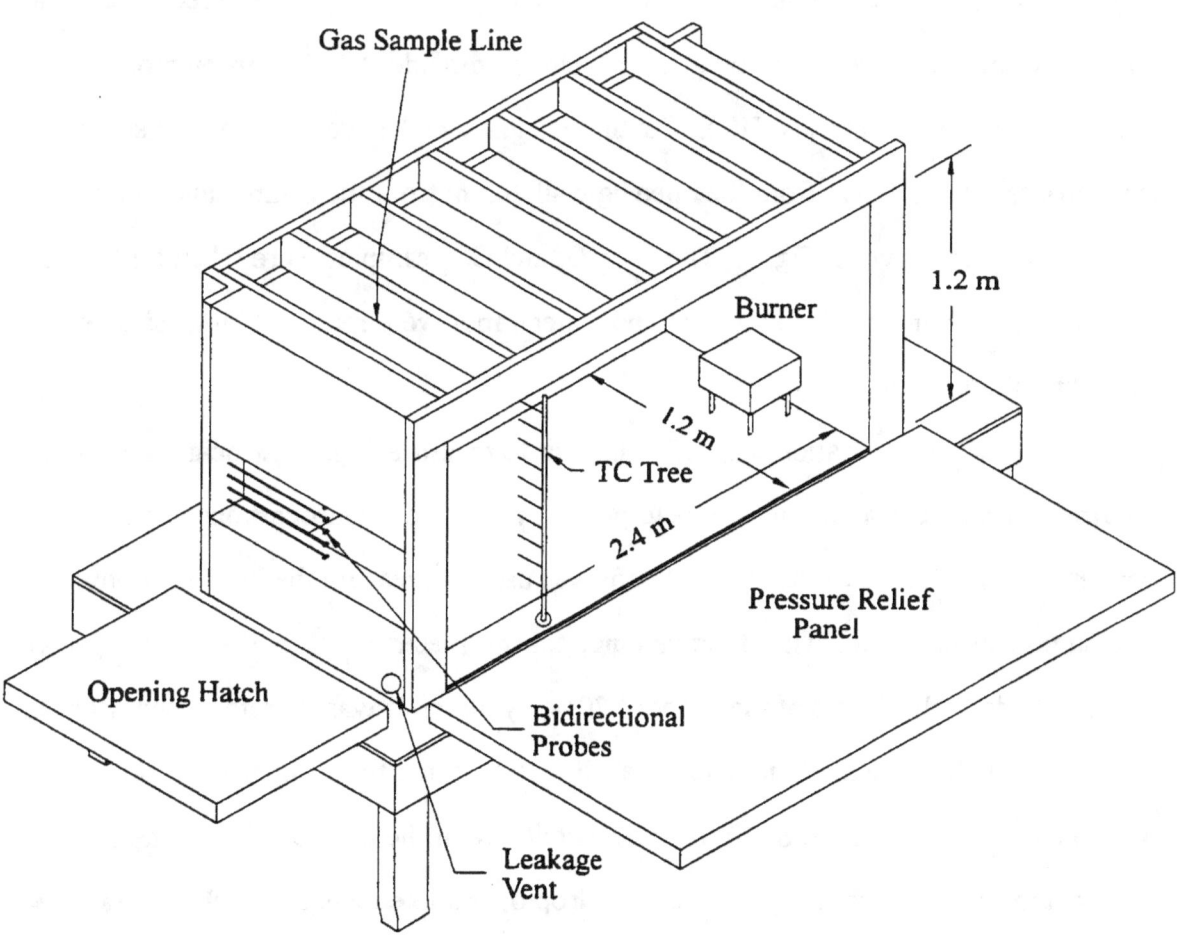

Figure 5.1 - Sketch of the half scale backdraft compartment showing important features of the apparatus.

Every effort was made to seal all construction holes to control leakage. The primary source of leakage into the compartment was found to be around the pressure relief panel and the opening hatch. Gaskets made from the refractory fiber blanket were compressed around the edges of these opening to reduce the leakage. A small 0.1 m diameter pressure relief vent was placed at the floor level to relieve the pressure from the initial burner ignition. Without this vent, a pressure rise sufficient to activate the pressure relief panel was produced. A computer controlled cover closed over this vent 15 s after ignition. Additional details of the apparatus can be found in Appendix A.

5.2.2 Species Concentration: In order to characterize the compartment conditions prior to a backdraft, the species concentration histories in the upper layer were recorded. Gas concentrations measured were: oxygen (O_2), carbon dioxide (CO_2), carbon monoxide (CO), and total hydrocarbons (HC). Continuous gas samples were taken with stainless steel probes located 0.6 m from the opening wall, 0.2 m from the ceiling, and 0.6 m from the side wall, as shown in Fig. 1. The O_2, CO, and CO_2 samples were taken through an unheated sample line in which the soot and water vapor were removed by glass fiber and desiccant filters, respectively.

The HC concentration was more difficult to measure. The hydrocarbon sample required a separate heated sample line to prevent the loss of hydrocarbons due to condensation. A flame ionization detector was used to measure the hydrocarbon (methane) concentration. The effective range for this meter was 0 to 1%. The expected range of hydrocarbons was of the order of 20%, by mass. It was therefore, necessary to dilute the sample. The dilution system was designed to mix the compartment sample with heated ambient air in a ratio of 20 to 1. The flow rate of the dilution air and the sample were determined by measuring the pressure drop over a fixed length of tubing. Sample flow rates were monitored continuously during the experiment. Typical dilution ratios would vary from 22 to 25 depending on the compartment gas temperature. A hot water jacket around the hydrocarbon sample line and dilution air line kept the gases over 60°C, well above the maximum calculated dew point of 44°C. To obtain the final species concentrations in the upper layer it was necessary to calculate the concentration of H_2O that was in the upper layer since there was no direct measurement of the H_2O. Detailed species balances were performed and can be found in Appendix B.

5.2.3 Temperatures: A vertical thermocouple tree was placed 0.6 m from the opening wall and 0.2 m from the pressure relief panel, as shown in Fig. 5.1. The thermocouples were made from 0.5 mm type K thermocouple wire with a stainless steel overbraid. The average bead diameter was 1.1 mm. The ten thermocouples were located at 0.10 m intervals, with the highest thermocouple at 0.15 m below the ceiling. The temperatures reported here are uncorrected values.

The thermal interface height history was calculated from the time dependent temperature profiles recorded from the thermocouple tree. The profiles were converted into unsteady average upper and lower layer temperatures using the method Quintiere et. al.[8] applied to steady state temperature profiles:

$$\int_0^{h_1}\left(\frac{1}{T}\right)dx = [h_1 - h_L]/T^{UL} + h_L/T^{LL}, \tag{5.1}$$

$$\int_0^{h_1} T\,dx = [h_1 - h_L]T^{UL} + h_L T^{LL}, \tag{5.2}$$

where T^{UL} and T^{LL} are the upper and lower layer temperatures, and h_1 and h_L are the heights of the compartment and the layer interface, respectively. Equation (5.1) is a mass balance and Eq. (5.2) retains the same mean temperature as in the data. Assuming that the lower layer temperature was the arithmetic average of the two lowest thermocouples allowed the upper layer temperature and thermal interface location to be calculated from Eqs. (5.1 and 5.2).

5.2.4 Compartment Pressure: The compartment pressure history was recorded using an electronic pressure transducer with an effective range of 0 to 1250 Pa. The pressure port was mounted in the stationary wall opposite the pressure relief panel at floor level. The ambient pressure reference was taken outside the building.

5.2.5 Hatch Flow: The flow in and out of the compartment after the hatch was opened was recorded using six bidirectional probes in the h/3 centered slot. The probes were located in the horizontal center of the opening and 65 mm apart. The outer probes were 43 mm from the soffit and sill. Probe velocities were calculated using the relationship given by McCaffrey and Heskestad:[9]

$$v = C(Re)\sqrt{\frac{2\Delta p}{\rho}}, \qquad (5.3)$$

where v is the velocity at the probe, Δp is the pressure difference measured across the probe, ρ is the density at the probe, and $C(Re)$ is an empirical calibration constant which is a function of the Reynolds number. Typically, $C(Re)$ can be taken as a constant at 0.926, which gives a maximum error of 7% for $Re > 520$. In these experiments $Re < 520$ was recorded and a Re correction was applied[12].

The temperature was recorded at each bidirectional probe in the opening using a bare bead 24 gauge Type K thermocouple with high temperature glass insulation. Four aspirated thermocouples, designed according to Newman et. al.,[10] were installed to correct the temperatures for radiation. Data from the aspirated thermocouples revealed that no radiation correction was required.

The mass flow was calculated by integrating the velocity and density profiles over the height of the opening as:

$$\dot{m}_l = C \int_{h_b}^{h_n} \rho v b \, dy, \qquad (5.4)$$

where $\dot{m}_l$ is mass flow in the lower part of vent, C is the flow coefficient taken as 0.68, b is width of opening, and h_b and h_n are the height of the sill, and neutral axis, respectively.

5.2.6 Data Acquisition System: Data from each sensor was recorded using a HP VECTRA 80486-33 computer with a 8 channel multifunction analog and digital input/output board. Two 32 channel analog input muliplexors were connected to this system. A total of 31 thermocouple and 17 voltage channels were used. To increase the scan rate, data was written to a RAM drive and then down loaded to a file on the hard disk immediately following the experiment. The system was capable of recording each channel 50 times a second. For experiments greater than 600 s the data was collected at a rate of 10 scan/s until 20s before opening when the rate was automatically increased to 50 scan/s. The reduced scan rate for the initial period of the experiment was done to reduce the data file size. Files for these experiments ranged in size from 4 to 10 Mbytes.

In addition to recording the data, the computer also controlled the experimental procedures using solid state relays activated by a digital input/output board also installed on the computer bus. The computer controlled systems included, experimental clock, burner pilot light, fuel flow, vent cover, spark ignitors, opening hatch, and still camera. Each system had a manual override.

5.2.7 Procedure: Before each experiment a 60 s baseline was taken to record the initial conditions. A pilot flame was ignited at the burner 5 s before the start of the experiment. At 0 s a solenoid was opened on the methane flow to the burner and the clock was reset to zero. The burner was left on for a predetermined time period. Gas flow to the burner was terminated 5 s before the hatch was opened. In some of the experiments, a spark was left on the entire time, while in other experiments the spark was not turned on until the hatch was opened. When the spark was left on, a dancing flame would often appear and consume some of the available hydrocarbons making consistent results difficult to attain. In later experiments the spark remained off until the hatch was opened. The dancing flame phenomenon is discussed in chapter 2.

5.3 Experimental Results

5.3.1 70 kW Fuel Flow Rate: The experimental parameters for the results presented in Figure 5.2 are: gas flow rate of 70 kW, burner flow time of 775 s, hatch opening at 780 s, and backdraft ignition above the burner. Figure 5.2a shows the upper layer species mass fraction histories for O_2, CO, CO_2, and HC. Idealized layer temperatures and height calculated from Eqs. (5.1 and 5.2) are shown in Fig. 5.2b. For the first 80 s, the temperature in the compartment rises as the layer within the compartment descends over the fire. The O_2 concentration is dropping as the fire consumes the available O_2 and the CO_2 concentration increases as a result of the combustion. Over the time period of 80 s to 170 s the temperature in the compartment drops as the fire is diminished. Over this same period the HC concentration is starting to rise as the O_2 and CO_2 level off. Waviness seen in the upper layer temperature history over the range of 140s to 170 s are the result of the flames pulsing before extinction. After 170 s the fire is completely out and the compartment begins to cool as seen in the exponential decay in the temperature profile. After 240 s the hydrocarbon concentration is steadily increasing. The oxygen concentration is slightly increasing as air leaks into the compartment and the CO_2 concentration declines as compartment gases are lost by leakage. At 720 s the HC analyzer became saturated and the slope of HC concentration approaches zero. At 780 s the hatch is opened and a the gravity current enters the compartment. Once the gravity current reaches the ignition source, a flame travels through the mixed region, stirs the compartment, drives combustible gases out the hatch, and culminates in a large external fireball approximately 4 m in diameter. The spikes in the temperature shown in Fig. 5.2b are caused by the wave propagation through the compartment. Gas concentrations after 780 s are unreliable due to the highly transient effects of the backdraft.

The idealized two zone approximation is compared with the temperature data from the thermocouple tree at 80 s and 780 s in Fig. 5.2c. At 80 s the compartment temperature is at its maximum and there is a substantial temperature gradient in the upper layer. The layer is located just above the burner surface. At opening, 780s, the layer is still near the top of the burner and the compartment temperature is nearly uniform vertically indicating that a one zone approximation is reasonable at this time. Figure 5.2d is a photograph from this experiment taken 1.5 s after ignition. Notice the flame burning along the top of the gravity current.

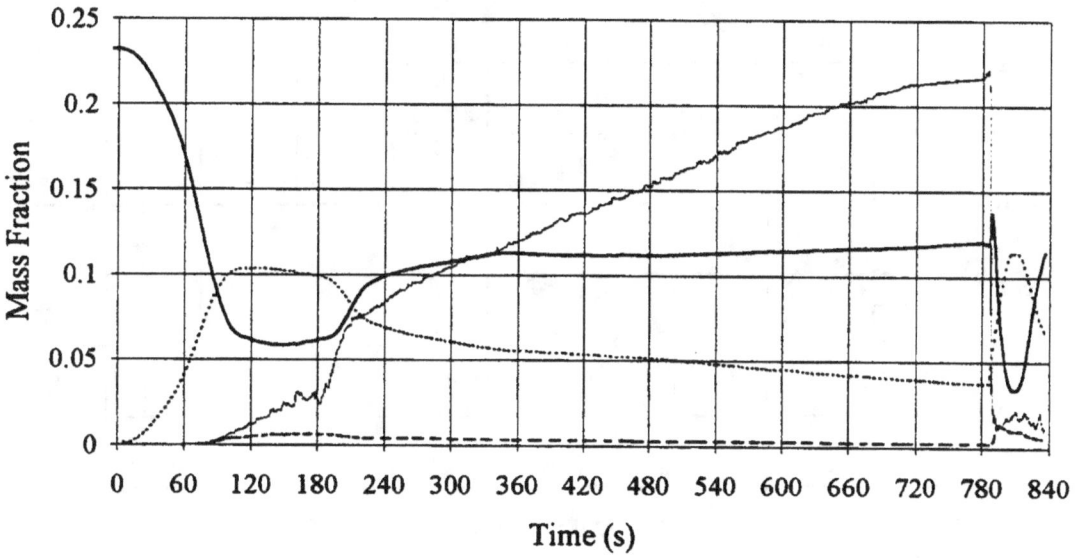

Figure 5.2a - Species concentration histories from backdraft experiment for O_2 (———), CO(– – –), CO_2 (·····), and HC (———) for the 70 kW (12th row in Table 5.1) fire source.

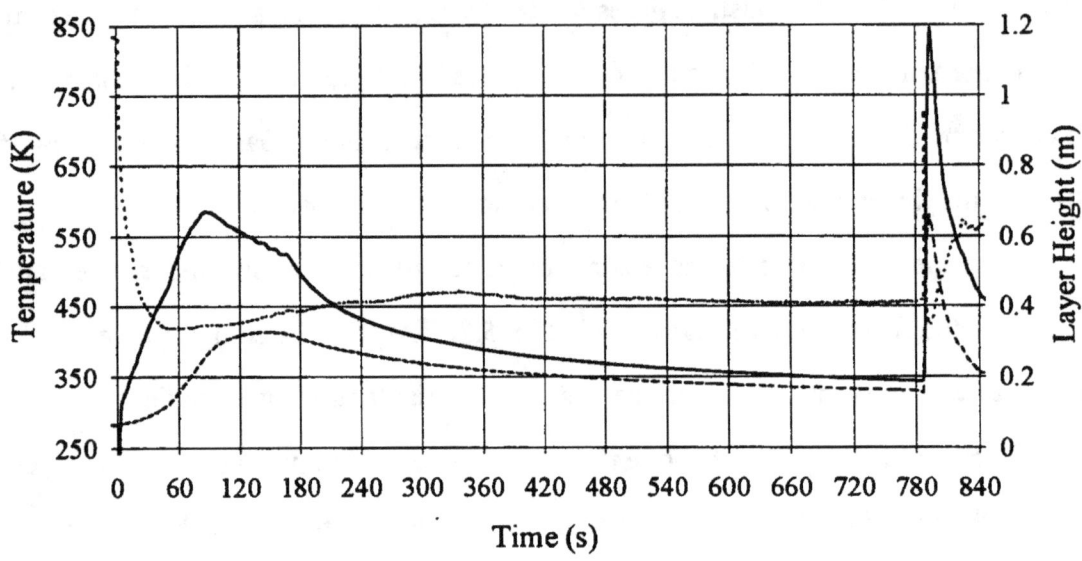

Figure 5.2b - Idealized two zone upper layer (———), lower layer (– – –), and layer height (······) histories from a 70 kW (12th row in Table 5.1) backdraft experiment.

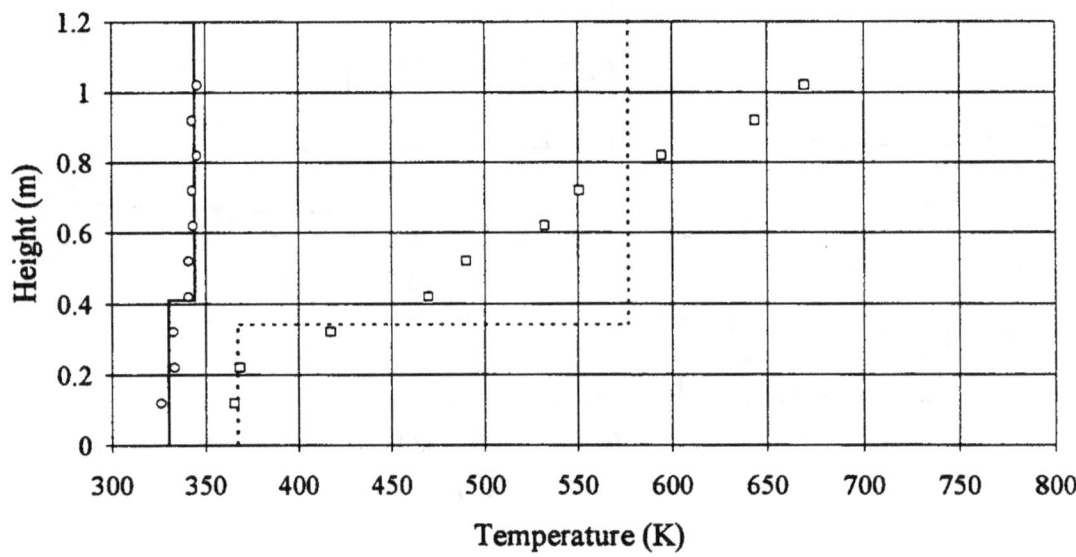

Figure 5.2c - Idealized two zone approximation compared with temperature data at 80 s (-----,□) and 780 s (———,○).

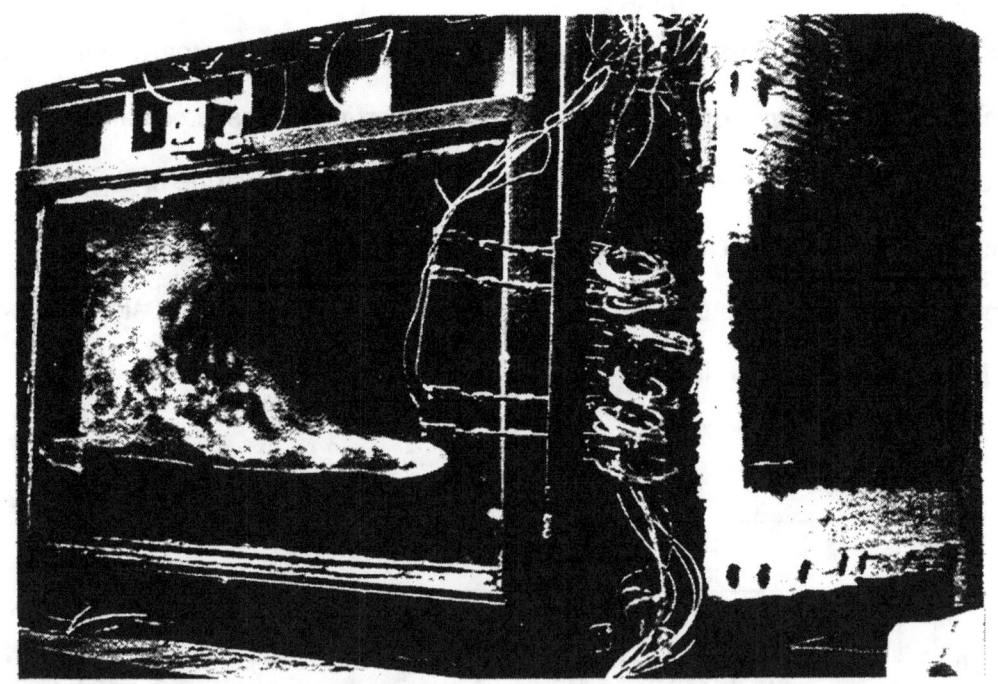

Figure 5.2d - Photograph showing flame structure for the same 70 kW fire source backdraft experiment. Photograph taken 1.5 s after ignition of the backdraft.

Figure 5.2e shows the effects of a backdraft, i.e., short histories for the compartment pressure and the total mass flow into through opening. The time starts at opening, 780 s and shows a 20 s period during which the gravity current enters the compartment, a backdraft is ignited and a large fire ball exits the compartment and conditions relax to quasisteady equilibrium. At 780 s the compartment pressure drops as the hatch falls open. At ~ 781.5 s the hatch strikes the table and causes a tremor in the pressure data. Ignition occurs at 785.6 s, marked with a 1 in Fig. 5.2d. Flames exit the compartment at 787.5 s as the pressure reaches the first peak, marked with a 2 in Fig 5.2e. The second peak is a result of the large fireball which exits the compartment. Even with the large indoor facility and approximately 11 m^2 of vent area, the pressure rises significantly as a large fireball erupts outside the compartment. The large drop after the spike is a repercussion of the large fireball. The total mass flow into the compartment is calculated from Eq. 4. Negative values indicate flow into the compartment. After opening and prior to ignition, the total mass which flows into the compartment shows a steady increase. After ignition, the flow into the compartment is reversed and all of the flow is out of the compartment as indicated by the period of zero slope in Fig. 5.2e. Once the flame has left the compartment and the fireball has subsided, the mass flow rate into the compartment reaches a quasisteady state as indicated by the constant slope in the last 10 s of Fig. 5.2e.

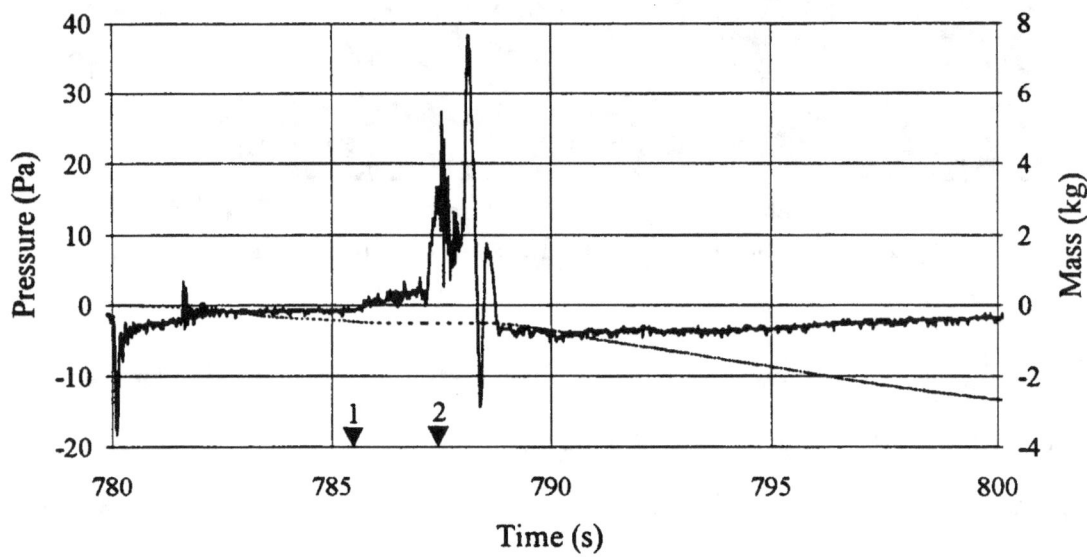

Figure 5.2e - Compartment pressure (———) and total mass inflow (------) histories for the same 70 kW fire source backdraft. Arrow 1 indicates ignition of the backdraft and arrow 2 indicates flame out the opening.

5.3.2 200 kW Fuel Flow: Typical results from a 200 kW experiment are shown in Fig. 5.3 a-e with the following experimental parameters: burn time of 175 s, hatch opened at 180 s, and backdraft ignition at the spark above the burner by a continuous spark. Figure 5.3a shows the species mass fraction histories for the O_2, CO, CO_2, and HC in the upper layer. Idealized layer temperatures and height calculated from Eqs. (5.1 and 5.2) are shown in Fig. 5.3b. As the temperature rises in the first 30 s, the O_2 concentration drops and the CO_2 concentration increases. After 30 s the burning is reduced as the O_2 reaches a minimum, the affects of the reduced burning can be seen in the declining CO_2 concentration and temperature. The HC concentration starts to increase after 30 s and increases to approximately 12% at 113 s when a dancing flame ignites at the spark and moves around the floor of the compartment consuming hydrocarbons in the upper layer. The temperatures in the upper layer rise due to the dancing flame. The slow response time of the thermocouple tree causes the temperature rise to occur at 120 s instead of 113 s when the dancing starts. The dancing flame stops at 130 s when the oxygen in the lower layer is consumed and the hydrocarbons start to build again.

At 180 s the hatch is opened and a the gravity current enters the compartment. At ignition the flame appears to propagate more through the main body of the gravity current rather than along the shear interface as seen in the 70 kW case. Then the flame exits the compartment and a large fireball approximately 2 m in diameter, considerably smaller than the 70 kW fireball, is produced. The spikes in the temperature shown in Fig. 5.3b are caused by the flame propagation through the compartment. Gas concentrations after 180 s are unreliable due to the highly transient backdraft.

The idealized two zone approximation is compared with the temperature data from the thermocouple tree at 25 s and 180 s in Fig. 5.3c. At 25 s the compartment temperature is at its maximum and there is a substantial temperature gradient in the upper layer. The layer is located just above the burner surface. At opening, 180s, the layer is still near the top of the burner and the compartment temperature gradient is considerably reduced. Figure 5.3d is a photograph showing the flame structure 1.5 s after ignition.

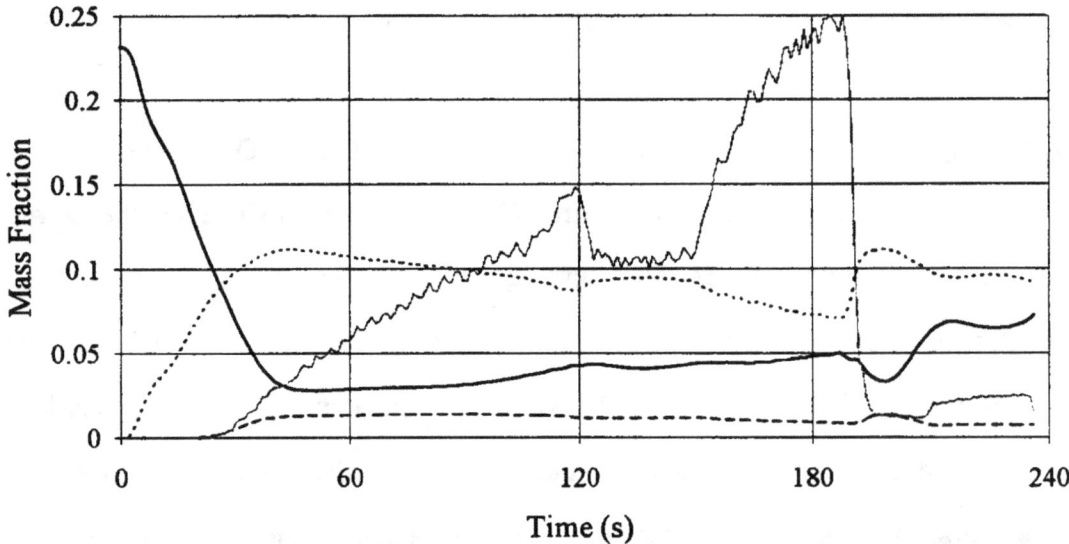

Figure 5.3a - Species concentration histories from backdraft experiment for O_2 (———), CO (– – –), CO_2 (······), and HC (——) for the 200 kW (15th row in Table 5.1) fire source.

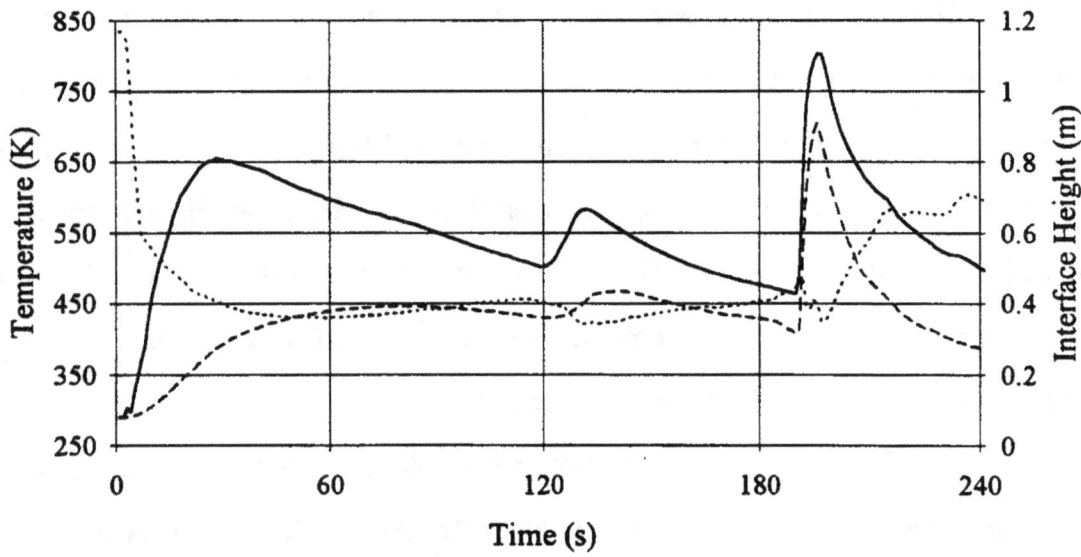

Figure 5.3b - Idealized two zone upper layer (———), lower layer (– – –), and layer height (······) histories from a 200 kW (15th row in Table 5.1) backdraft experiment.

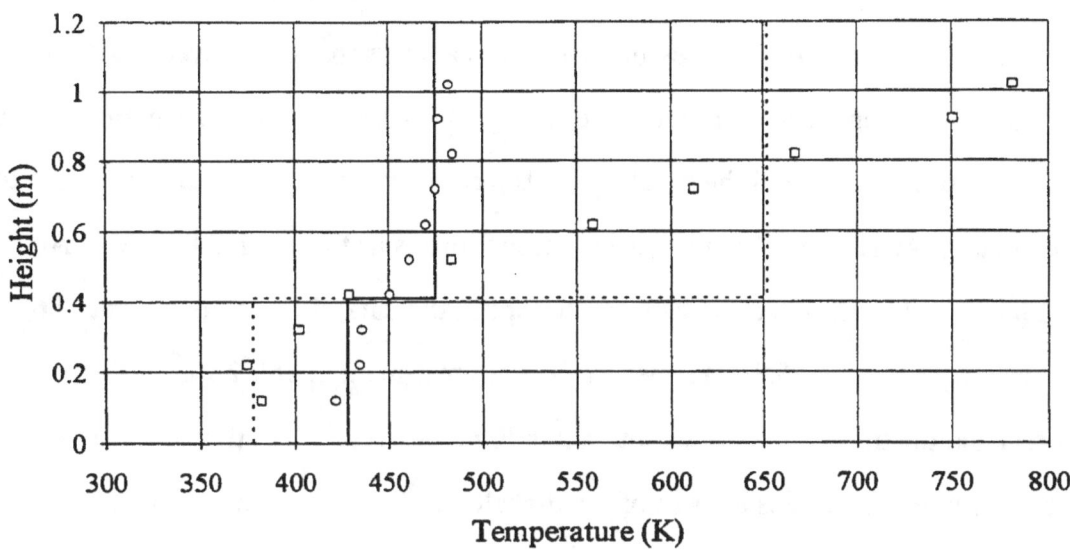

Figure 5.3c - Idealized two zone approximation compared with temperature data at 25 s (-----,□) and 180 s (——— ,o) for the same 200 kW (15th row in Table 5.1) backdraft experiment.

Figure 5.3d - Photograph showing flame structure for the 200 kW (15th row in Table 5.1) fire source backdraft experiment. Photograph taken 1.5 s after ignition of the backdraft.

Figure 5.3e shows the effects of a backdraft for the 200 kW fire source, i.e., histories for the compartment pressure and the total mass that has flowed in through the hatch since opening. The time starts at opening, 180 s and shows a 20 s period in which the backdraft is ignited and the fireball exits the compartment and conditions return to a quasisteady. At 180 s the compartment pressure drops as the hatch falls open. At ~181.5 s the hatch strikes the table and causes a tremor in the data. Ignition occurs at 186.7 s marked with a 1 in Fig. 5.3e. Flames exit the compartment at 189.2 s after the pressure has reached the first peak marked with a 2 in Fig 5.3e. As seen in the 70 kW case, the second peak in Fig. 5.3e is a result of the fireball which exits the compartment.

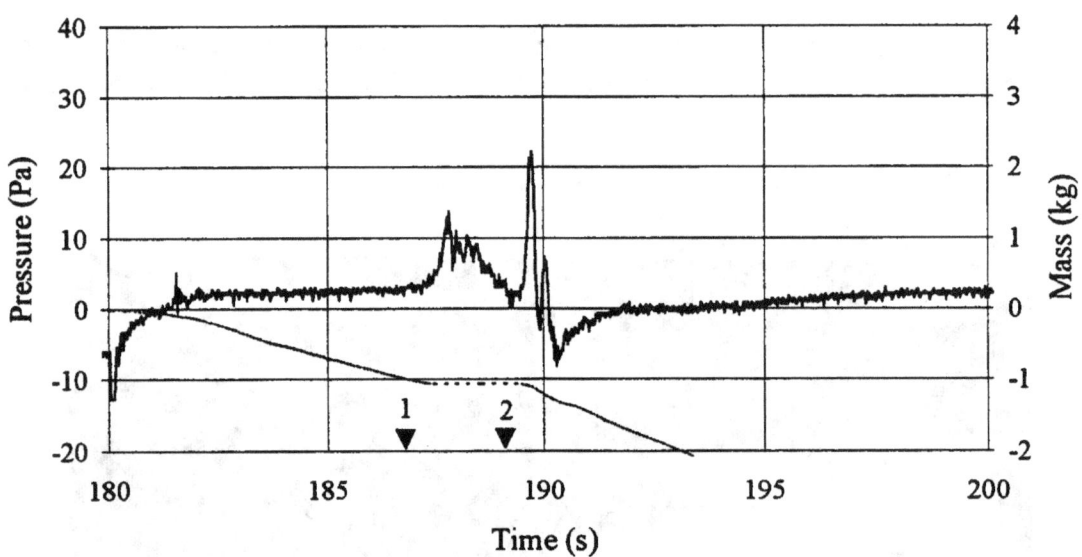

Figure 5.3e - Compartment pressure (———) and total mass inflow (······) histories for the 200 kW Figure 5.3d - Photograph showing flame structure for the 200 kW (15th row in Table 5.1) fire source backdraft experiment. Arrow 1 indicates ignition of the backdraft and arrow 2 indicates flame out the opening.

5.3.3 Summary: Table 5.1 is a summary of the 17 experiments reported here. columns 1 and 2 are the burner characteristics, i.e., the burner flow rate and the time the burner gas is flowing. Columns 3 - 6 are the compartment species concentrations at opening for O_2, CO, CO_2, and HC, respectively. Columns 7 -9 are results calculated from Eqs. (5.1 and

5.2) for the lower layer temperature, the upper layer temperature, and the layer height at opening. Column 10 is the observed ignition delay time. Values shown in parentheses were ignited at the burner spark, all other backdrafts were ignited at the spark above the burner. Column 11 is the peak pressure measured inside the compartment. Column 12 is the total mass carried into the compartment after hatch opening. Column 13 is an estimate of the diameter of the fireball which burns outside the compartment. Column 14 is the run name which corresponds to the data given in appendix C.

In order to obtain a quiescent environment within the compartment, the gas flow rate was set at the low value of 70 kW and the spark ignitors were left off until the hatch was opened. The flow rate was more difficult to control at the 70 kW rate and some minor fluctuation is seen in the data reported in column 1. Burn times ranged from 295 s to 775 s. Times greater than 775 s were felt to be too hazardous to attempt safely. The species concentrations are nearly constant for O_2, CO, and CO_2 indicating the repeatability of the system. The HC concentrations histories are similar to Fig. 5.2a with a long gradual build up, although the slope of the curve changed slightly depending on the burner flow rate. The idealized layer temperatures decreased as the burner times increased due to the energy loss to the boundaries. Layer height maybe taken as constant throughout the 70 kW experiments considering the calculation method used. The maximum pressure and size of the fireball can be considered as gross measurement of the intensity of the backdraft and are seen to increase with the HC concentration as expected.

Experiments were also conducted using a 200 kW fire source and varying the burn times from 115 s to 235 s. In all five experiments, the spark ignitor was left on throughout the experiment and a dancing flame was observed in all but one experiments. The dancing flames causes large thermal instabilities within the compartment and increased mixing between the upper and lower layer. The increased turbulence makes it difficult to obtain repeatable conditions.

Table 5.1 - Summary data from backdraft experiments showing burner characteristics, Species concentrations at opening, idealized layer temperature, and height, ignition delay time, peak pressure, total mass flow into the compartment before ignition, and fireball diameter.

Fuel Flow (kW)	Burner Time (s)	Y_{O_2}	Y_{CO}	Y_{CO_2}	Y_{HC}	T_{UL} (K)	T_{LL} (K)	h_L (m)	Ignition Delay (s)	p_{max} (Pa)	m_{IN} (kg)	Fire Ball (m)	Run Name
72	295	0.09	0.005	0.07	0.10	417	378	0.43	5.27	4	0.72	0	P3EXP38
72	355	0.11	0.004	0.06	0.12	390	361	0.43	5.36	6	1.05	2	P3EXP36
72	415	0.11	0.004	0.06	0.14	379	353	0.42	5.43	9	0.78	2	P3EXP35
72	475	0.11	0.003	0.05	0.16	362	339	0.42	6.63	28	0.91	3	P3EXP34
69	535	0.11	0.003	0.05	0.16	377	356	0.43	5.47	43	0.61	3	P3EXP33
77	535	0.11	0.003	0.05	0.20	363	344	0.41	5.67	40	0.69	4	P3EXP45
69	555	0.11	0.003	0.05	0.19	359	340	0.42	(3.7)	73	0.46	4	P3EXP32
69	595	0.12	0.003	0.04	0.19	363	346	0.41	5.40	43	0.62	4	P3EXP42
73	655	0.12	0.003	0.04	0.21	350	331	0.43	(3.63)	50	0.44	4	P3EXP39
71	715	0.11	0.003	0.04	0.20	348	332	0.44	5.63	33	0.66	4	P3EXP40
68	715	0.12	0.003	0.04	0.22	347	332	0.42	5.70	49	0.62	4	P3EXP43
70	775	0.12	0.002	0.04	0.22	344	330	0.41	6.60	39	0.52	4	P3EXP41
200	115	0.04	0.012	0.09	0.13	517	445	0.44	4.27	14	0.56	0	P3EXP26
200	145	0.04	0.012	0.10	0.10	570	475	0.34	6.27	8	0.93	1	P3EXP30
200	175	0.05	0.009	0.07	0.24	474	427	0.41	6.73	22	1.07	2	P3EXP27
200	205	0.06	0.008	0.06	0.29	447	408	0.43	(4.33)	8	0.64	0	P3EXP28
200	235	0.06	0.007	0.06	0.29	433	400	0.43	5.27	36	0.96	4	P3EXP29

5.4 Conclusions

The results presented here show that the HC concentration must be >10% in order for a backdraft to occur. When the HC concentration is < 10% the flame travel is slow and the compartment overpressure is much lower. As the HC concentration increases the compartment overpressure increases and the backdraft becomes more severe. The 70 kW burner flow rate experiments produced excellent backdrafts. The species concentrations show long slow changes. The results presented here suggest that the 200 kW source was too large for the compartment. Experiments with this large source were highly transient and compartment conditions are too unstable to interpret trends. The 200 kW experiments were further complicated by dancing flames due to the ignitors being on throughout the experiment.

Future work should concentrate on designing and building a full scale apparatus using a controlled fire source and possibly more realistic fuels. Improved gas analysis would also be useful to determine what hydrocarbons are present in the real fuel experiments. Actual opening geometries could also be used to investigate the effects of typical doors and windows. Openings in the ceiling could also be incorporated to study the effects of firefighter ventilation tactics.

References - Chapter 5

[1] Croft, W.M., "Fires Involving Explosions", Fire Safety Journal, 3, 3-24, 1980/81.

[2] Steward, P. D. C., "Dust and Smoke Explosions", NFPA Quarterly, 7, 424-428, 1914.

[3] Fire Ventilation Practices, 5th Edition, International Fire Service Training Association, Oklahoma, 32-33, 1970.

[4] Dunn, V., "Beating the Backdraft", Fire Engineering, 44-48, 1988.

[5] Zalosh, R.G., "Explosion Protection," in SFPE Handbook of Fire Protection Engineering, (P.J. DiNenno, ed.) 2-88 to 2-105, National Fire Protection Association, Quincy, MA, 1988.

[6] Factory Mutual Engineering Corporation, "Damage-Limiting Construction," Loss Prevention Data Sheet 1-44, Norwood, MA, 1991.

[7] "Neoceram Super Heat-Resistant Glass for Industrial Use," Nippon Electric Glass Co. Ltd.

[8] Quintiere, J.G., Steckler, K., and Corley, D., "An Assessment of Fire Induced Flows in Compartments" Fire Science and Technology, 4, 1-14, 1984.

[9] McCaffrey, B.J. and Heskestad, G., "A Robust Bidirectional Low-Velocity Probe for Flame and Fire Application", Combustion and Flame, 26 125-127, 1976.

[10] Newman, J.S. and Croce, P.A., " A Simple Aspirated Thermocouple For Use In Fires", Fire & Flammability, 10, 326-336, 1979.

CHAPTER 6

BACKDRAFT EXPERIMENTS USING A SIMULATED WINDOW OPENING

6.1 Introduction

A backdraft is defined as a rapid deflagration following the introduction of oxygen into a compartment filled with accumulated unburned fuels. The dangerous consequences of backdraft to both firefighters and civilians are well documented in numerous fire service publications and training manuals.[1,2,3,4] However, little research has been done on the fundamental phenomena underlying backdrafts.

Consider a fire in a closed compartment where the only ventilation provided is by leakage. As the fire heats the compartment, leaks in the compartment bounding surfaces permit outflows that minimize any pressure differential[5]. As the hot layer descends over the fire, the available oxygen is reduced and the combustion efficiency decreases. Excess pyrolyzates[6] accumulate upper layer forming a fuel rich upper layer. Suddenly, a new ventilation source is provided by a window breaking or door opening. Cold, oxygen rich air enters the compartment and propagates across the floor as a gravity current. If an ignition source is not immediately available, the gravity current will reflect off the rear wall and propagate back to the opening. Once the gravity current has reached the opening, the compartment becomes reservoir in a reservoir filling problem. A new lower layer made up of a mixture of fuel rich upper layer and oxygen rich incoming air continues to grow. Once an ignition source is available, e.g. smoldering ember can be or small flame, the backdraft ignites and a deflagration wave propagates through the compartment driving the unburned fuel out the opening where it combusts in the spectacular fireball commonly associated with backdraft.

Experimental results are presented from a series of 11 experiments conducted in the half room scale apparatus shown in Fig. 6.1. Here, where the opening geometry is a 0.4 m square window centered, on a short wall, a dramatic horizontal flame jet exits the opening prior to the formation of the fireball. Experimental variables include fuel flow rate, burn time, ignition location, ignition delay time, burner height, species sample location, and opening size. The fuel source was methane from a 0.3 m square sand burner centered at 0.3 m height along the wall opposite the opening. The ignition source for the backdraft was two sparks located 0.15 m in front of the burner. Data are presented which characterize the conditions in the compartment prior to backdraft and quantify the severity of the deflagration.

6.2 Experimental Design & Procedures

6.2.1 Apparatus: Experiments were conducted in a special compartment designed to safely control the dangerous overpressures expected in backdrafts. The experimental apparatus dimensions were limited to half a small residential room to minimize the expected hazards. It also allowed the experiments to be conducted inside a 900 m^3 building at the Richmond Field Station of the University of California at Berkeley. Figure 6.1 shows a schematic of the apparatus giving the internal dimensions of the compartment and the instrumentation locations. In order to control the overpressure hazard, one long wall was a pressure relief panel[7,8]. A 0.9 m high by 1.5 m wide observation window of Neoceram[9] was installed in the wall opposite the pressure relief panel.

To simulate a window breaking due to thermal stress, a 0.4 m high by 0.4 m wide opening was centered in the short wall opposite the burner, as shown in Fig. 6.1. This opening was covered with a computer activated hatch which was opened after the fire had been burning for several minutes. The burner was placed against the wall opposite the opening, as seen in Fig. 6.1. The ignition source for the backdraft was two spark ignitors 0.15 m in front of the burner and 0.15 m and 0.3 m above the floor. A 10,000 volt transformer was used for each spark ignitor to produce an arc between two 3 mm diameter 308 stainless steel electrodes spaced 5 mm apart.

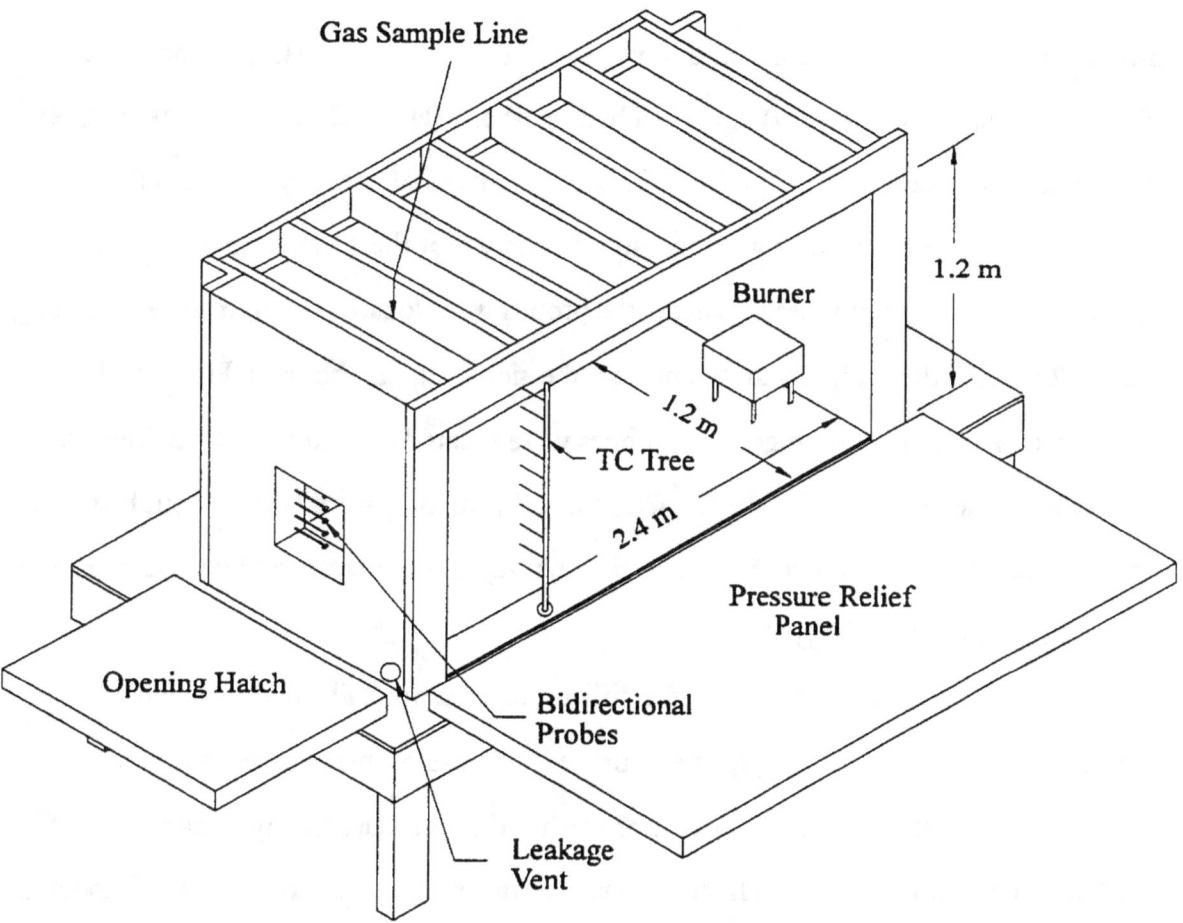

Figure 6.1 -Sketch of the half-scale backdraft compartment showing important features of the apparatus.

Every effort was made to seal all construction holes to control leakage. The primary source of leakage into the compartment was around the pressure relief panel and the opening hatch. A small 0.1 m diameter pressure relief vent was placed at the floor level to relieve the pressure from the initial burner ignition. Without this vent, the initial pressure rise sufficed to activate the pressure relief panel. A computer controlled cover closed over this vent 15 s after ignition. Additional apparatus description can be found in refs. 5 and 7.

6.2.2 Species Concentration: In order to characterize the compartment conditions prior to a backdraft, the species concentration histories in the upper and lower layers were recorded. Gas concentrations measured were: oxygen (O_2), carbon dioxide (CO_2), carbon monoxide (CO), and total hydrocarbons (HC). Continuous gas samples were taken with stainless steel probes. Only one analyzer of each type was available so the lower layer concentrations were measured in only 3 of the 11 experiments. For the upper layer concentration measurements the probes were located 0.6 m from the opening wall, 0.2 m from the ceiling, and 0.6 m from the side wall, as shown in Fig. 6.1. For the lower layer species measurements, the probes were located 0.15 m above the floor and 0.76 m from the burner wall. The O_2, CO, and CO_2 samples were taken through an unheated sample line in which the soot and water vapor were removed by glass fiber and desiccant filters, respectively.

The HC concentration was more difficult to measure. The hydrocarbon sample required a separate heated sample line to prevent the loss of hydrocarbons due to condensation. A flame ionization detector was used to measure the hydrocarbon (methane) concentration. The effective range for this meter was 0 to 1%. The expected range of hydrocarbons was of the order of 25%, by volume. It was therefore, necessary to dilute the sample. The dilution system was designed to mix the compartment sample with heated ambient air in a ratio of 25 to 1. The flow rate of the dilution air and the sample were determined by measuring the pressure drop over a fixed length of tubing, 1.7 mm ID. Sample flow rates were monitored continuously during the experiment. Typical dilution ratios would vary from 22 to 25 depending on the compartment gas temperature and the soot build up in the sample line. A hot water jacket around the hydrocarbon sample line and the dilution air line kept the gases over 60°C, well above the maximum calculated dew point of 44°C. To obtain the final species concentrations in the upper layer it was necessary to calculate the upper layer H$_2$O concentration, since there was no direct measurement of H$_2$O. Detailed species balances were performed[7].

6.2.3 Temperatures: A vertical thermocouple tree was placed 0.6 m from the opening wall and 0.2 m from the pressure relief panel, as shown in Fig. 6.1. The thermocouples were made from 0.5 mm type K thermocouple wire with a stainless steel overbraid. The average bead diameter was 1.1 mm. The ten thermocouples were located at 0.10 m intervals, with the highest thermocouple at 0.15 m below the ceiling. The temperatures reported here are uncorrected values.

The thermal interface height history was calculated from the time dependent temperature profiles recorded from the thermocouple tree. The profiles were converted into unsteady idealized upper and lower layer temperatures using the method Quintiere et. al.[10] applied to steady state temperature profiles:

$$\int_0^{h_1} \left(\frac{1}{T}\right) dx = [h_1 - h_L]/T^{UL} + h_L/T^{LL}, \qquad (6.1)$$

$$\text{and} \quad \int_0^{h_1} T dx = [h_1 - h_L]T^{UL} + h_L T^{LL}, \qquad (6.2)$$

where T^{UL} and T^{LL} are the upper and lower layer temperatures, and h_1 and h_L are the heights of the compartment and the layer interface, respectively. Equation (1) is a mass balance and Eq. (2) retains the same mean temperature as in the data. Assuming that the lower layer temperature was the arithmetic average of the two lowest thermocouples allowed the upper layer temperature and thermal interface location to be calculated from Eqs. (1 and 2).

6.2.4 Compartment Pressure: The compartment pressure history was recorded using an electronic pressure transducer with an effective range of 0 to 1250 Pa. The pressure port was mounted in the stationary wall opposite the pressure relief panel at floor level. The ambient pressure reference was taken outside the building.

6.2.5 Opening Flow: The flow in and out of the compartment after the hatch was opened was recorded using six bidirectional probes in the opening. The probes were located in the horizontal center of the opening and 65 mm apart. The top and bottom probes were 43 mm from the soffit and sill. Probe velocities were calculated using the relationship given by McCaffrey and Heskestad[11]:

$$v = C(Re)\sqrt{\frac{2\Delta p}{\rho}}, \qquad (6.3)$$

where v is the velocity at the probe, Δp is the pressure difference measured across the probe, ρ is the density at the probe, and $C(Re)$ is an empirical calibration constant which is a function of the Reynolds number.

The temperature was recorded at each bidirectional probe in the opening using a bare bead 0.05 mm Type K thermocouple with high temperature glass insulation. Four aspirated thermocouples, designed according to Newman et. al.,[12] were installed to correct the temperatures for radiation. Data from the aspirated thermocouples revealed that no radiation correction was required.

The total mass flow was calculated by integrating the velocity and density profiles over the height of the opening as:

$$\dot{m}_{in} = C\int_{h_b}^{h_n}[\rho(y,t)\cdot v(y,t)\cdot b]dy, \qquad (6.4)$$

$$\dot{m}_{out} = C\int_{h_n}^{h_t}[\rho(y,t)\cdot v(y,t)\cdot b]dy \qquad (6.5)$$

where $\dot{m}_{in}$ and $\dot{m}_{out}$ are the mass flow in the lower portion and the mass flow out the upper portion of the vent opening, C is the flow coefficient taken as 0.68, b is width of opening, and h_b, h_t, and h_n are the height of the sill, soffit, and neutral axis, respectively. h_n was determined empirically from the probe velocity data. The density was corrected for temperature using the thermocouple measurements. When flames reach the opening the thermocouples cannot respond fast enough to accurately measure the temperature and the mass flow calculations are terminated.

6.2.6 Data Acquisition System: Data from each sensor were recorded using a HP VECTRA 80486-33 computer with a 8 channel multifunction analog and digital input/output board. Two 32 channel analog input muliplexors were connected to this system. A total of 31 thermocouple and 17 voltage channels were used. To increase the scan rate, data were written to a RAM drive and then downloaded to a file on the hard disk immediately following the experiment. The system was capable of recording each channel 50 times a second. For experiments greater than 600 s the data was collected at a rate of 10 scan/s until 20s before opening when the rate was automatically increased to 50 scan/s. The reduced scan rate for the initial period of the experiment simply to reduce the data file size. Files for these experiments ranged from 4 to 10 Mbytes.

In addition to recording the data, the computer also controlled the experimental procedures using solid state relays activated by a digital input/output board also installed on the computer bus. The computer controlled systems included: experimental clock, burner pilot light, fuel flow, vent cover, spark ignitors, opening hatch, and still camera. Each system had a manual override.

6.2.7 Procedures: Before each experiment a 60 s baseline was taken to record the initial conditions. A pilot flame was ignited at the burner 5 s before the start of the experiment. At 0 s a solenoid was opened on the methane flow to the burner and the clock was reset to zero. The burner was left on for a predetermined time period. Gas flow to the burner was terminated 5 s before the hatch was opened. After opening there was a predetermined time delay ranging from 15-30 s before the spark ignitors were activated.

6.3 Experimental Results

6.3.1 Exemplar Data

The experimental parameters for the results presented in Figs. 6.2 - 6.5 are: gas flow rate of 70 kW, burner flow time of 535 s, hatch opening at 540 s and backdraft ignition delay time of 20 s. Figure 6.2a shows the upper layer species mass fraction histories for O_2, CO, CO_2, and HC. Lower layer species concentrations are given in Fig. 6.2b. Idealized layer temperatures and height calculated from Eqs. (6.1 and 6.2) are shown in Fig. 6.3a. For the first 70 s, the temperatures in the compartment rise as the layer within the compartment descends over the fire. The upper layer O_2 concentration is dropping as the fire consumes the available O_2 and the CO_2 concentration increases as a result of the combustion. Species concentrations in the lower layer are relatively unaffected during this early period. Over the time period of 70 s to 150 s the temperature in the compartment drops as the fire oscillates and then dies. After 150 s the fire is completely out and the compartment begins to cool as seen in the exponential decay in the temperature profile. Over this same period the upper layer HC concentration rises as the O_2 and CO_2 level off. The O_2 concentration in the lower layer drops considerably as the CO_2 and HC concentrations increase. After 240 s the hydrocarbon concentration in both the upper and lower layers is steadily increasing with the lower layer at ~20% that of the upper layer. The oxygen concentration slightly increases as air leaks into the compartment and the CO_2 concentration declines as compartment gases are lost due to leakage. At 480 s the HC analyzer became saturated and the slope of HC concentration approaches zero. A linear extrapolation would give a more accurate HC concentration in the last 60 s. At 540 s the hatch is opened and a gravity current enters the compartment.

The gravity current travels across the floor and is reflected back to the opening. After the gravity current reaches the opening oxygen rich air continues to pour into a new lower layer as the upper layer gases continue to exit the compartment. At 560 s the sparks are turned on and ignite a flame which travels through the compartment, driving the flammable gases out the opening and culminates in a large external fireball, ~5 m diameter. The spikes in the temperature shown in Fig. 6.3a are caused by the wave propagation through the compartment. Gas concentrations after 560 s are unreliable due to the highly transient effects of the backdraft.

An idealized two zone approximation is compared with the temperature data from the thermocouple tree at 60 s and 540 s in Fig. 6.3b. At 60 s the compartment temperature is near the peak temperature and there is a substantial temperature gradient in the upper layer. The layer interface is located just above the burner surface. At opening, 540s, the layer interface is still near the top of the burner and the compartment temperature is nearly uniform vertically indicating that a one zone approximation is reasonable at that time.

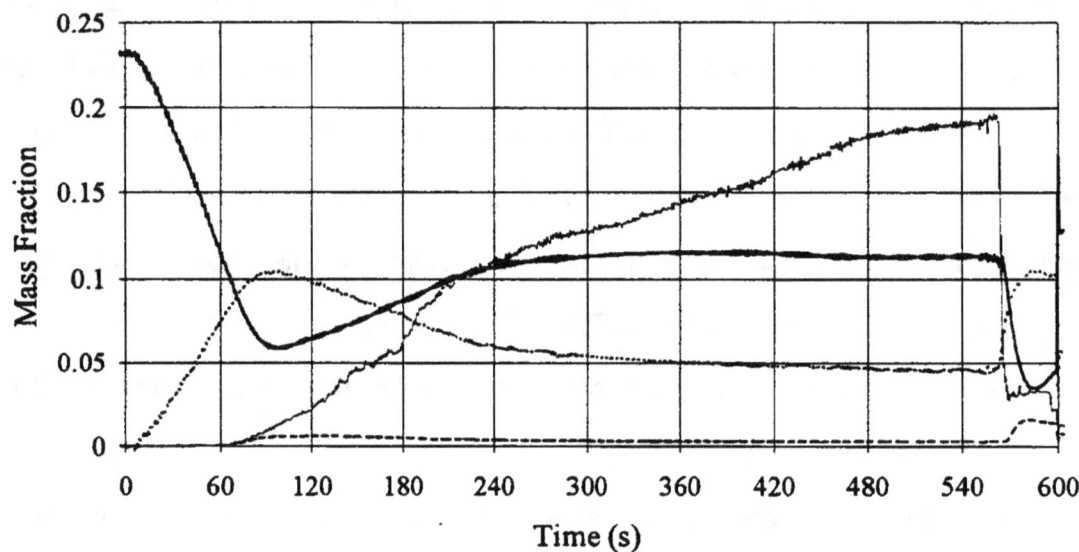

Figure 6.2a - Upper layer species concentration histories from backdraft experiment for O_2 (——), CO (– – –), CO_2 (·····), and HC (——). Run 9 in Table 6.1 and 6.2.

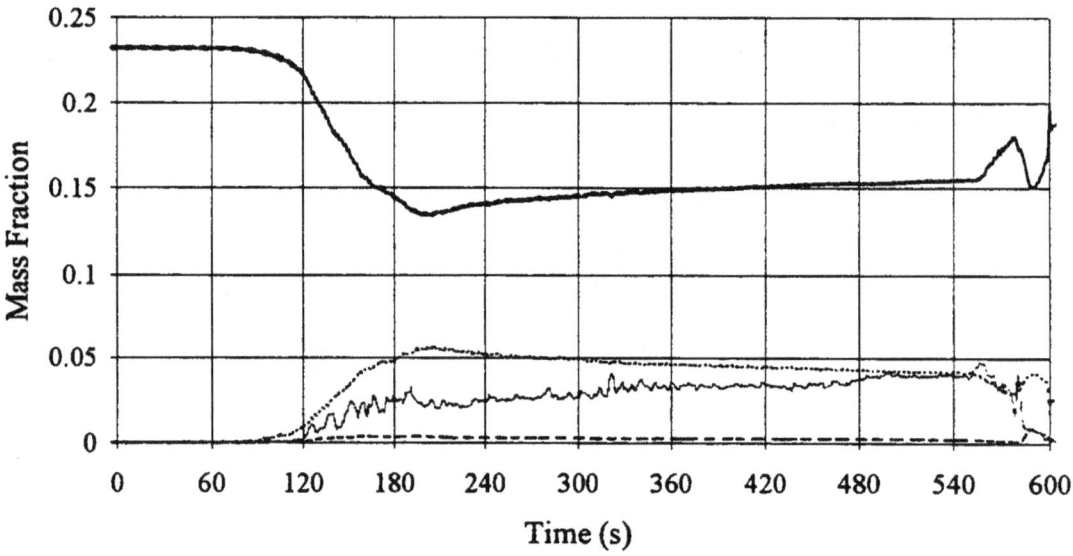

Figure 6.2b - Lower layer species concentration histories from backdraft experiment for O_2 (——), CO (– – –), CO_2 (·····), and HC (——). Run 9 in Table 6.1 and 6.2.

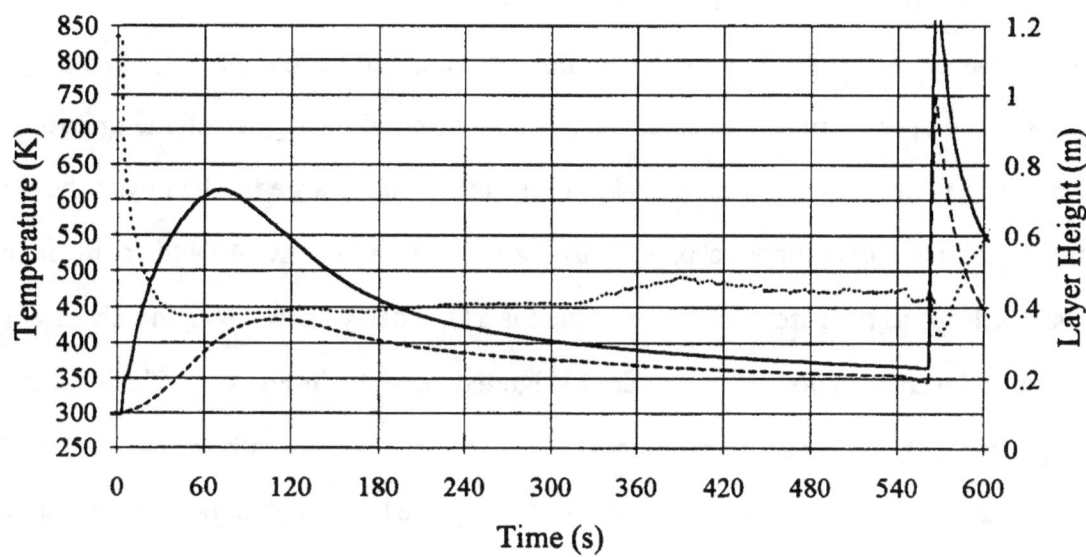

Figure 6.3a - Idealized two zone upper layer (———), lower layer (– – –), and layer height (······). Run 9 in Table 6.1 and 6.2.

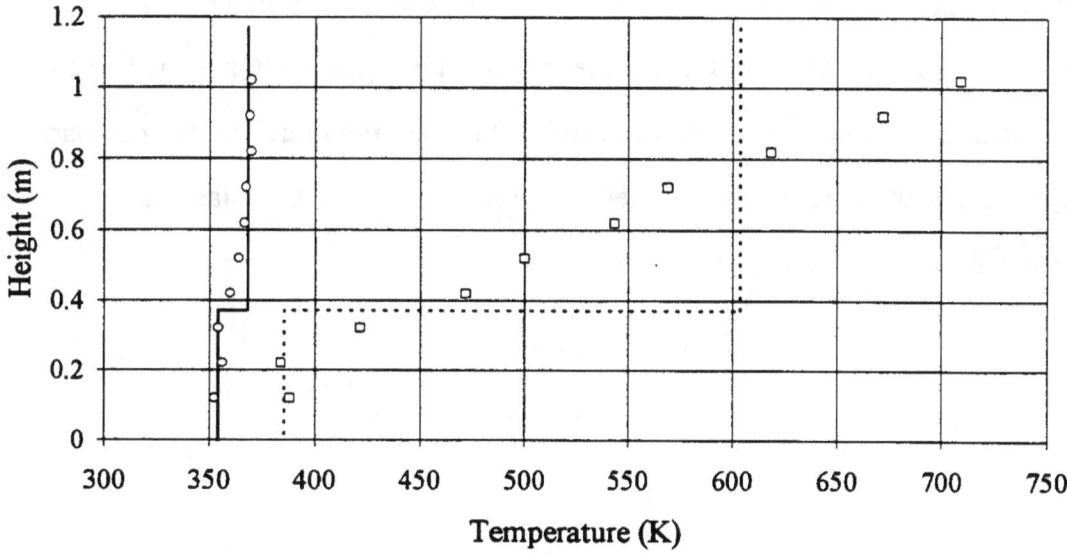

Figure 6.3b - Idealized two zone approximation compared with temperature data at 60 s (······,□) and 540 s (———,○). Run 9 in Table 6.1 and 6.2.

Figure 6.4 shows the effects of a backdraft, i.e., short histories for the compartment pressure and the total mass flow into and out of the opening. The graph begins at the opening time, 540 s, and shows a 30 s period during which the gravity current enters the compartment, a backdraft is ignited and a large fire ball exits the compartment and conditions relax to quasisteady equilibrium. At 540 s the compartment pressure drops as the hatch falls open. There is a 20 s delay as the oxygen rich air enters the compartment. At 560 s the backdraft is ignited, arrow 1 in Fig. 6.4. Flames exit the compartment at 560.9 s as the pressure reaches 160 Pa, marked with arrow 2 in Fig. 6.4. The pressure spike in the compartment is caused by the large fireball erupting outside the compartment in the large, 900 m^3, indoor facility with approximately 11 m^2 of vent area. The total mass flow into and out of the compartment is calculated from Eqs. (6.4 and 6.5). Negative values indicate flow into the compartment. After opening and prior to ignition, the total mass which flows into and out of the compartment steadily increases. After ignition, the flow into the compartment is reversed and all of the flow is out of the compartment as indicated by the period of zero slope on the inflow curve and the steep increase in the slope of the outflow curve, as shown in Fig. 6.4. Once the flame leaves the compartment and the fireball has subsided, the mass flow rate into the compartment reaches a quasisteady state as indicated by the constant slope in the last 8 s of Fig. 6.4, which is flow into the compartment as it cools.

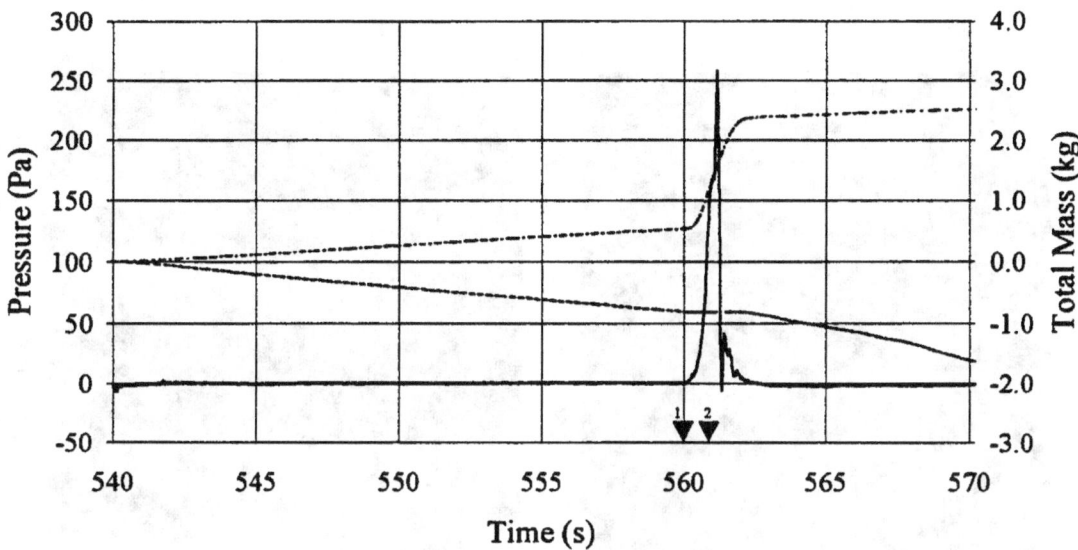

Figure 6.4 - Compartment pressure (———) and total mass inflow (······) and total mass outflow (– – –) histories. Arrow 1 indicates ignition of the backdraft and arrow 2 indicates flame out the opening. Run 9 in Table 6.1 and 6.2.

Figure 6.5 is a series of video images from this experiment taken at 0.2 s intervals after ignition. The flame is initially spherical in shape and then stretches toward the opening, exiting the compartment 0.9 s after ignition.

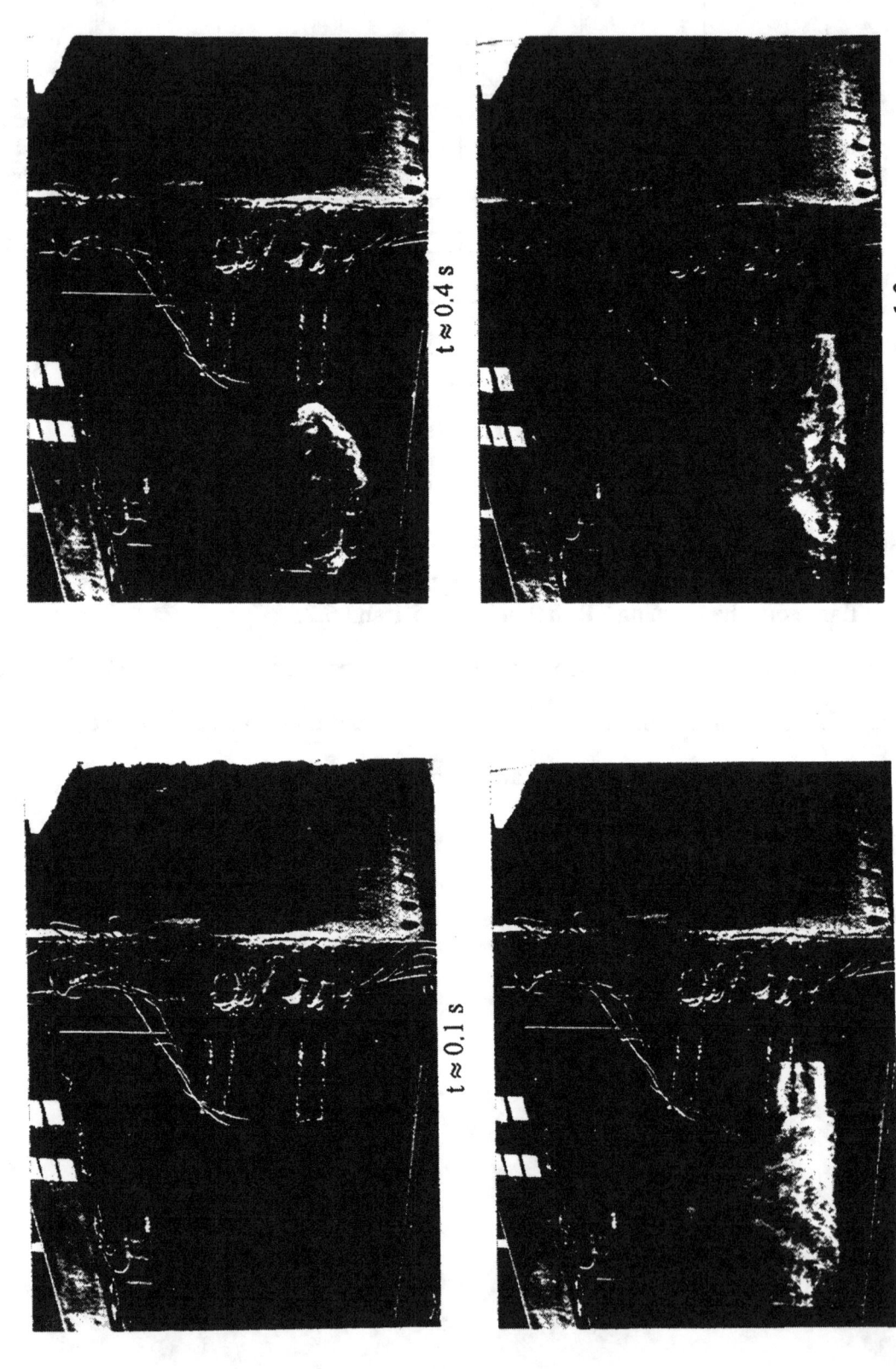

Figure 6.5a-d - Series of photographs showing flame structure of the backdraft. Photographs are taken 0.2 s apart. Run 9 in Table 6.1 and 6.2.

6.3.2 Summary: Table 6.1 is a summary of the 11 experiments reported here. Column 1 is the experiment number. Columns 2 and 3 are the burner characteristics, i.e., the burner flow rate and the time the burner gas is flowing. Columns 4 - 6 are results calculated from Eqs. (1 and 2) for the lower layer temperature, the upper layer temperature, and the layer height at opening. Column 7 is the observed ignition delay time between compartment opening and ignition of the backdraft. Column 8 is the peak pressure measured inside the compartment. Column 9 - 10 are the total mass which flows into and out of the compartment after hatch opening and prior to ignition, respectively. Column 11 - 12 are the total mass which flows into and out of the compartment after opening and prior to flames exiting the compartment. Column 13 is the run name which corresponds to the data given in appendix C.

Table 6.1 Summary of the 11 window experiments reporting the burner characteristics, layer temperatures and height, ignition delay, peak compartment pressure, and mass flow into and out of the compartment.

Run Number	Fuel Flow (kW)	Burner Time (s)	T_{UL} (K)	T_{LL} (K)	h_L (m)	$t_{ig} - t_o$ (s)	p_{max} (Pa)	$m_{in}^{t=t_i}$ (kg)	$m_{out}^{t=t_i}$ (kg)	$m_{in}^{t=t_{fo}}$ (kg)	$m_{out}^{t=t_{fo}}$ (kg)	Run Number Appendix C
1	70	415	382	360	0.48	10.33	2	0.35	0.27	*	*	P3EXP82
2	72	415	384	364	0.45	15.00	13	0.62	0.40	0.62	1.57	P3EXP79
3	72	415	381	362	0.46	20.47	9	0.74	0.55	0.75	1.63	P3EXP77
4	70	415	384	363	0.48	25.50	14	0.93	0.68	0.94	1.67	P3EXP85
5	72	475	375	359	0.46	15.03	22	0.57	0.39	0.57	1.15	P3EXP81
6	75	475	374	356	0.44	20.13	82	0.79	0.53	0.80	1.56	P3EXP76
7	70	475	378	361	0.46	25.00	115	0.96	0.66	0.96	1.62	P3EXP86
8	72	535	369	355	0.45	15.00	189	0.62	0.40	0.62	1.70	P3EXP83
9	74	535	368	354	0.44	20.00	258	0.81	0.54	0.81	1.20	P3EXP78
10	73	535	371	357	0.46	25.27	213	0.95	0.65	0.95	1.31	P3EXP80
11	70	535	368	354	0.46	30.40	102	1.13	0.72	1.13	1.47	P3EXP84

The burner flow rate was difficult to control and some fluctuation is seen in the data reported in column 2. Burn times ranged from 420 s to 540 s. Times greater than 540 s were considered too hazardous to attempt safely. The ignition delay ranged from 15 s to 30 s. The position of the gravity current within the compartment can be calculated as a function of time from salt water modeling assuming a constant Froude number discussed in chapter 3. Velocity current velocity for these experiments was approximately 0.5 m/s. For these experiments the gravity current reaches the rear wall approximately 7 s after opening and returns to the front wall approximately 14 s after opening. The species concentrations are nearly constant for O_2, CO, and CO_2 indicating the repeatability of the system. The HC concentrations histories are similar to Fig. 6.2a with a long gradual build up, although the slope of the curve changed slightly according to the burner flow rate. The idealized layer temperatures decreased as the burner times increased due to the energy loss to the boundaries and leakage of the hot layer gases. Layer interface height may be taken as approximately constant throughout the experiments. The maximum pressure and size of the fireball can be considered as a gross measurement of the intensity of the backdraft. They are seen to increase with the HC concentration as expected.

6.3.3 Chemical Energy Accounting: An energy budget on the compartment was prepared to determine the amount of energy available for the backdraft. The integral over time of the burner flow rate can be calculated to determine the total chemical energy released into the system:

$$E_B = \int_0^t (\Delta H_C \cdot \dot{m}_B) \, dt \tag{6.6}$$

where E_B is the total chemical energy released by the burner, to anytime t, ΔH_c is the heat of combustion and $\dot{m}_B$ is the burner flow rate. The burner energy is assumed to be either consumed in the burner flame, stored within the compartment, or lost due to leakage in the compartment boundary. The energy stored in the compartment at each time t is calculated assuming the compartment can be divided into two idealized layers as discussed above.

$$E_s = \Delta H_c (Y_{HC}^{UL} \cdot m^{UL} + Y_{HC}^{LL} \cdot m^{LL}), \qquad (6.7)$$

where E_s is the amount of chemical energy stored within the compartment, Y_{HC}^{UL} and Y_{HC}^{LL} are the mass fractions for the hydrocarbons in the upper layer and lower layer, respectively, and m^{UL} and m^{LL} are the total mass of the upper and lower layers, respectively. The total mass in each layer is calculated from the layer height and average layer density:

$$m^{UL} = \rho^{UL} \cdot w \cdot L \cdot (h_1 - h_L), \qquad (6.8)$$

$$m^{LL} = \rho^{LL} \cdot w \cdot L \cdot h_L, \quad \text{for } t < t_0, \qquad (6.9)$$

where t_0 is the time when the compartment hatch is opened, ρ^{UL} and ρ^{LL} are the densities of the upper and lower layers, respectively, w is the compartment width, L is the compartment length, and h_1 and h_L are the compartment height and the height of the layer above the floor. The energy lost by leakage is found by the difference, $E_L = E_B - E_S - E_C$, where E_L is the energy lost from leakage and E_C is the energy combusted at the burner. The leakage mass loss rate $\dot{m}_L$ is then estimated from:

$$E_L = \int_0^t \Delta H_C Y_{HC}^{UL} \dot{m}_L dt. \qquad (6.10)$$

The chemical energy histories for the compartment are shown in Fig. 6.6 for run 9 in Table 6.1 and 6.2. The dashed line indicates the total chemical energy which enters the compartment through the burner, (Eq. 6.6). The solid line indicates the chemical energy which is stored within the compartment as a function of time. The chemical energy supplied by the burner simply increases linearly due to the constant flow rate (74 kW) through the burner. At 535 s the burner is shut off and the total chemical energy supplied by the burner remains constant. The total stored chemical energy is initially zero since all of the chemical energy released by the burner is consumed by the fire. As the fire burns out, the hydrocarbons start to build as unburned fuel is stored within the compartment. The stored energy continues to increase to 480 s. The difference between the burner supplied energy and the stored energy is a function of the chemical energy consumed by the fire and the chemical energy lost due to leakage out of the compartment. The amount of chemical energy consumed by the fire is approximately $E_C \sim 7$ MJ. The chemical energy lost due to leakage is estimated at 9 MJ. From Eq. (6.10) that suggests a constant leakage rate of 5 gm/s. The total chemical energy history with that mass loss rate is shown by the dotted line in Fig. 6.6. With $E_B \sim 39$ MJ the energy available at the time of opening is $E_S \sim$ 19 MJ.

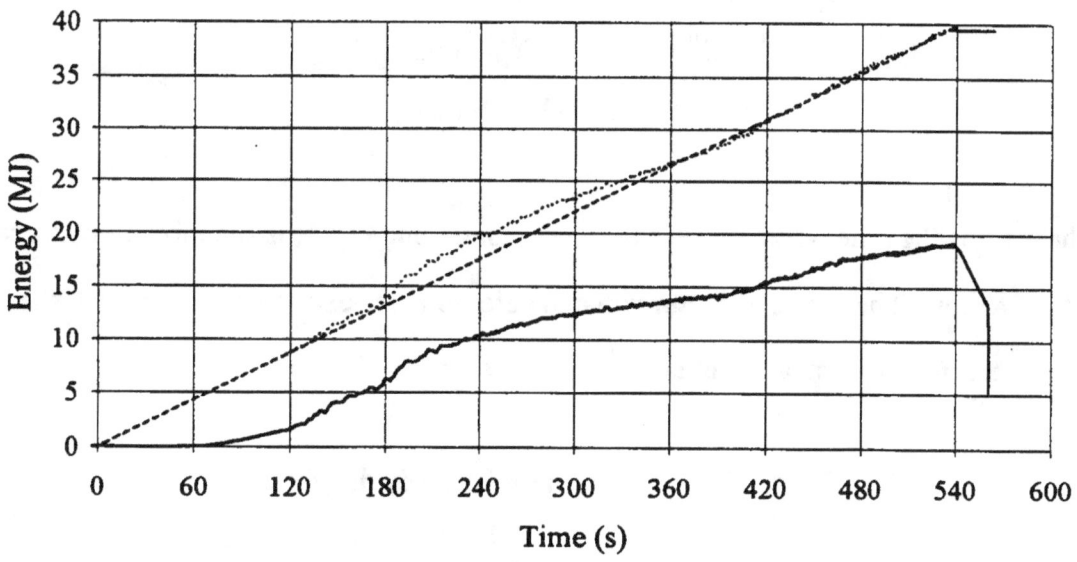

Figure 6.6 - Chemical energy histories for the energy within the compartment (——) and the chemical energy supplied by the burner (– – –) and the total energy history with an assumed constant mass leakage rate of 5 gm/s (·····). From run 9 in Table 6.1 and 6.2.

The available energy for backdraft is assumed to consist of: 1) the amount of energy which leaves the compartment prior to ignition, E_I, 2) the amount which leaves the compartment after ignition and before flame exits the opening, E_F, 3) the amount which is contained in the compartment when the flames reach the opening and not consumed there, E_N. Each of these will now be estimated.

After opening the compartment but prior to ignition of the backdraft, the mass flow out of the compartment is assumed to be out of the upper layer and the flow into the compartment is assumed to be into the lower layer. The unburned chemical energy that has flowed out of the compartment is:

$$E_I = \Delta H_C \ Y_{HC}^{UL} \ C \ b \int_{t_o}^{t_i} \int_{h_n}^{h_t} \rho(y,t) \, v(y,t) \, dy \, dt, \qquad \text{for } t_o > t > t_i, \quad (6.11)$$

where Y_{HC}^{UL} is assumed constant over that time interval.

After ignition of the backdraft, the compartment gases are assumed to be well mixed and the hydrocarbon mass fraction is taken as:

$$Y_{HC}^{Mix} = \frac{Y_{HC}^{UL}(m_{t=t_o}^{UL} - m_{Out}) + Y_{HC}^{LL} \cdot m_{t=t_o}^{LL}}{m_{t=t_o}^{UL} + m_{t=t_o}^{LL} + m_{In} - m_{Out}}, \qquad \text{for } t_i > t > t_{FO}, \quad (6.12)$$

where t_{FO} is the time when flames exit the compartment, Y_{HC}^{Mix}, is the mixture hydrocarbon mass fraction. Therefore, after ignition and before flames exit the compartment, the energy exiting the compartment is:

$$E_F = Y_{HC}^{Mix} \ C \ b \int_{t_o}^{t_{fo}} \int_{h_s}^{h_t} [\rho(y,t) \cdot v(y,t)] \, dy \, dt, \text{ for } t_i > t > t_{fo}. \quad (6.13)$$

To estimate the amount of chemical energy within the compartment which is not consumed it is assumed that all the oxygen within the compartment when flames reach the opening is burned. The amount of oxygen which is stored within the compartment at opening, $m_S^{O_2}$, is,

$$m_S^{O_2} = (Y_{O_2}^{UL} \cdot m^{UL} + Y_{O_2}^{LL} \cdot m^{LL}), \qquad (6.14)$$

where $Y_{O_2}^{UL}$ and $Y_{O_2}^{LL}$ are the mass fraction of oxygen in the upper and lower layer respectively. After opening, oxygen is gained from the incoming air and mass is lost in the out flow from the upper layer for a net change in the mass of oxygen, $\Delta m_I^{O_2}$, given by

$$\Delta m_I^{O_2} = Y_{O_2}^{Air} C b \int_{t_o}^{t_i} \int_0^{h_n} \rho(y,t) v(y,t) \, dy \, dt - Y_{O_2}^{UL} C b \int_{t_o}^{t_i} \int_{h_n}^{h_t} \rho(y,t) v(y,t) \, dy \, dt \qquad (6.15)$$

for $t_o > t > t_i$, and where $Y_{O_2}^{Air}$ is the oxygen mass fraction of air.

After ignition, the compartment gases are assumed to be well mixed as in Eq. (6.12) and the oxygen mass fraction is taken as:

$$Y_{O_2}^{Mix} = \frac{Y_{O_2}^{UL}(m_{t=t_o}^{UL} - m_{Out}) + Y_{O_2}^{LL} \cdot m_{t=t_o}^{LL}}{m_{t=t_o}^{UL} + m_{t=t_o}^{LL} + m_{In} - m_{Out}}, \qquad t_i > t > t_{fo}, \qquad (6.16)$$

where $Y_{O_2}^{Mix}$ is assumed constant over that time interval. The change in the stored oxygen concentration over the interval $t_I > t > t_{FO}$, $\Delta m_{FO}^{O_2}$, is then taken as:

$$\Delta m_{FO}^{O_2} = Y_{O_2}^{Air} C b \int_{t_I}^{t_{FO}} \int_0^{h_n} \rho(y,t) v(y,t) \, dy \, dt - Y_{O_2}^{Mix} C b \int_{t_I}^{t_{FO}} \int_{h_n}^{h_t} \rho(y,t) v(y,t) \, dy \, dt \qquad (6.17)$$

The amount of oxygen that is stored in the compartment when flames exit the compartment is then, $m_{FO}^{O_2}$, then:

$$m_{FO}^{O_2} = m_S^{O_2} + \Delta m_I^{O_2} + \Delta m_{FO}^{O_2}, \qquad (6.18)$$

The maximum amount of stored chemical energy at the time flames reach the opening which can burn in the compartment, E_O, is,

$$E_O = 12.5\, m_{FO}^{O_2}, \qquad (6.19)$$

where the 12.5 factor is the amount of energy released per unit mass of oxygen based on an oxygen-methane reaction. The energy is the compartment, but not consumed in,

$$E_N = E_S - E_I - E_F - E_O. \qquad (6\text{-}20)$$

The portion of the chemical energy stored in the compartment at opening that remains available for the fireball, E_{FB}, can now be estimated as:

$$E_{FB} = E_I + E_F + E_N = E_S - E_O, \qquad (6.21)$$

In Fig. 6.6 for run 9 in Tables 6.1 and 6.2 at 540 s the hatch covering the opening falls open and the upper layer gases start to exit the compartment. As these gases leave the compartment the stored energy, $E_S = 19.4$ MJ, drops as demonstrated by the steep negative slope seen in Fig. 6.6 after 540 s. At ignition, t = 560 s, $E_I = 5.1$ MJ and the slope of the stored chemical energy line becomes very large as the compartment gases are forced out of the compartment by the expanding flame, $E_F \sim 2.7$ MJ. Some of the remaining chemical energy stored within the compartment has been consumed there $E_O \sim 5.3$ MJ; $E_N \sim 6.3$ MJ was not. The chemical energy forced out of the compartment, $E_{FB} \sim 14.1$ MJ, may burn as a large fireball outside of the compartment. The combustion efficiency of the fireball remains unknown. Table 6.2 is a summary of the calculated chemical energy parameters. Column 1 is the experiment run number. Columns 2 - 5 are the compartment species concentrations at opening for O_2, CO, CO_2, and HC, respectively. Species concentrations printed in parenthesis indicate the sample was taken from the lower layer. The average hydrocarbon mass fraction for the compartment evaluated at backdraft ignition is shown in column 6. Column 7 is the chemical energies stored in the compartment when the compartment was opened. Column 8 is the chemical energy which flows out of the compartment prior to backdraft ignition. The chemical energy which flows out of the opening between compartment ignition and the flames exiting the opening is shown in column 9. Column 10 is the amount of energy burned in the compartment assuming all the available oxygen is consumed. Column 11 is the energy in the compartment which is not burned there. Column 12 is the amount of chemical energy which is available for the fireball. Column 13 is an estimate of the diameter of the fireball which burns outside the compartment. Column 14 is the run name which corresponds to the data given in appendix C.

The presence of a fireball is used as the indication of a backdraft in these experiments. When the fireball is not observed, the flame travel in the compartment is noticeably slower and less intense than in the experiments with a fireball. The fireball also gives a crude indication of the severity of the backdraft. If the fireball were contained within a volume scaled with the compartment, such as a corridor, the pressure rise within the compartment would be significantly higher, possibly enough to cause structural damage.

Table 6.1 Summary of chemical energy accounting reporting the species concentrations at compartment opening, the chemical energy stored in the compartment at opening, the chemical energy which exists in the compartment prior to ignition, the chemical energy that exists in the compartment after ignition and before flames exit the compartment, the chemical energy released in the compartment by the backdraft, the chemical energy that is available for the fireball, and an estimate of the fireball diameter.

Run Number	$Y_{O_2}^{(LL)}$	$Y_{CO}^{(LL)}$	$Y_{CO_2}^{(LL)}$	$Y_{HC}^{(LL)}$	Y_{HC}^{Mix}	E_S (MJ)	E_I (MJ)	E_F (MJ)	E_O (MJ)	E_{FB} (MJ)	Fire Ball (m)	Run Number Appendix C
1	0.11	0.003	0.05	0.15	0.08	14.43	2.03	0.00	6.22	8.21	0	P3EXP82
2	0.11	0.003	0.05	0.17	0.08	16.59	3.40	4.86	4.03	12.55	0	P3EXP79
3	0.11	0.003	0.05	0.18	0.08	17.47	4.95	4.27	5.30	12.17	0	P3EXP77
4	(0.17)	(0.002)	(0.03)	(0.02)	0.06	15.26	5.44	3.02	4.81	10.45	0	P3EXP85
5	0.11	0.003	0.05	0.18	0.09	17.28	3.51	6.67	3.26	14.02	0	P3EXP81
6	0.11	0.003	0.05	0.19	0.08	18.72	5.04	4.36	4.57	14.15	4	P3EXP76
7	(0.16)	(0.002)	(0.04)	(0.03)	0.07	18.13	6.27	3.50	4.96	13.18	4	P3EXP86
8	0.11	0.003	0.05	0.18	0.09	17.45	3.60	5.67	3.79	13.66	4	P3EXP83
9	0.11	0.003	0.05	0.20	0.08	19.39	5.08	2.72	5.28	14.11	5	P3EXP78
10	0.11	0.003	0.05	0.20	0.08	18.99	6.50	2.53	5.54	13.45	5	P3EXP80
11	(0.15)	(0.002)	(0.04)	(0.04)	0.08	19.84	7.92	2.85	5.77	14.08	4	P3EXP84

6.4 Future Research

Future work should concentrate on designing and building a full scale apparatus using a controlled fire source and possibly more realistic fuels. Improved gas analysis would also be useful to determine what hydrocarbons are present in the real fuel experiments. Additional opening geometries using different vent/wall area ratios and locations could also be used to investigate the effects vent flow has on the backdraft severity and chance of occurrence.

References - Chapter 6

[1] Croft, W.M., "Fires Involving Explosions", Fire Safety Journal, 3, 3-24, 1980/81.

[2] Steward, P. D. C., "Dust and Smoke Explosions", NFPA Quarterly, 7, 424-428, 1914.

[3] Fire Ventilation Practices, 5th Edition, International Fire Service Training Association, Oklahoma, 32-33, 1970.

[4] Dunn, V., "Beating the Backdraft", Fire Engineering, 44-48, 1988.

[5] Emmons, H.W., "The Calculation of a Fire in a Large Building," ASME Paper No. 81-HT-2, American Society for Mechanical Engineers, New York, 1981.

[6] Pagni, P. J., and Shih, T. M., "Excess Pyrolyzates," 16th Symposium (Int'l) on Combustion, pp. 1329-1343, The Combustion Institute, Pittsburgh, PA, 1976.

[7] Zalosh, R.G., "Explosion Protection," in SFPE Handbook of Fire Protection Engineering, (P.J. DiNenno, ed.) pp. 2-88 to 2-105, National Fire Protection Association, Quincy, MA, 1988.

[8] Factory Mutual Engineering Corporation, "Damage-Limiting Construction," Loss Prevention Data Sheet 1-44, Norwood, MA, 1991.

[9] "Neoceram Super Heat-Resistant Glass for Industrial Use," Nippon Electric Glass Co. Ltd.

[10] Quintiere, J.G., Steckler, K., and Corley, D., "An Assessment of Fire Induced Flows in Compartments" Fire Science and Technology, 4, 1-14, 1984.

[11] McCaffrey, B.J. and Heskestad, G., "A Robust Bidirectional Low-Velocity Probe for Flame and Fire Application", Combustion and Flame, 26 125-127, 1976.

[12] Newman, J.S. and Croce, P.A., "A Simple Aspirated Thermocouple For Use In Fires", Fire & Flammability, 10 326-336, 1979.

CHAPTER 7

CONCLUSIONS

7.1 Summary

The backdraft scenario presented here accurately describes the phenomena known as backdraft. Experiments show that the ignition of the backdraft does not occur immediately upon opening the compartment. It is this time delay caused by the gravity current propagation which creates a hazard to firefighters who enter a compartment and become trapped in the backdraft process.

The gravity current entering a compartment is both qualitatively and quantitatively similar to other naturally occurring gravity currents. The nondimensional velocity for the full opening compares well with the perfect fluid theory. The structure of the entering gravity current head for the full open condition shows a shallow mixed region riding in the current near the interface. It also has the detailed features reported for steady state gravity currents, i.e. billows, lobes, and clefts. Entering currents for the slot, window, and door openings show a different gravity current structure with the mixed region occupying nearly the entire current due to the enhanced mixing near the opening. Similar detailed features of the steady state gravity current appear in these currents.

The values of v^* and h^* obtained here for a variety of opening geometries, are independent of the density difference ratio, β. The exiting gravity current is also independent of β, and has a v^* approximately equal to the entering current. These results can be applied to predict the time to ignition for a backdraft with compartment and opening geometries similar to the conditions reported here.

The two dimensional numerical simulation performed at the National Institute of Standards and Technology accurately predicts the compartment gravity current for the fully open condition. The two dimensional computational density profile shows that the main body of the current is made up ambient fluid while a mixed layer made up of large vortical structures travels along the shear interface between the two fluids. Such structure is also observed in the salt water experiments although the individual vortices are not visible due to the averaging inherent in of the flow visualization techniques. For the slot opening, the results of the two dimensional simulation also compare favorably with the experimental results. The computed density profile shows that the structure of the gravity current for the $h_1/3$ centered slot opening is significantly different from the fully open condition. For the slot opening, the entire gravity current is filled with a complex structure of large scale vortices. This large scale mixing for the $h_1/3$ centered slot opening results from the rearward facing step, forced by the lower edge of the slot. Two dimensional transit times predicted by the numerical simulation compare well with the salt water experiments although the computed times are slightly longer for the slot opening. Comparing the computational transit time with the time to ignition for the backdraft experiments gravity current also shows good agreement. Numerical velocity profile predictions in the opening compared well with backdraft experiment results and fall well within the experimental error bounds.

The results presented for the quantitative backdraft experiments show that the hydrocarbon concentration must be >10% in order for a backdraft to occur. When the hydrocarbon concentration is < 10% the flame travel is slow and the compartment overpressure is much lower. As the hydrocarbon concentration increases, the compartment overpressure increases and the backdraft becomes more severe. Hydrocarbon concentrations > 15% exhibit large fire balls outside the compartment and would result in significantly higher pressures if contained within an adjacent corridor or room with a volume of the order of the experimental compartment.

7.2 Future Research

The salt water modeling was limited to only one compartment aspect ratio and limited opening geometries. Future work should focus on different compartment aspect ratios, smaller openings, and openings at the floor level. Improved flow visualization techniques including slit lighting to obtain a two dimensional image of the gravity current are recommended. Time dependent concentrations should be measured within the gravity current for comparison with computational results.

Future backdraft experiments should concentrate on two areas. The first should focus on designing and building a full scale apparatus i.e. a minimum $h_1 = 2.4$ m. Full scale experiments should focus on quantifying the effects of backdraft and how to prevent or reduced the impact of backdraft. Suggested areas to investigate include: 1) Can vertical ventilation prevent backdrafts? 2) Effects of horizontal ventilation when it is not possible to vent the ceiling of the structure. 3) Impact of closing the door after entering a compartment. 4) Effect of containing the fireball in an adjacent space of the same order as the backdraft compartment.

Further research is also necessary on common fuels which may contribute to backdraft. Results presented here indicate that high unburned fuel concentrations are necessary in order for a backdraft to occur. Experimental work is required to determine the effects of ventilation on the pyrolysis rate of the fuel and to quantify the amount of pyrolyzates consumed in the flame as a function of the available oxygen concentration. Gas analysis capable of determining the nature of the hydrocarbons present would also be helpful in calculating flammability limits.

APPENDIX A

SPECIES CONCENTRATIONS

In order to evaluate the composition of the upper layer of the compartment it is necessary to determine all of the major species present, specifically O_2, CO_2, CO, HC, H_2O, and N_2. Direct measurement of species concentrations were only possible for only O_2, CO_2, CO, and HC. Concentrations of H_2O and N_2 had to be calculated using species balances on O_2 and total C. The following assumptions were necessary for the analysis: 1) that the upper layer is well stirred and 2) that the overall reaction was:

$$CH_4 + a(O_2 + 3.77N_2) + bH_2O \rightarrow cCO_2 + dCO + eH_2O + fN_2 + gCH_4 \quad (A\text{-}1)$$

Using these assumptions, which are reasonable since the fire source was a gas burner flowing methane, an overall balance on the oxygen and total carbon will yield the two concentrations, H_2O and N_2.

The first step is to determine the amount of N_2 which is present in the system. This is done by a simple O_2 balance. Starting with the dry sample which has had all of the water scrubbed out using a desiccant filter the total amount of oxygen is calculated:

$$n^A_{O_2} = n^A (X^A_{O_2} + X^A_{CO_2} + \frac{1}{2} X^A_{CO}) \quad (A\text{-}2)$$

Symbols follow standard chemical notation when possible and a complete listing appears in the nomenclature. The oxygen is not only in the analyzed gases but also in the water which has been scrubbed. The amount of H_2O which is present is a result of the combustion of CH_4 and can be estimated from the following relationship derived from a C balance,

$$n^C_{H_2O} = 2(n^A_{CO_2} + n^A_{CO}) \Rightarrow 2n^A (X^A_{CO_2} + X^A_{CO}) \quad (A\text{-}3)$$

Combining Eqs. (A-2 and A-3), multiplying by 3.77, i.e. the molar ratio of N_2/O_2, and assuming $n_D=1$ gives the total amount of N_2:

$$n^L_{N_2} = 3.77 n^L_{O_2} \Rightarrow 3.77[X^A_{O_2} + 2X^A_{CO_2} + \frac{3}{2}X^A_{CO}] \tag{A-4}$$

Knowing the amount of N_2, the remaining unknown is the concentration of H_2O. There are two sources of H_2O, that produced by combustion and the amount present in the air. The amount of H_2O produced by combustion has already been estimated in eq. (A-3). The amount of H_2O present in the air is obtained from the following definition of relative humidity:

$$w = \left[0.622 \frac{\phi p_{sat}}{(101,325 - \phi p_{SAT})}\right] \tag{A-5}$$

where w is the specific humidity, ϕ is the relative humidity, and p_{sat} is the saturation pressure (Pa). The relative humidity was recorded before each experiment and p_{sat} can be calculated from the following relationship taken from ref. 1:

$$p_{SAT}(T) = CT^{-B/R_v} \exp\left(\frac{-A}{R_v T}\right), \tag{A-6}$$

where $A=3.18 \times 10^6$ J kg^{-1}, $B=2470$ J kg^{-1} K^{-1}, and $C=6.05 \times 10^{26}$ N m^{-2}. Knowing the relative humidity and temperature the amount of H_2O in the ambient air can be calculated from the following relationship:

$$n^A_{H_2O} = \frac{w \cdot m^D_{Air}}{M_{H_2O}}. \tag{A-7}$$

The mass of the dry air is calculated from the:

$$m_{Air}^A = n^A \left[M_{O_2}\left(X_{O_2}^A + 2X_{CO_2}^A + \frac{3}{2}X_{CO}^A \right) + M_{N_2} X_{N_2}^A \right] \qquad \text{(A-8)}$$

Using eqs. (A-3 and A-7) and the assumption that $n_D=1$, the total number of moles in the system can be calculated as:

$$n^L = n^A + n^C + n_{H_2O}^{Air} \Rightarrow 1 + n^C + n_{H_2O}^{Air} \qquad \text{(A-9)}$$

Applying eq. (A-9) the mole fractions for the upper layer can be calculated as,

$$X_{O_2}^L = \frac{n^A X_{O_2}^A}{n^L}$$

$$X_{CO_2}^L = \frac{n^A X_{CO_2}^A}{n^L}$$

$$X_{CO}^L = \frac{n^A X_{CO}^A}{n^L}$$

$$X_{N_2}^L = \frac{n^A X_{N_2}}{n^L}$$

$$X_{H_2O}^L = \frac{n^C + n^{Air}}{n^L}$$

The final result are reported as mass fractions. To convert the mole fractions the molecular weight of the upper layer sample is calculated,

$$M^L = X_{O_2}^L M_{O_2} + X_{CO_2}^L M_{CO_2} + X_{CO}^L M_{CO} + X_{N_2}^L M_{N_2} + X_{H_2O}^L M_{H_2O}. \qquad \text{(A-10)}$$

Using the definition of mass fraction we can convert mole fractions to mass fractions with:

$$Y_i^L = \frac{X_i^L M_i}{n^L M^L} \qquad \text{(A-11)}$$

References - Appendix A

[1] Sahota, M.S. and Pagni, P.J. "Heat and Mass Transfer In Porous Media Subject to Fires", <u>International Journal of Heat and Mass Transfer,</u> **22**, 1069-1081, (1979).

APPENDIX B

COMPARTMENT DETAILS

Pressure Relief Panel Detail

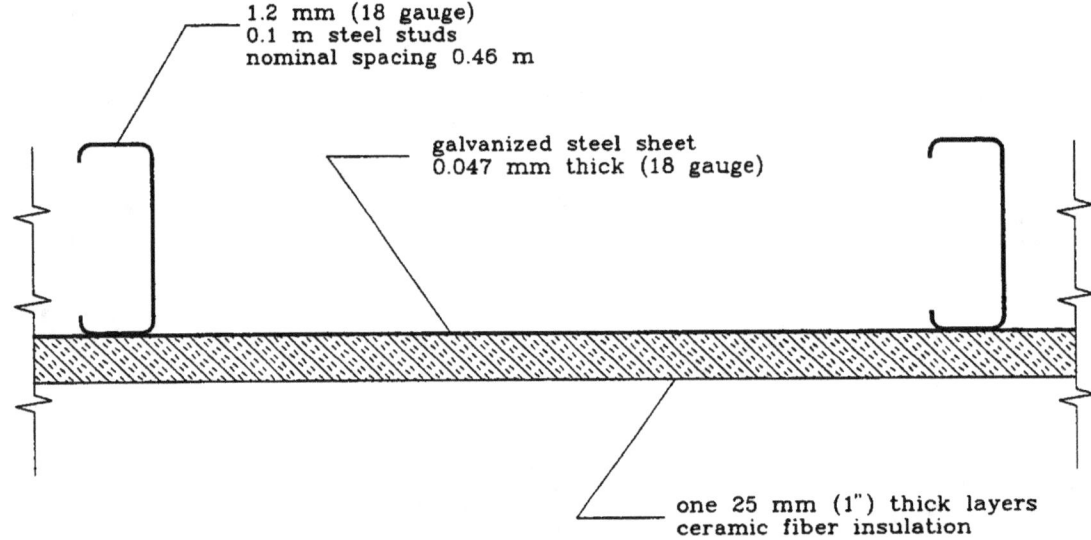

Wall Detail

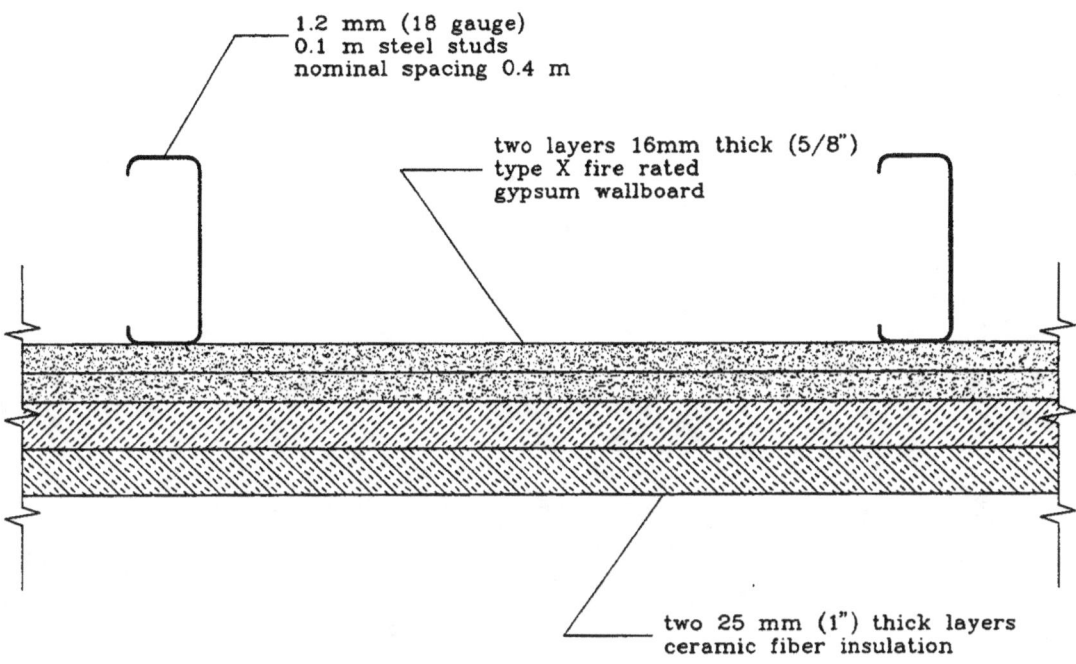

Ceiling Detail

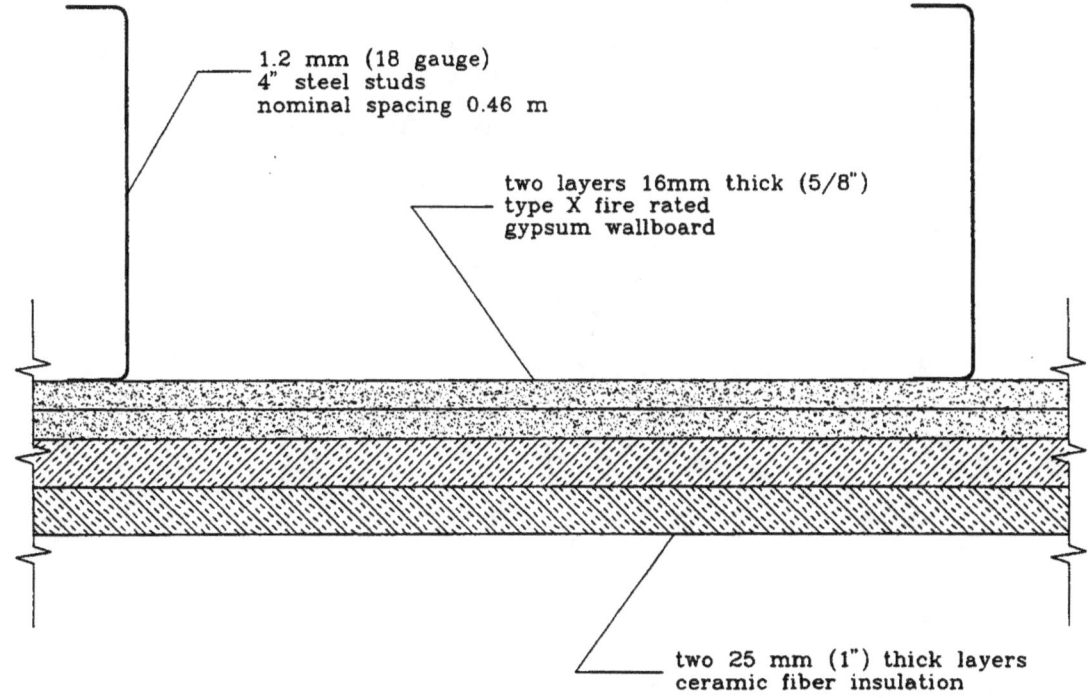

APPENDIX C
BACKDRAFT EXPERIMENTAL DATA

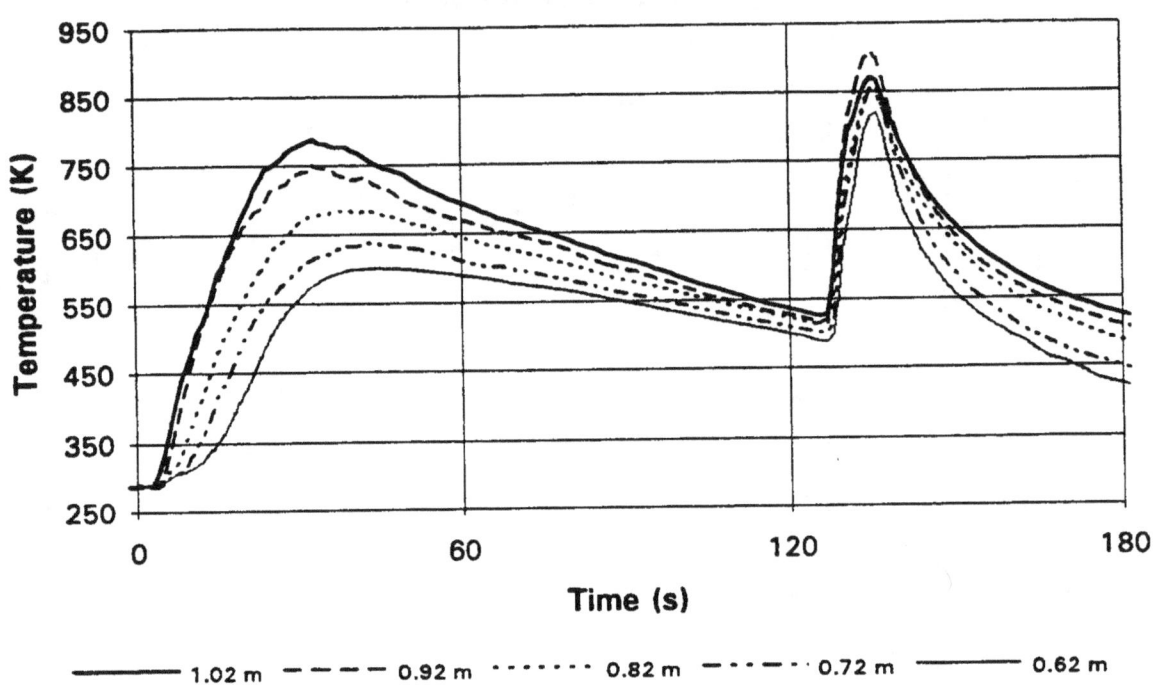

Layer Temperature History - TC Tree Data

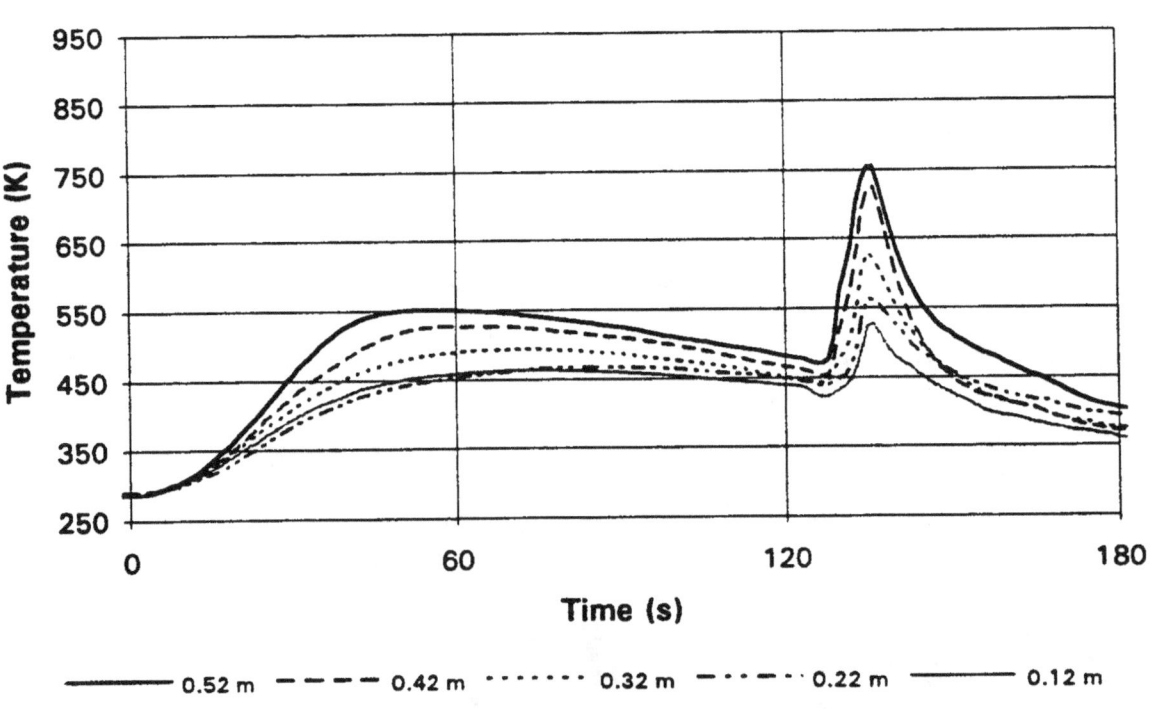

Layer Temperature History - TC Tree Data

Run: P3EXP26

Two Zone Approximation - Temperature & Interface Histories

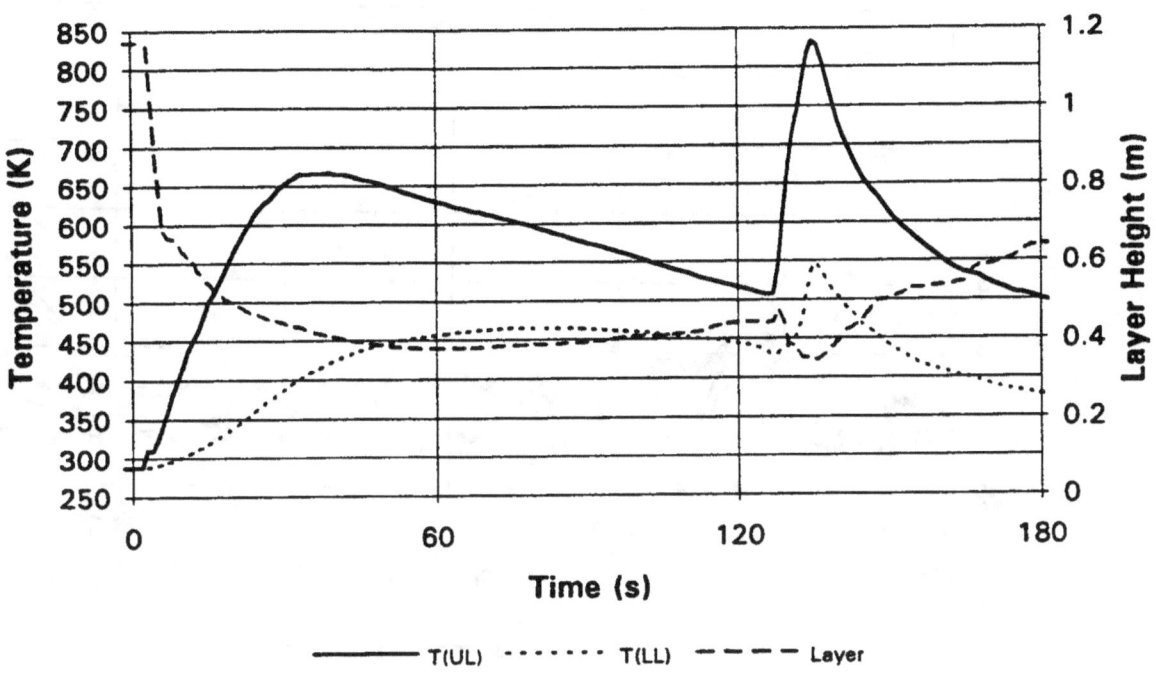

Mass Fraction History - Compartment Gases

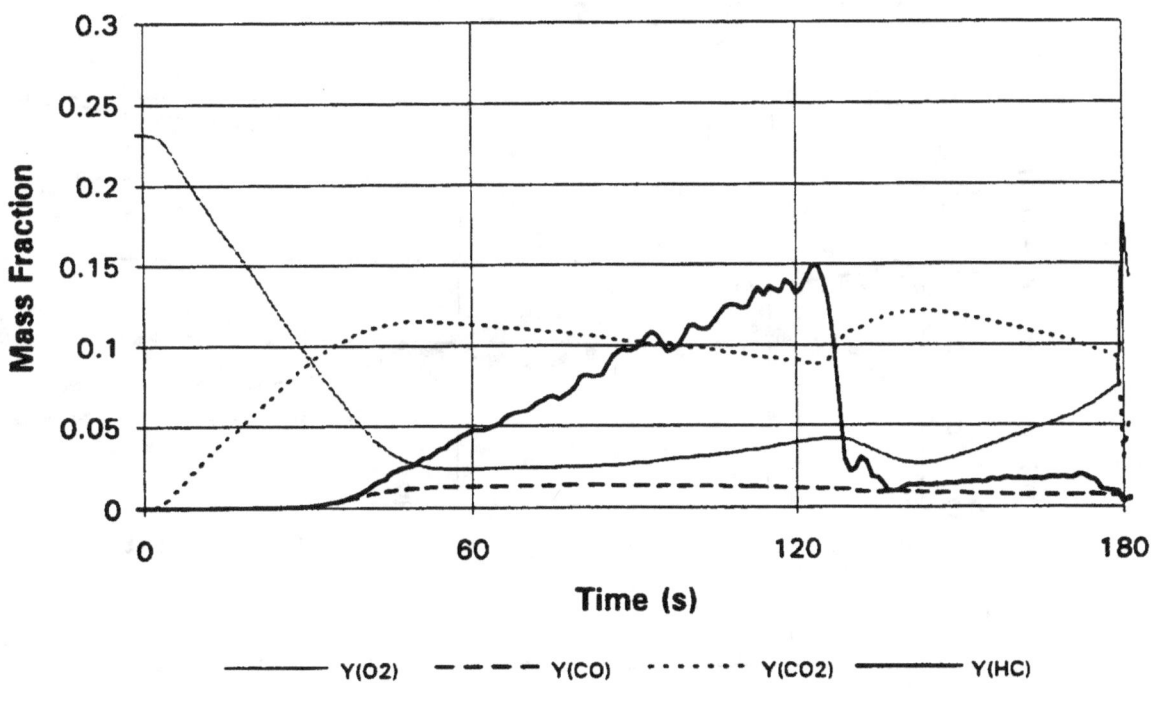

Run: P3EXP26

Bidirectional Probe Velocity History

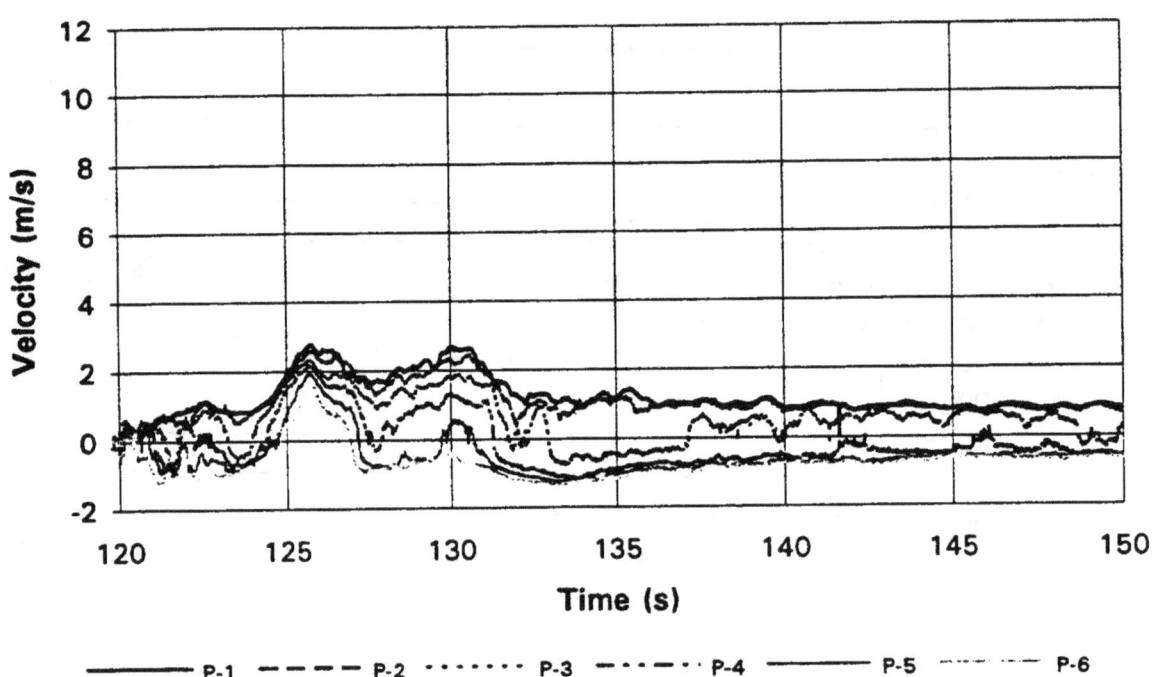

Pressure & Total Mass Out Histories

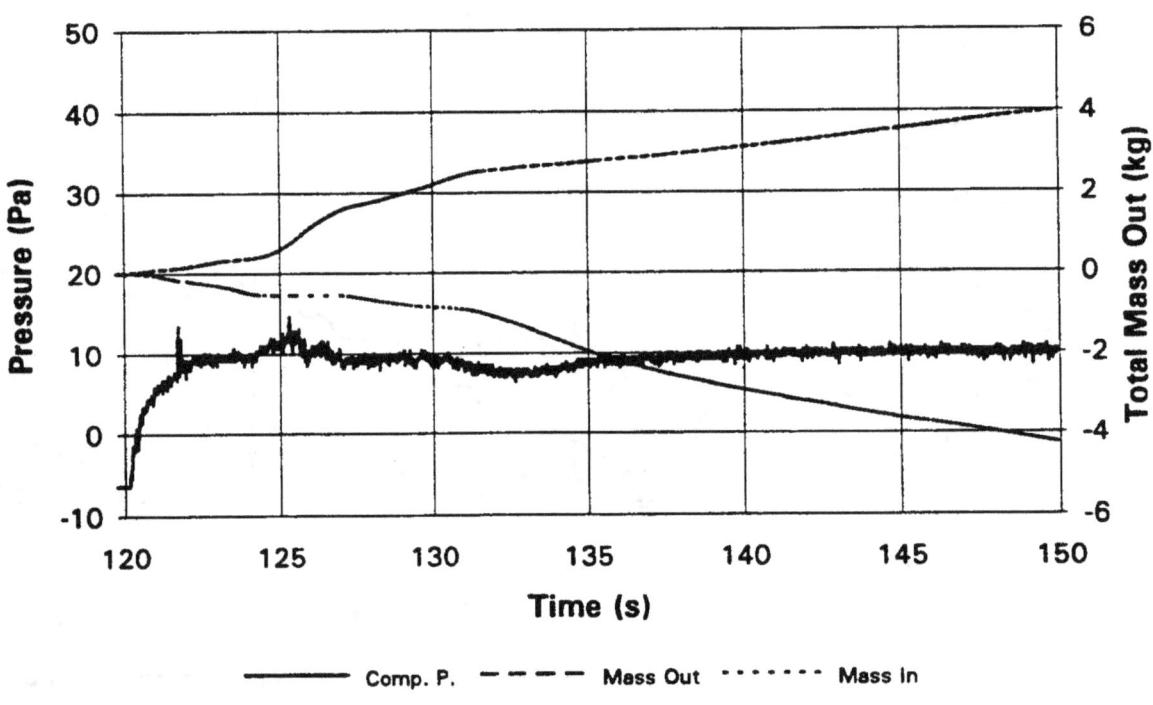

Run: P3EXP26

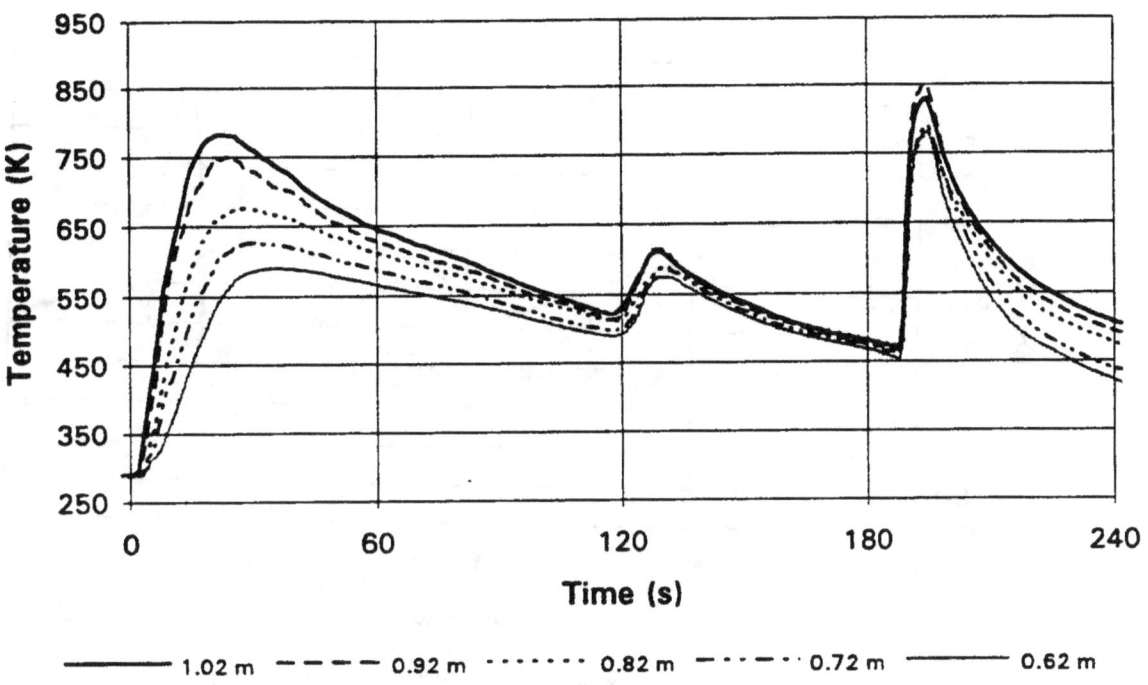

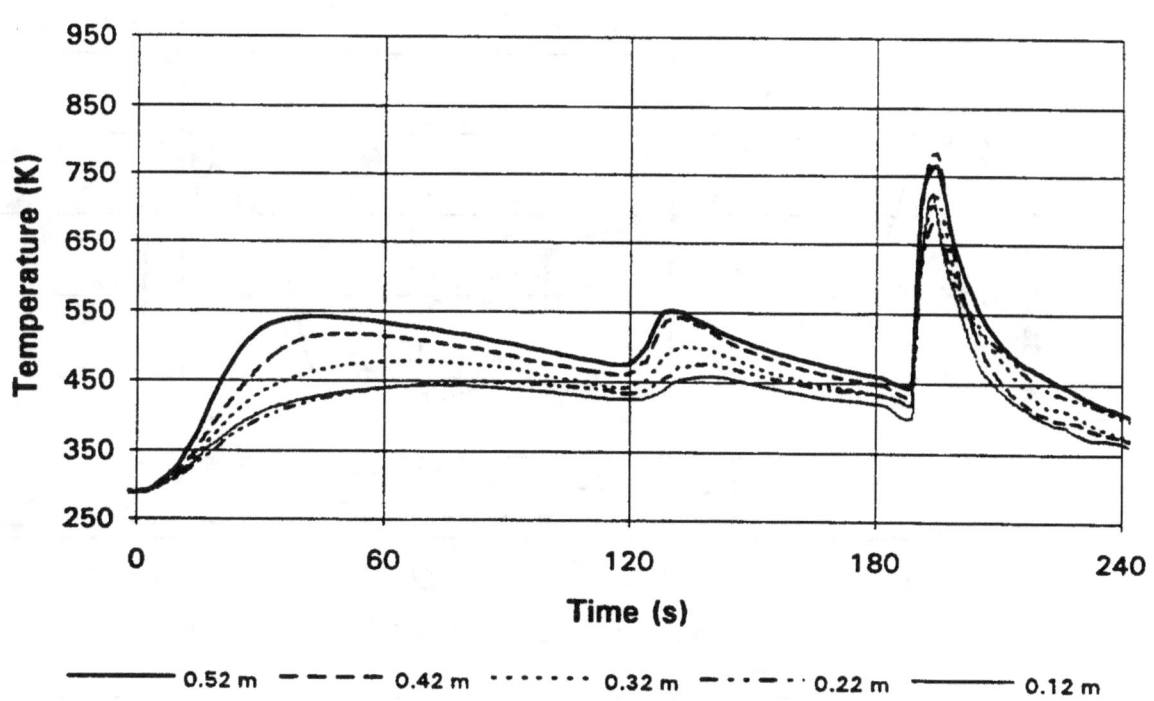

Run: P3EXP27

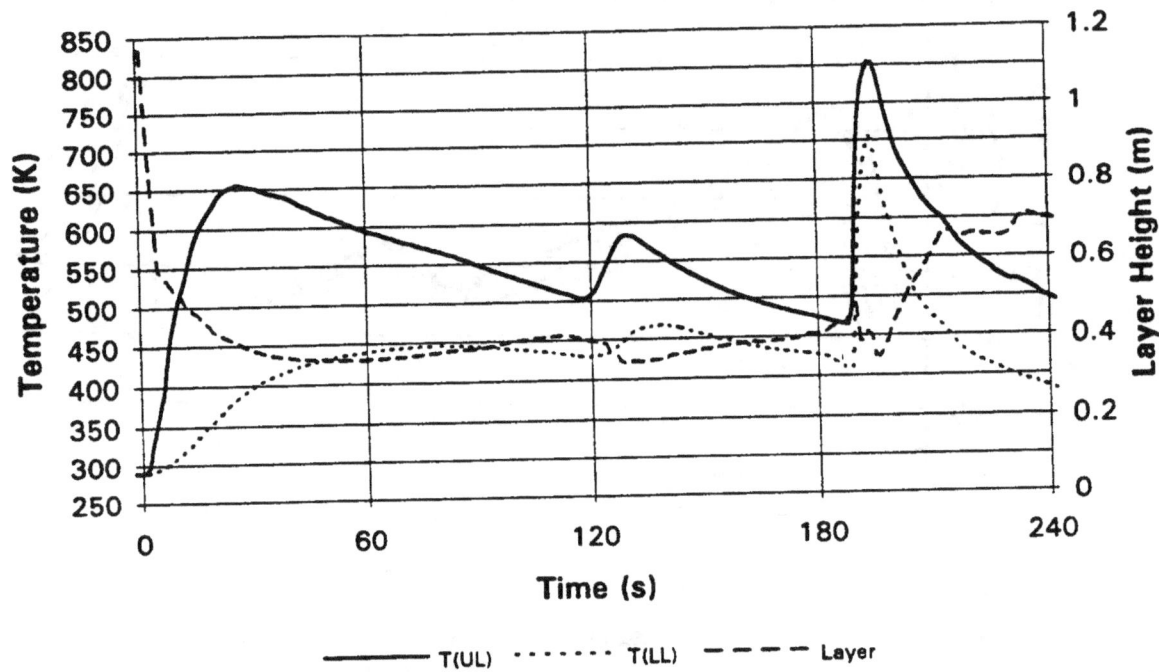

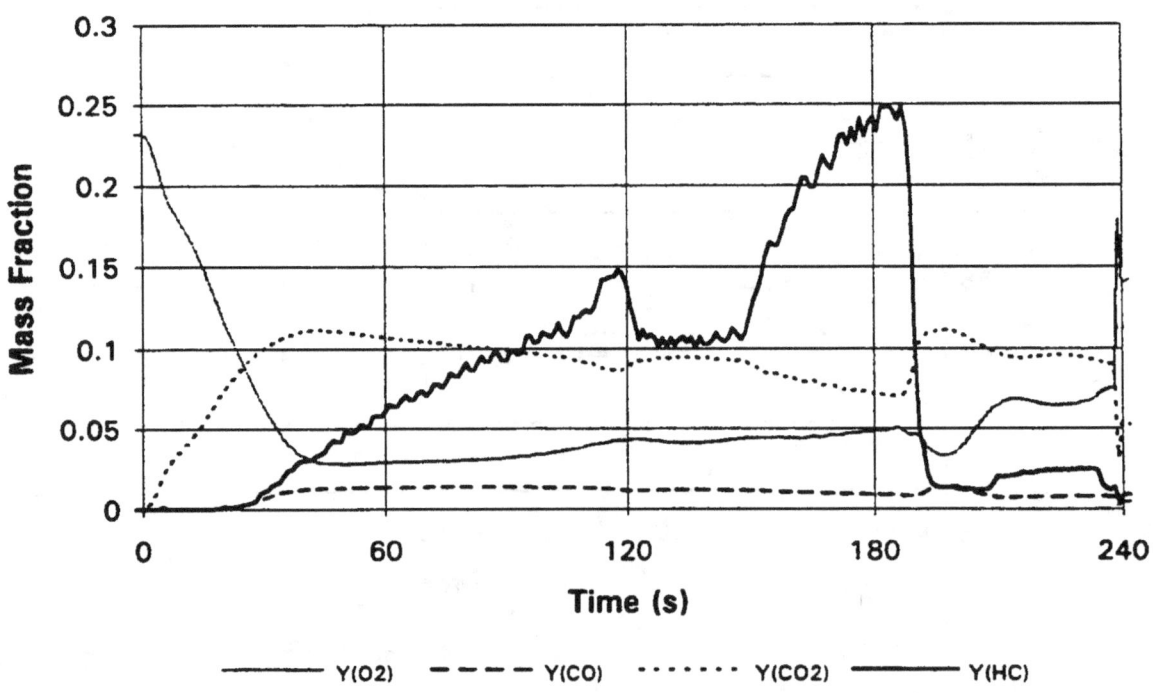

Run: P3EXP27

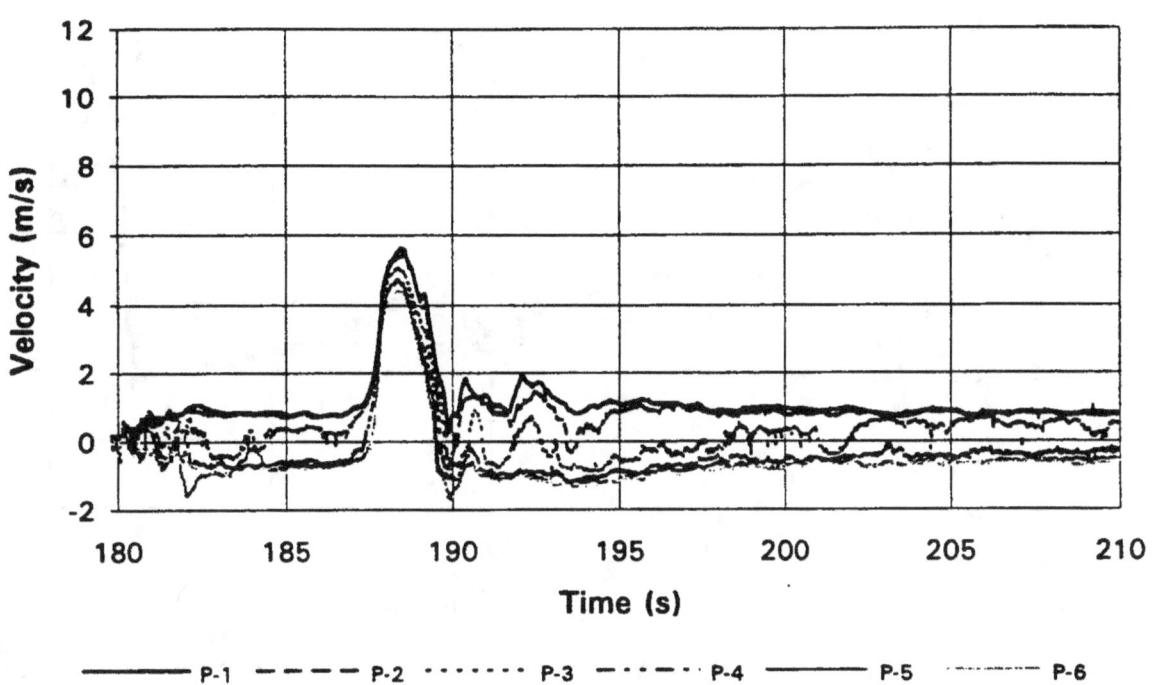

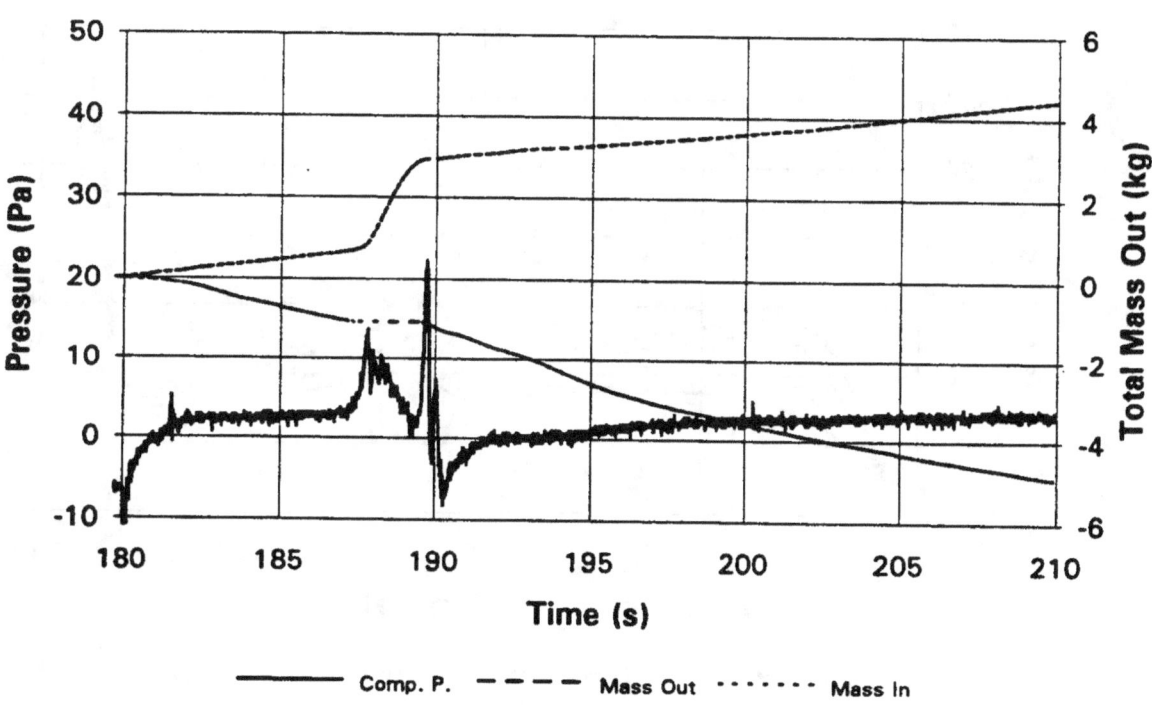

Run: P3EXP27

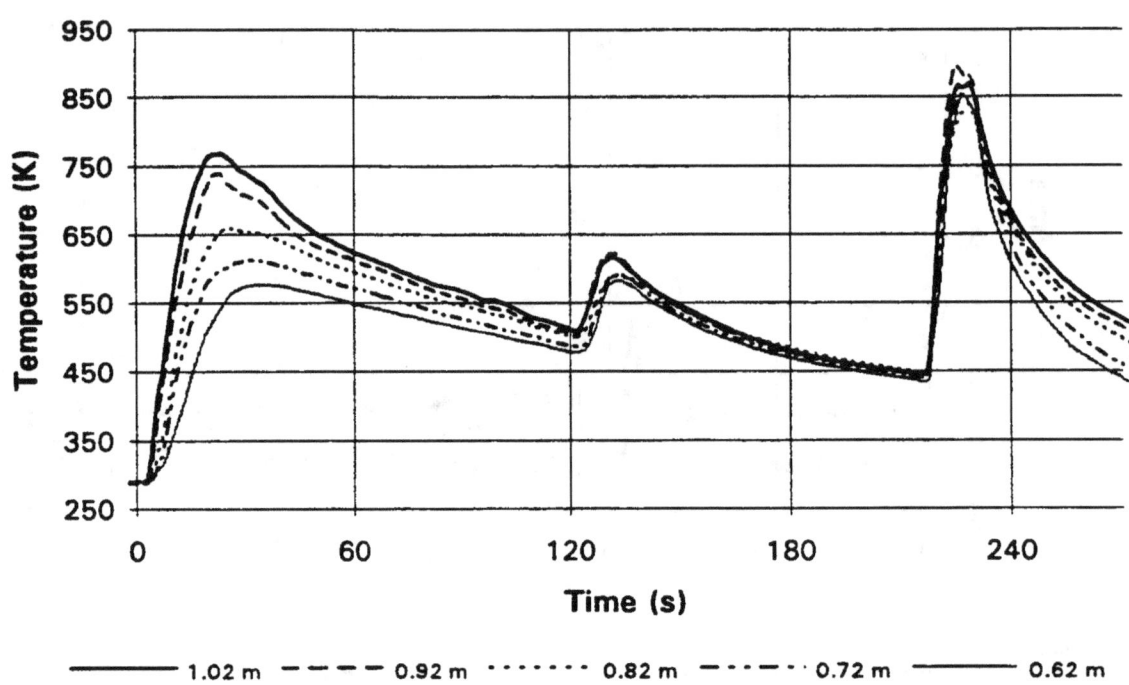

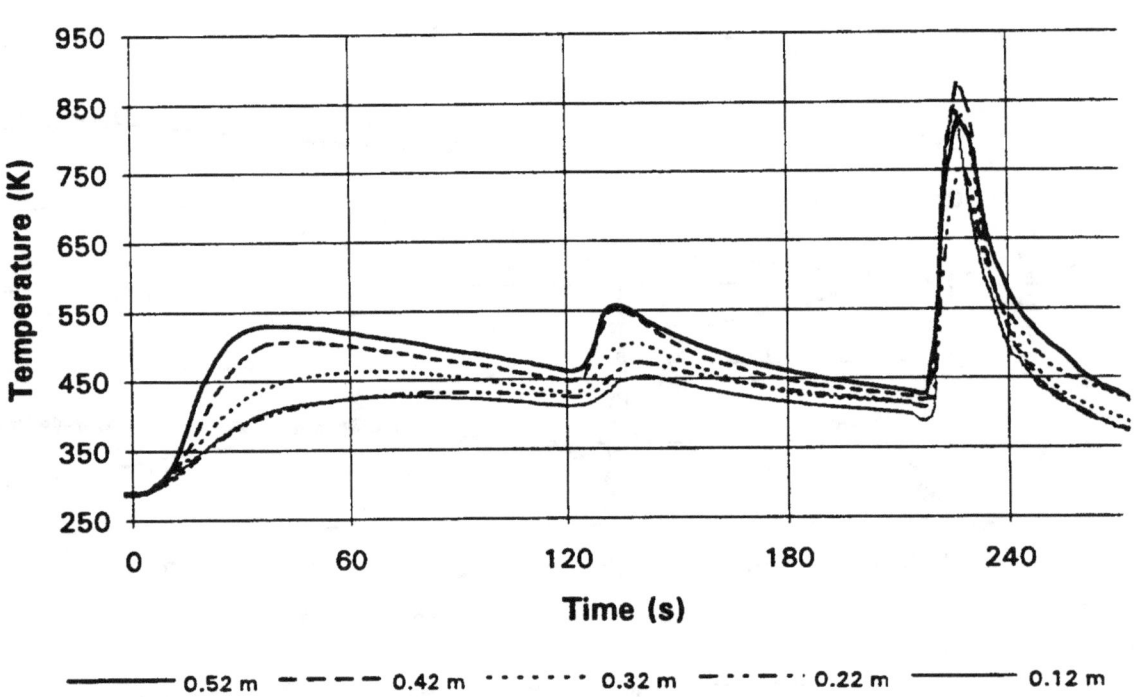

Run: P3EXP28

Two Zone Approximation - Temperature & Interface Histories

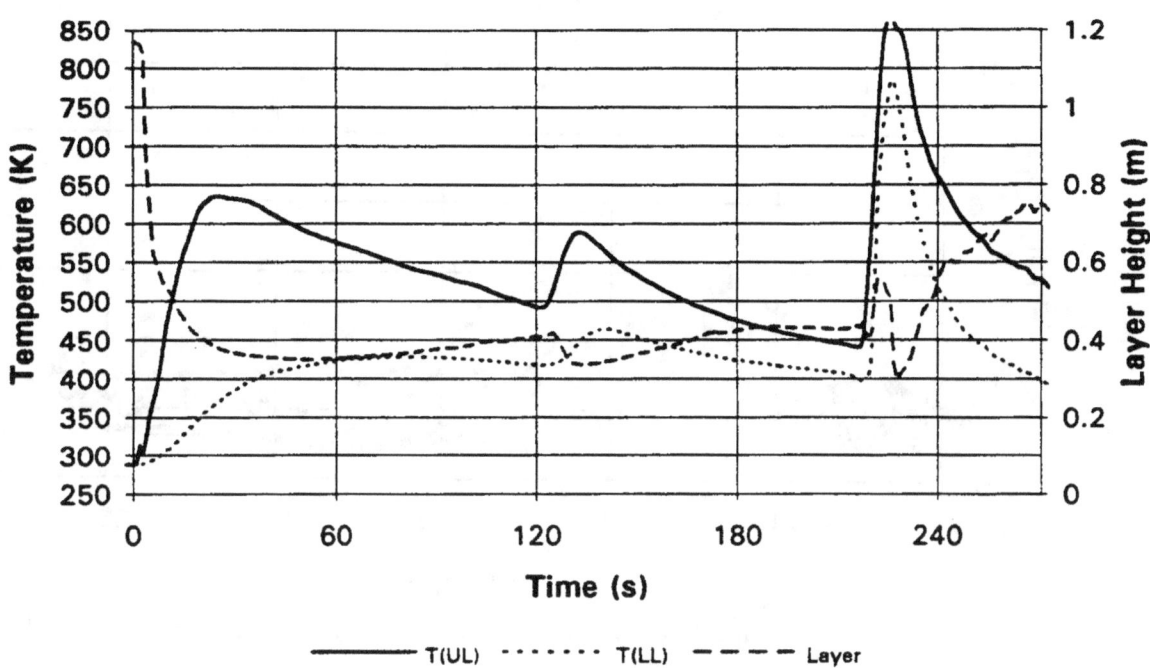

Mass Fraction History - Compartment Gases

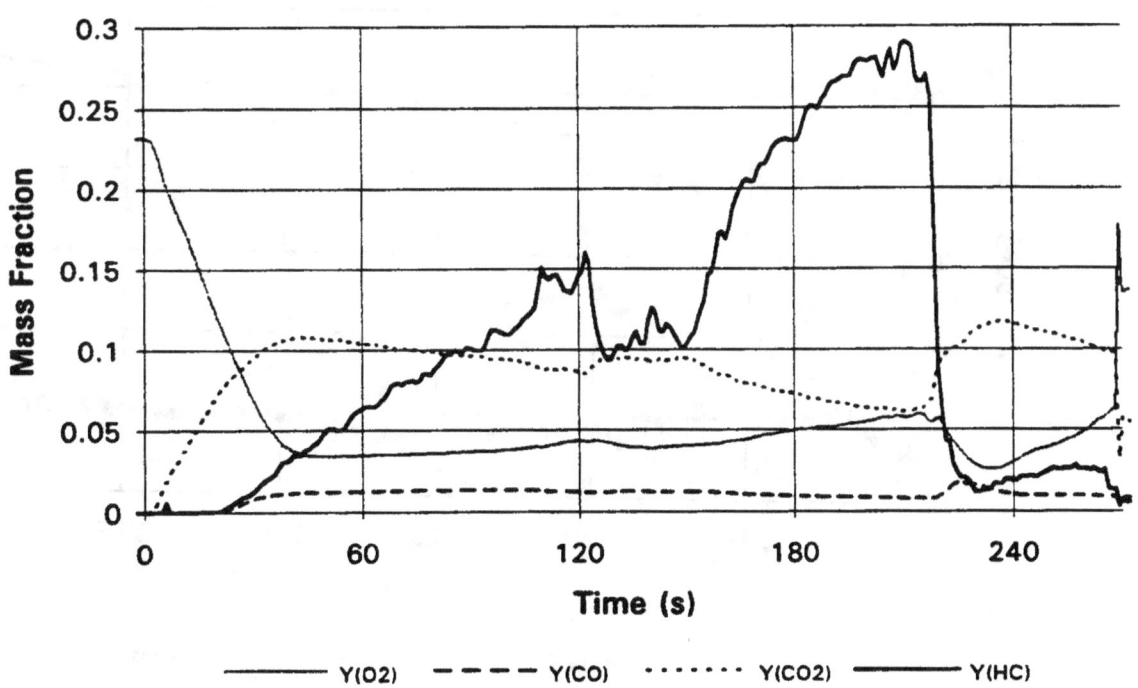

Run: P3EXP28

Bidirectional Probe Velocity History

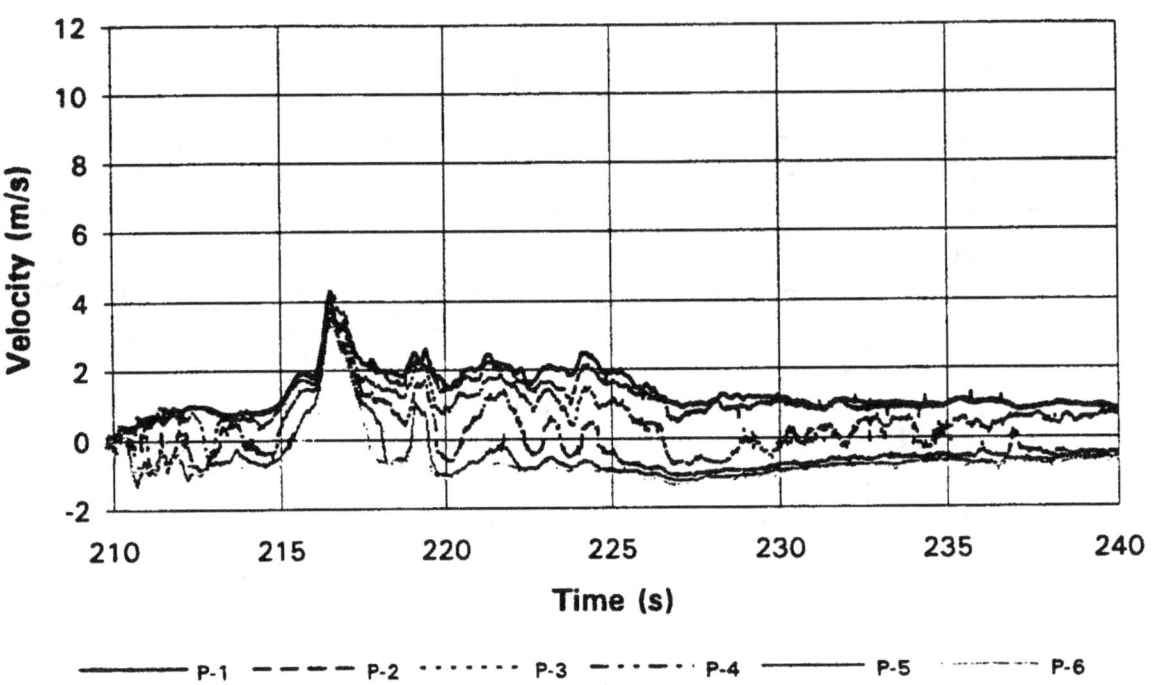

Pressure & Total Mass Out Histories

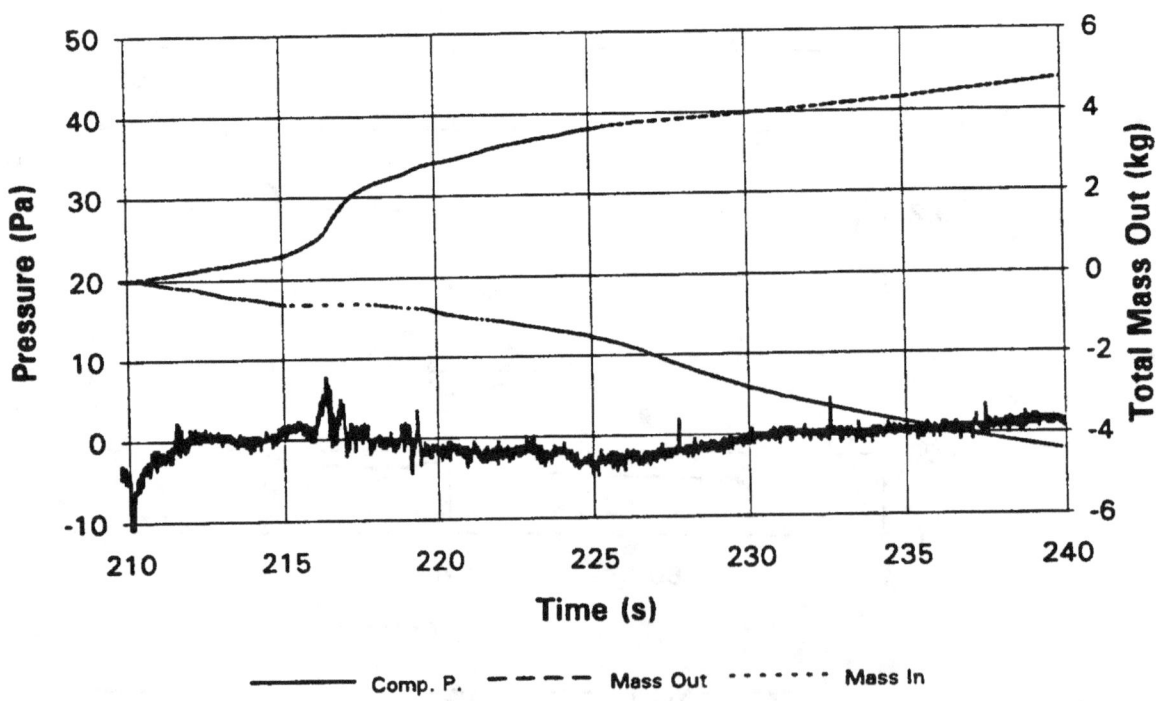

Run: P3EXP28

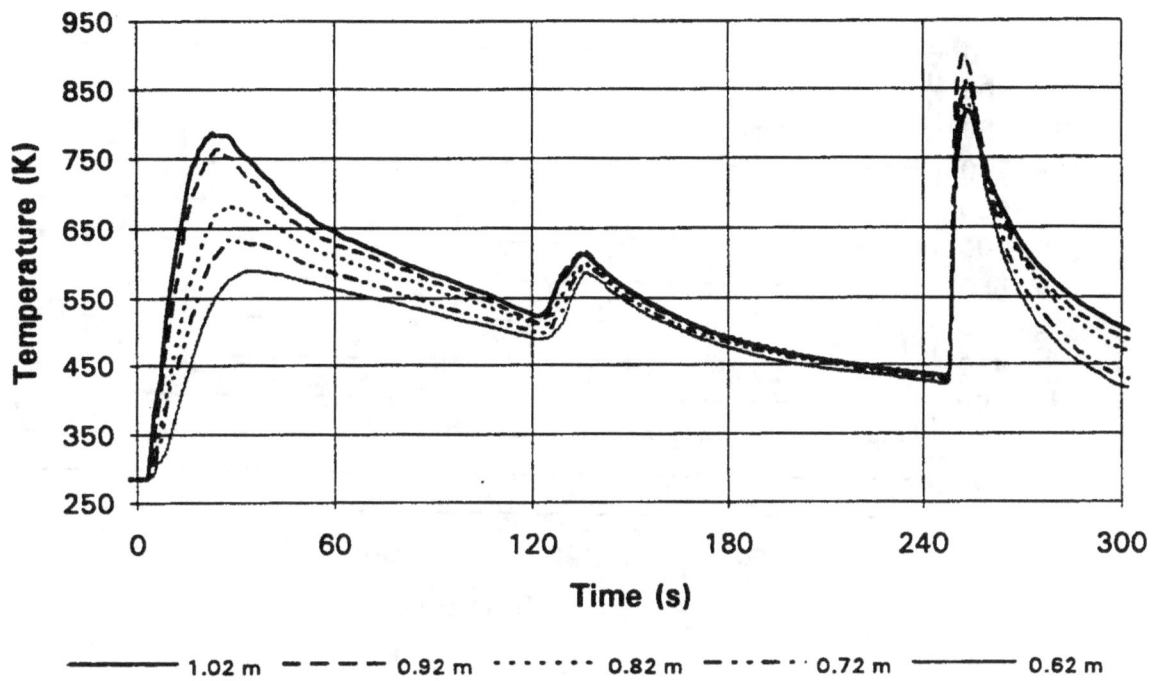

Layer Temperature History - TC Tree Data

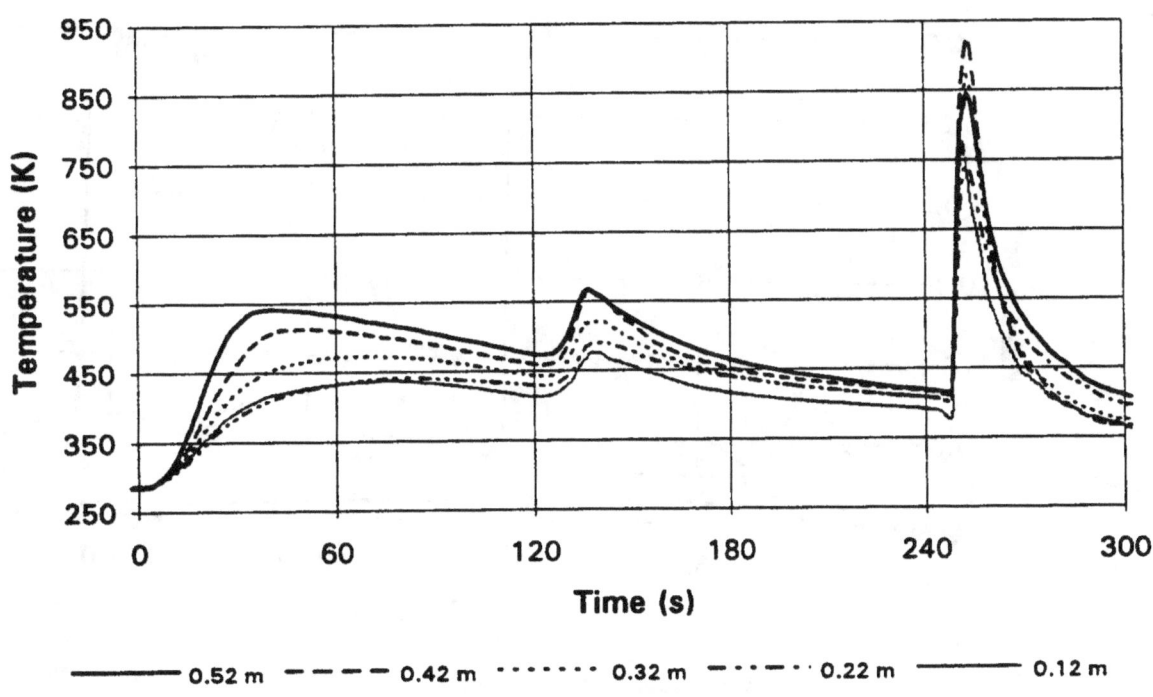

Layer Temperature History - TC Tree Data

Run: P3EXP29

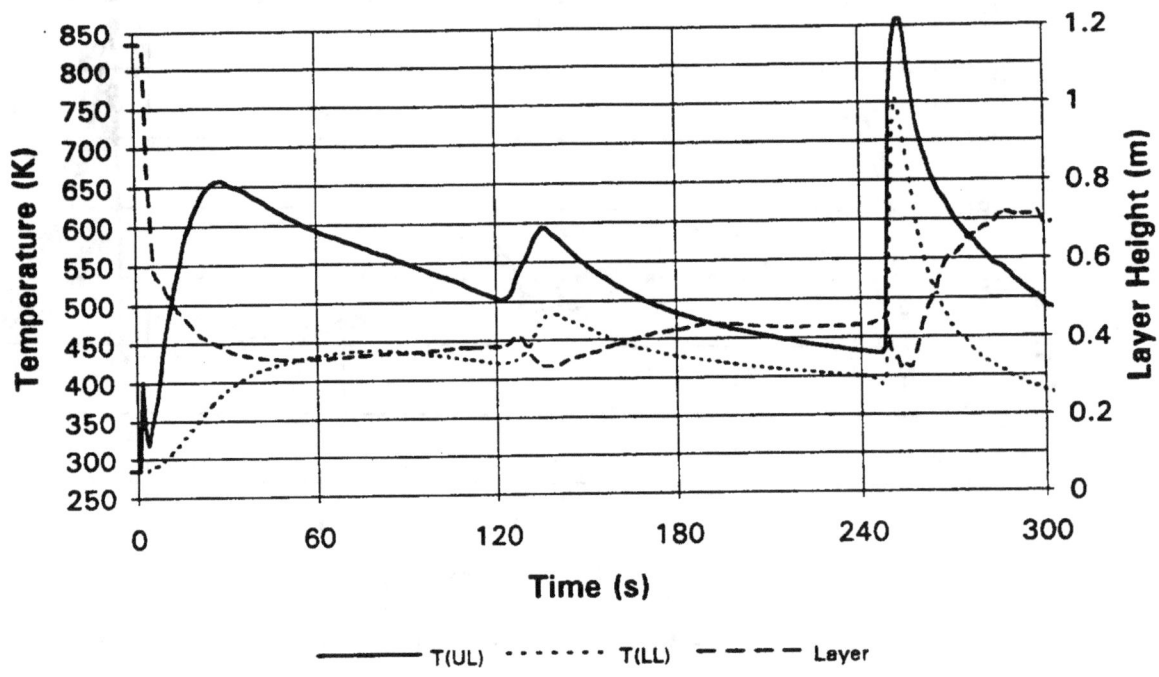

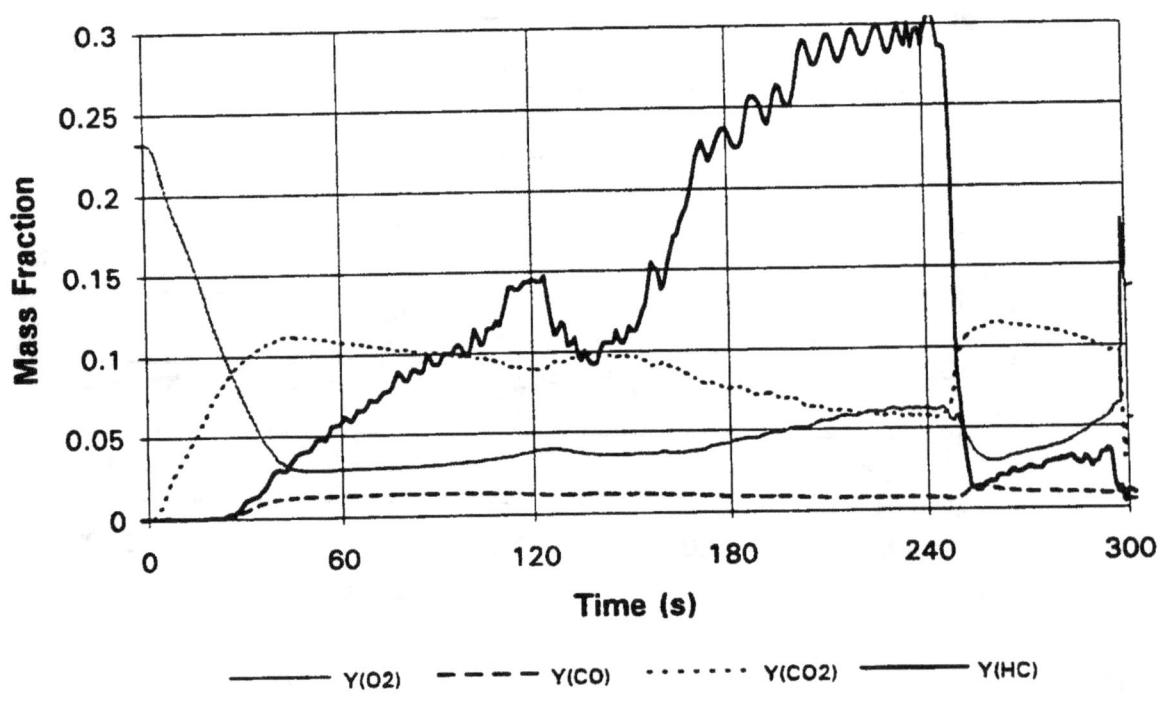

Run: P3EXP29

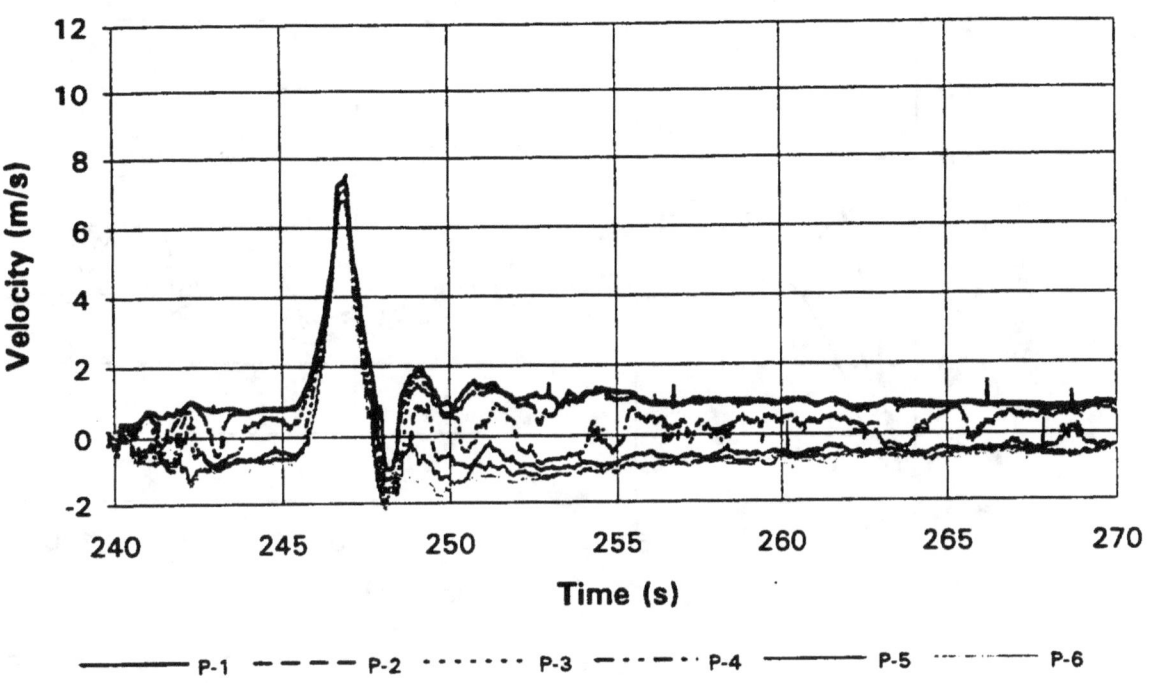

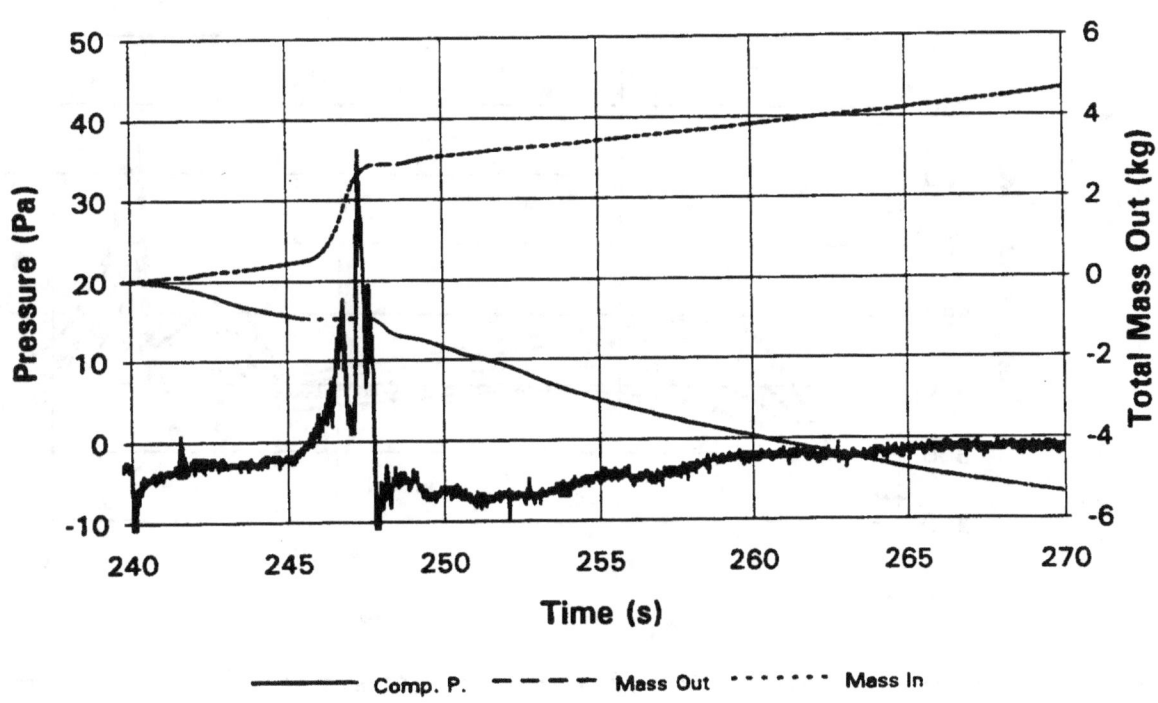

Run: P3EXP29

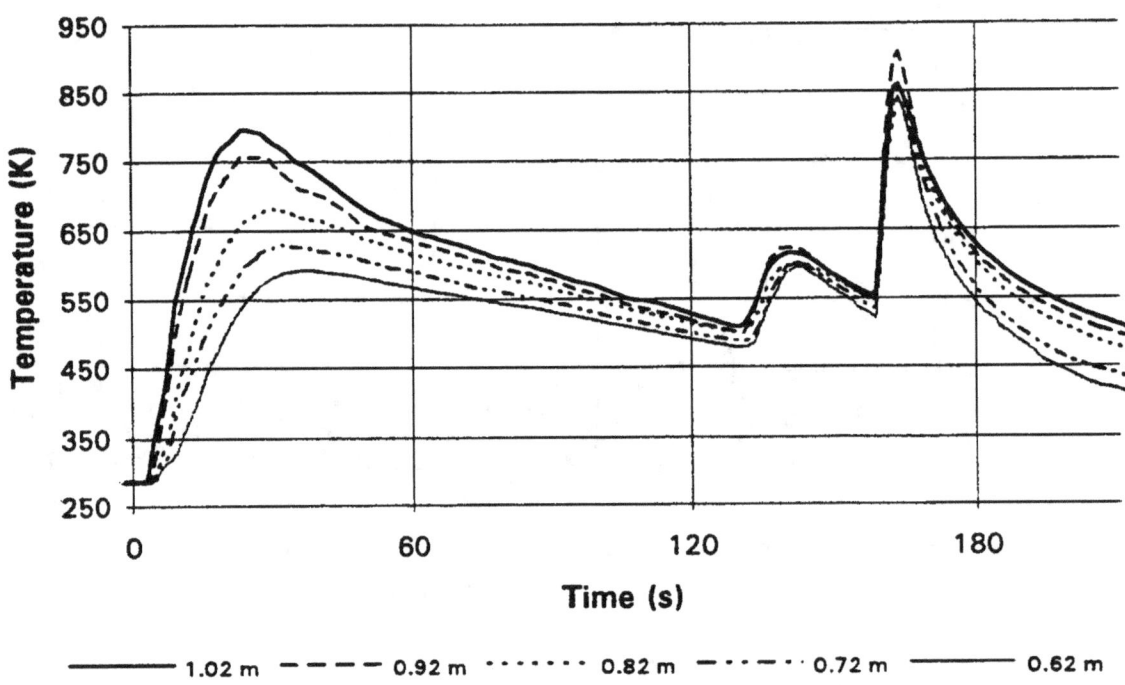

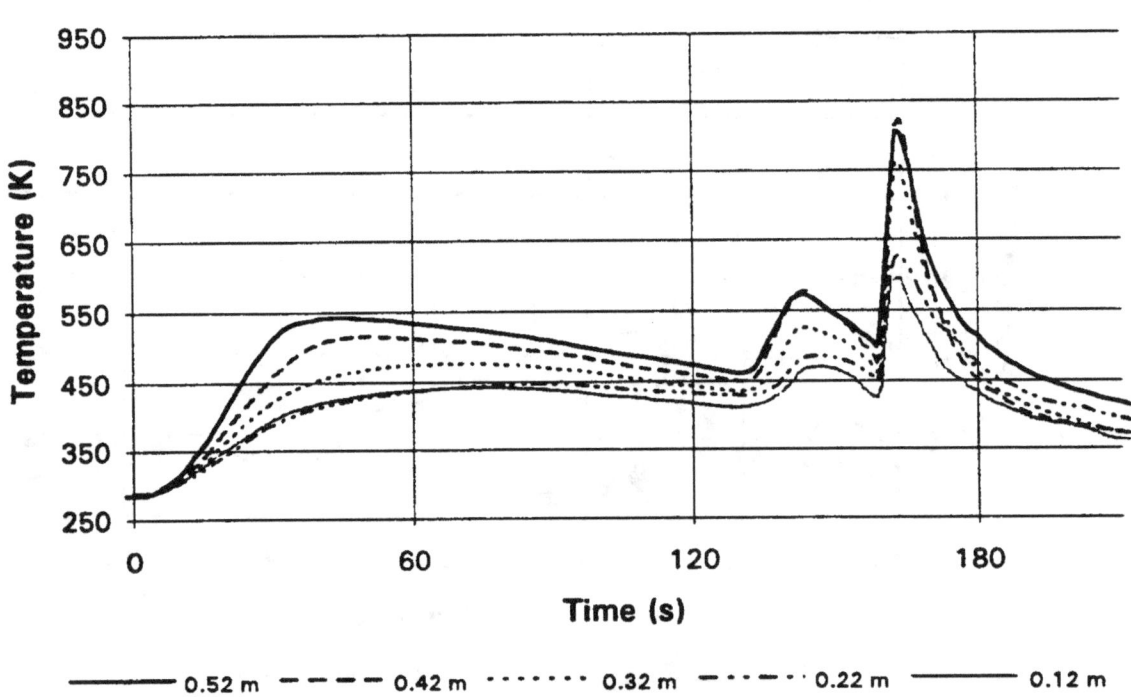

Run: P3EXP30

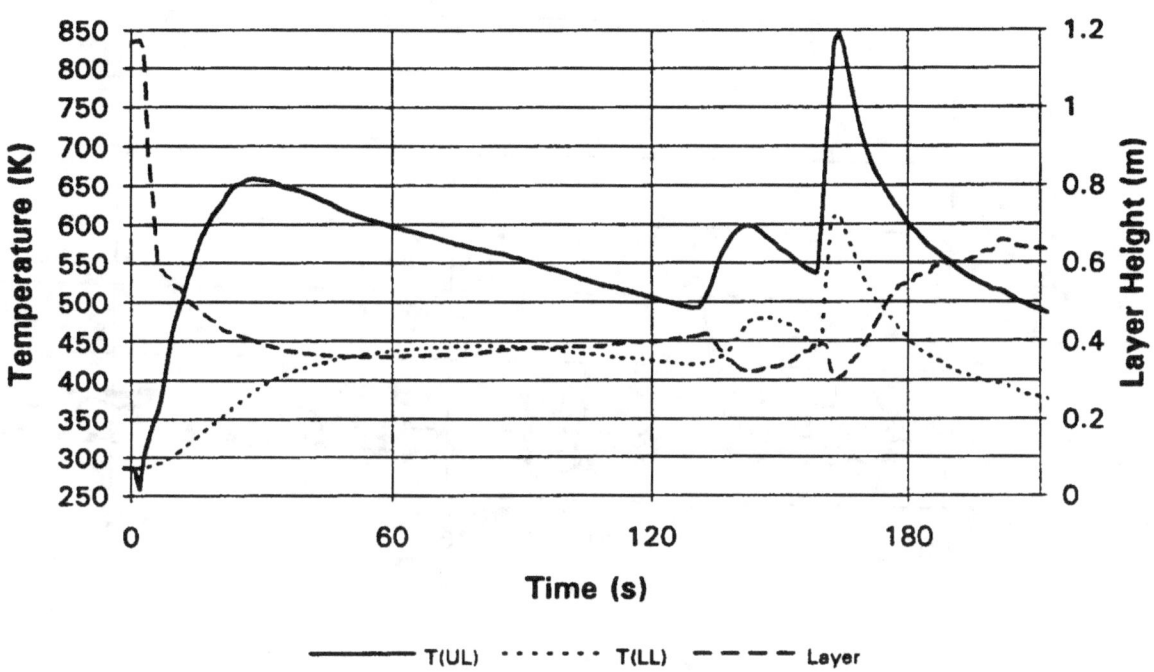

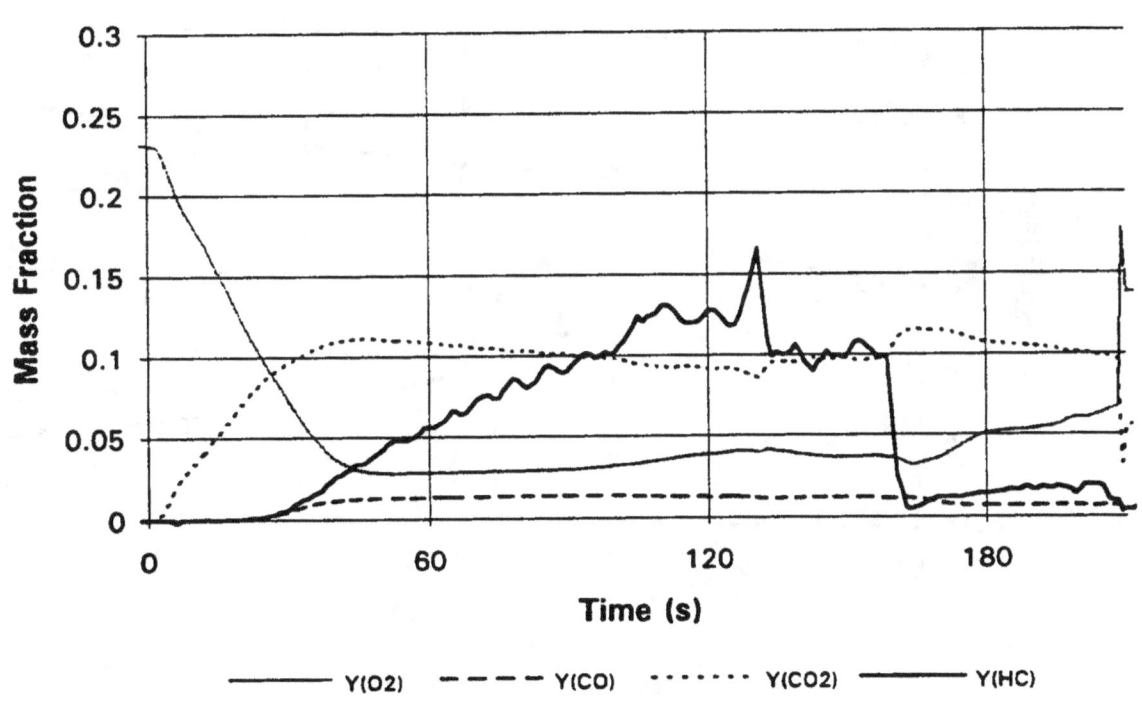

Run: P3EXP30

Bidirectional Probe Velocity History

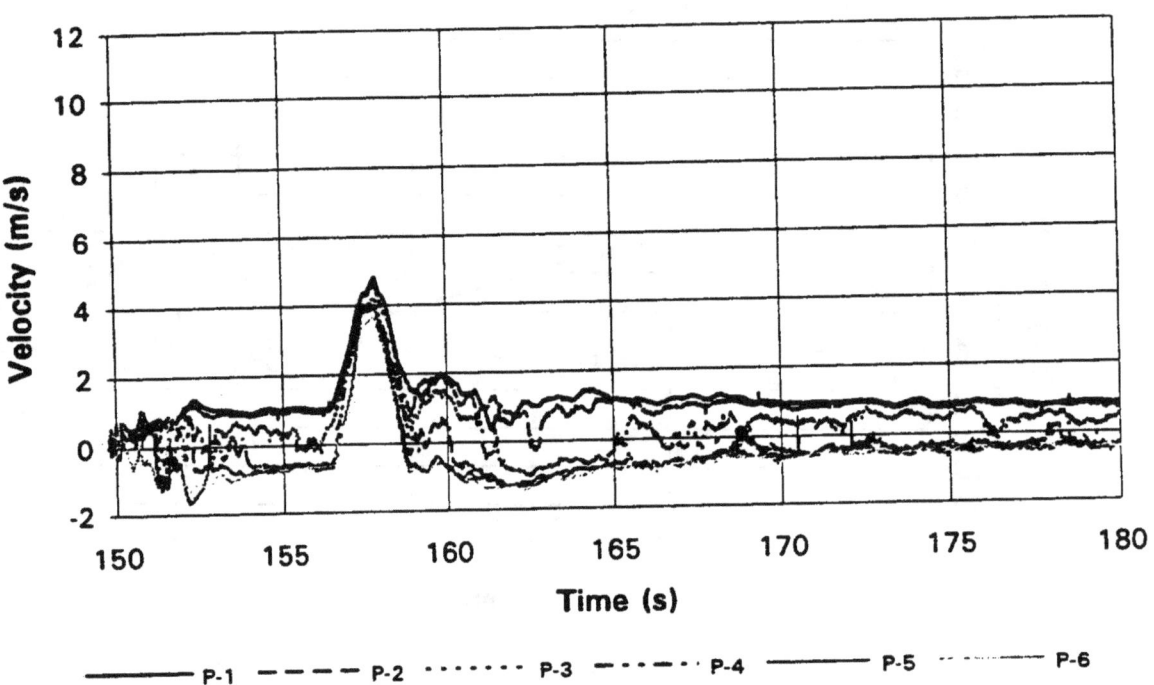

Pressure & Total Mass Out Histories

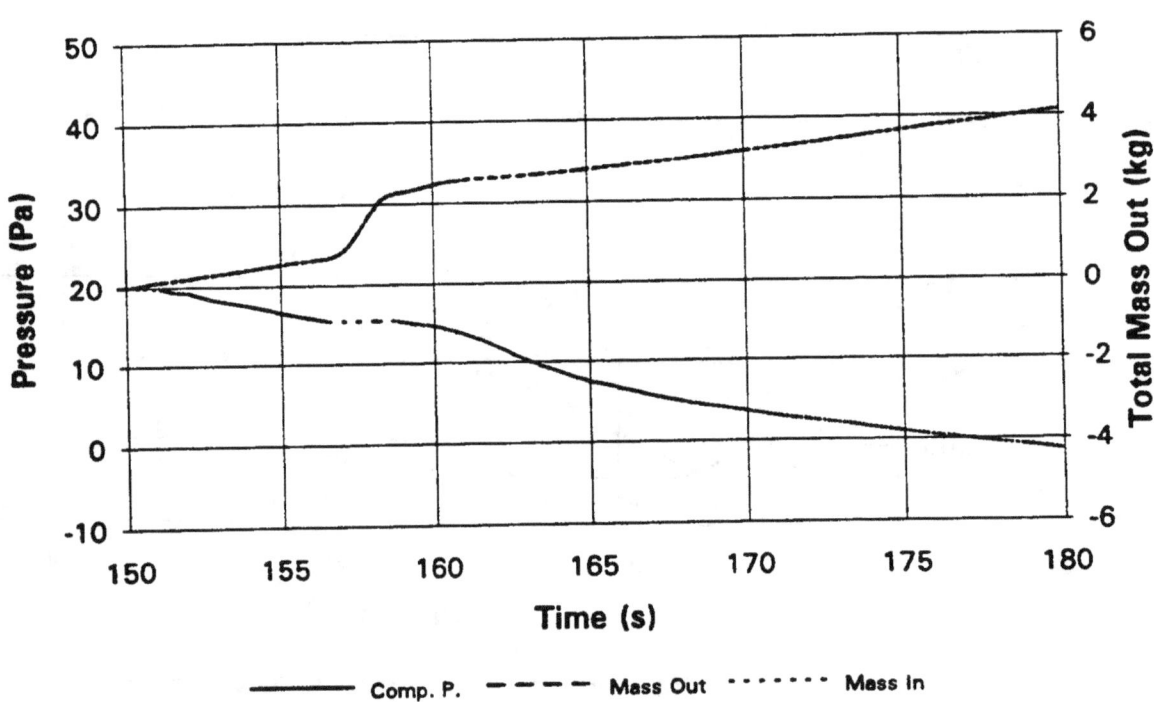

Run: P3EXP30

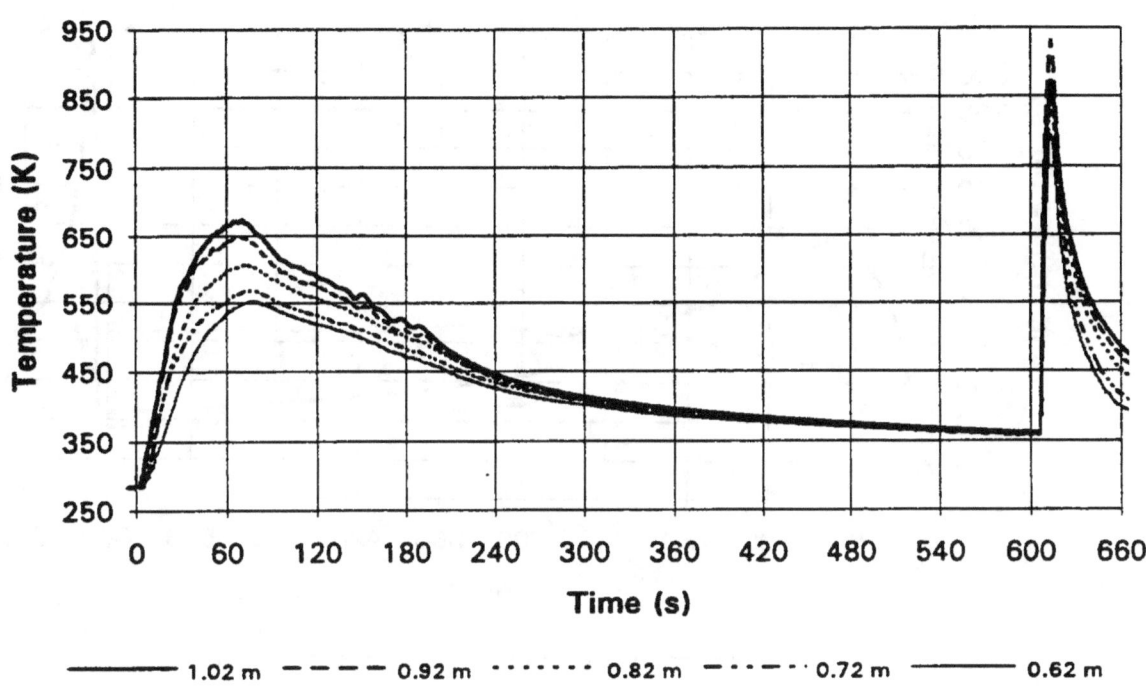

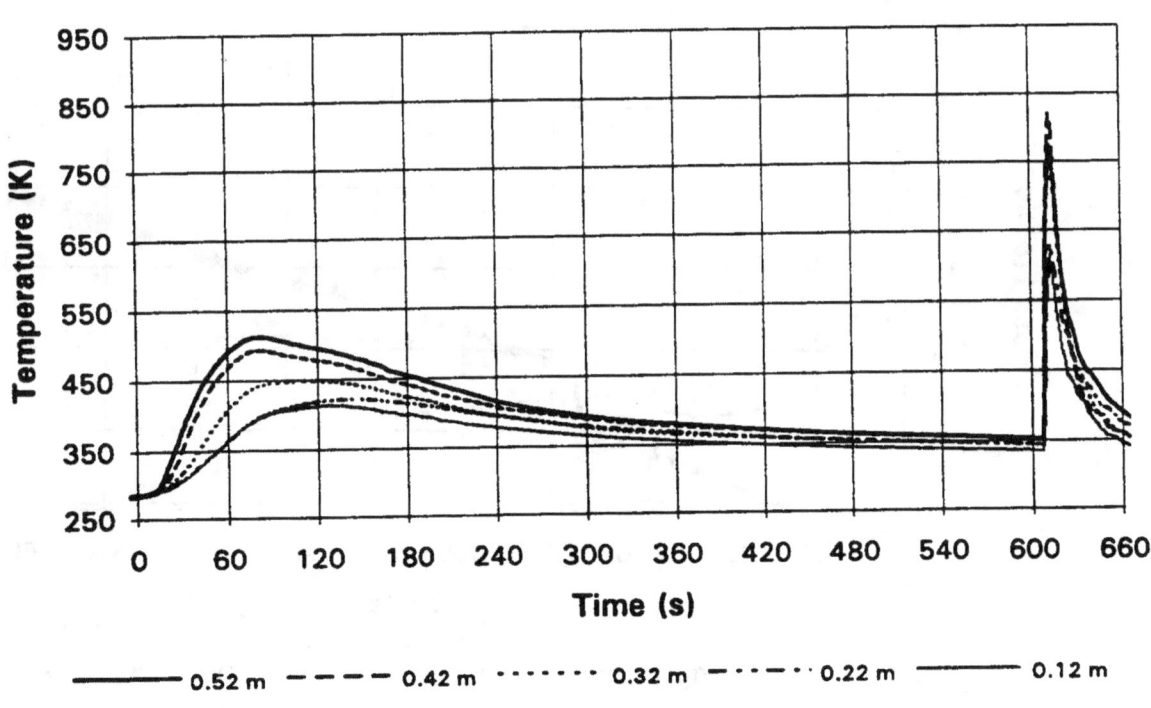

Run: P3EXP32

Two Zone Approximation - Temperature & Interface Histories

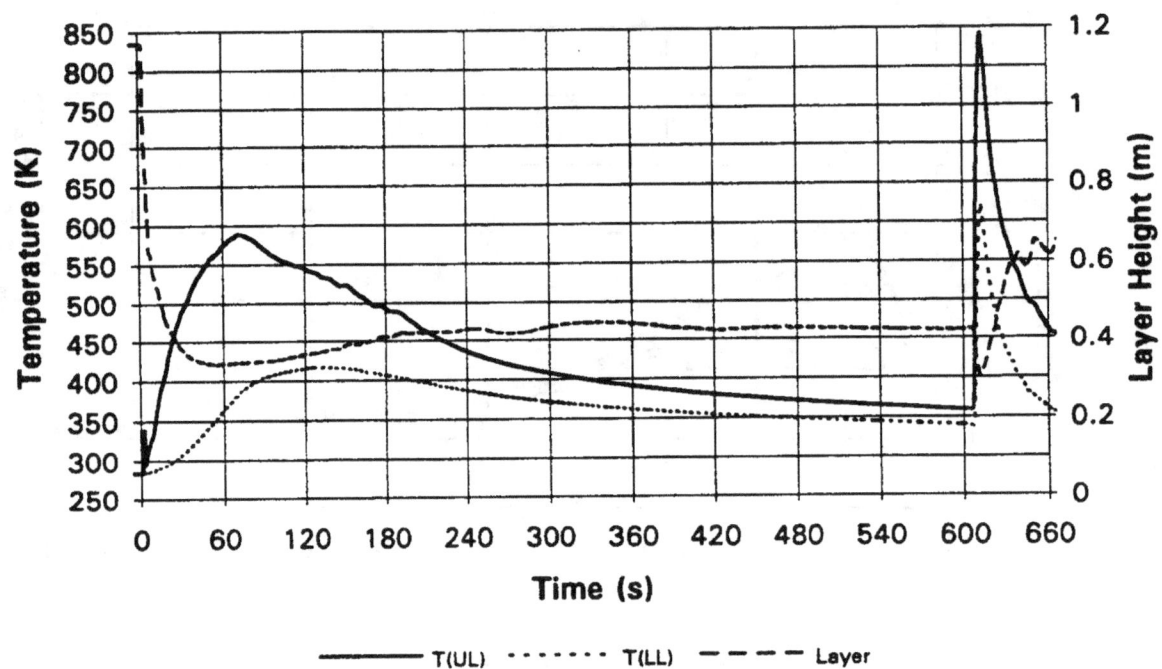

Mass Fraction History - Compartment Gases

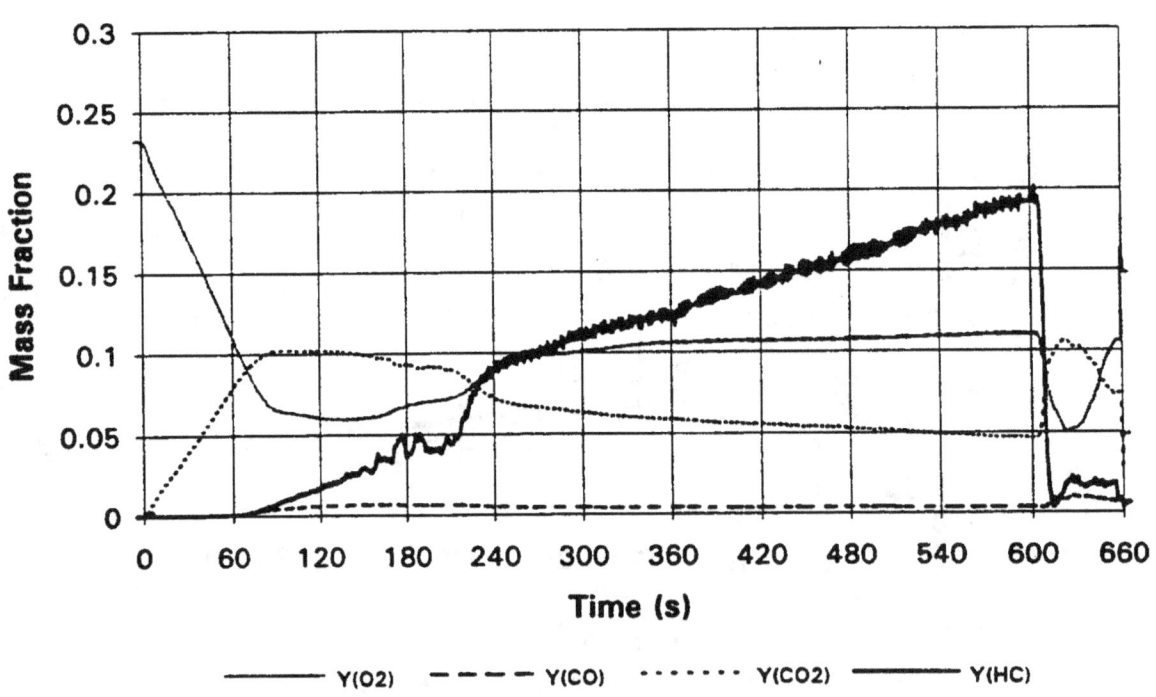

Run: P3EXP32

Bidirectional Probe Velocity History

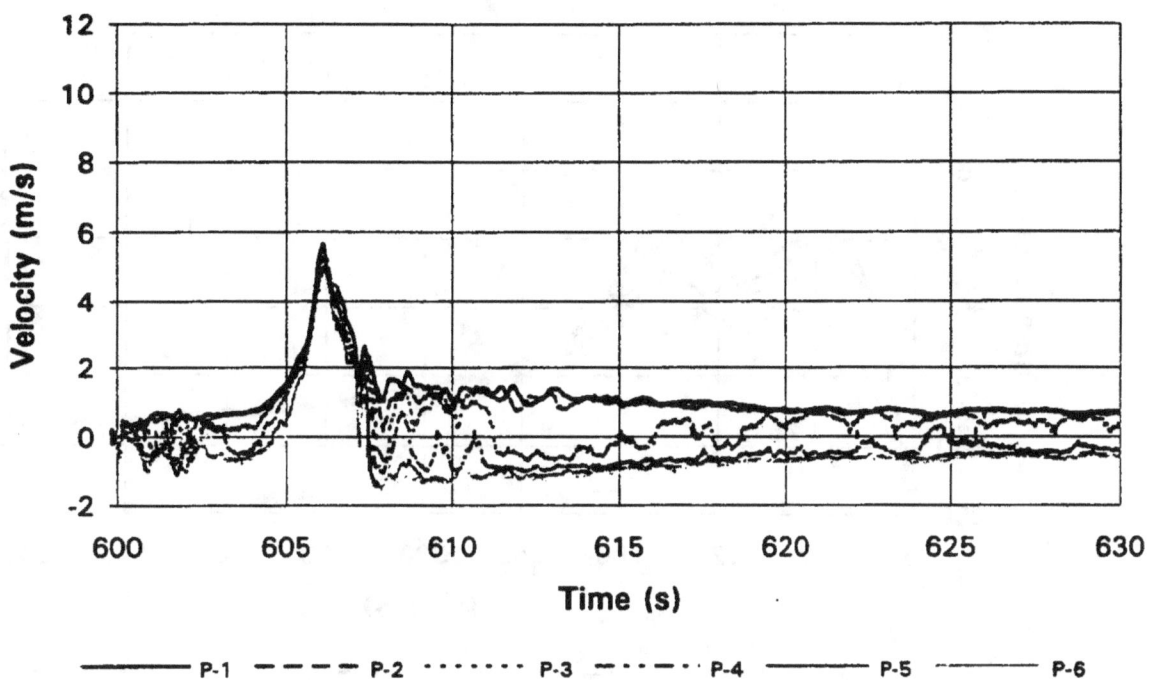

Pressure & Total Mass Out Histories

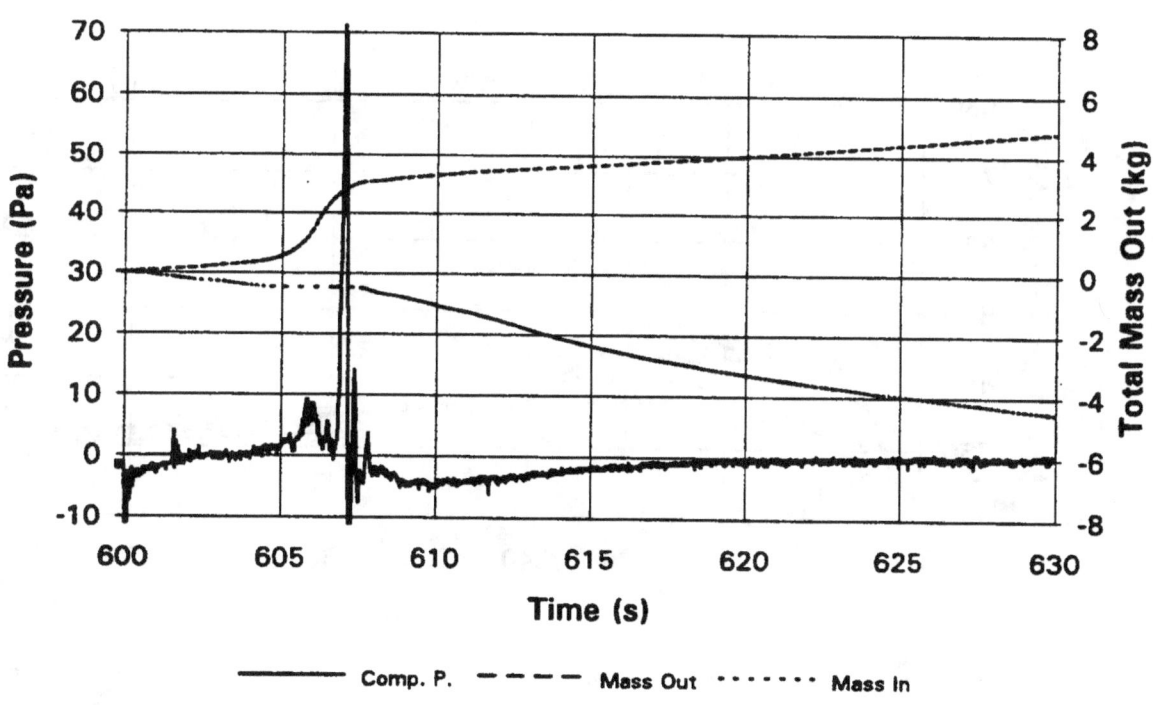

Run: P3EXP32

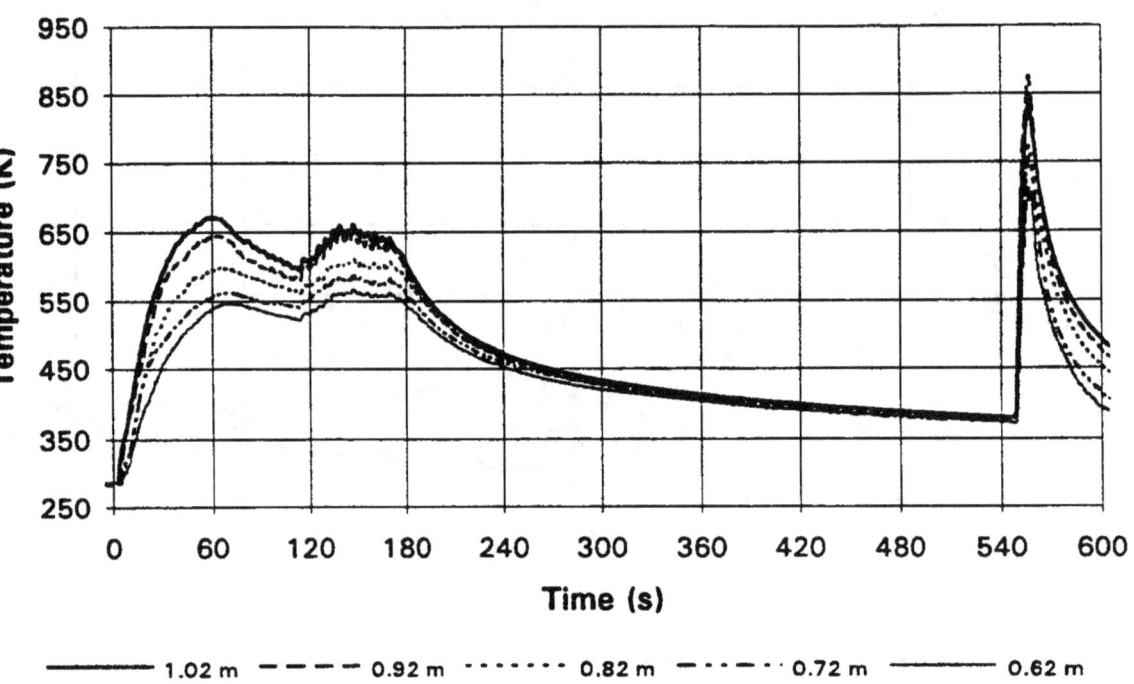

Layer Temperature History - TC Tree Data

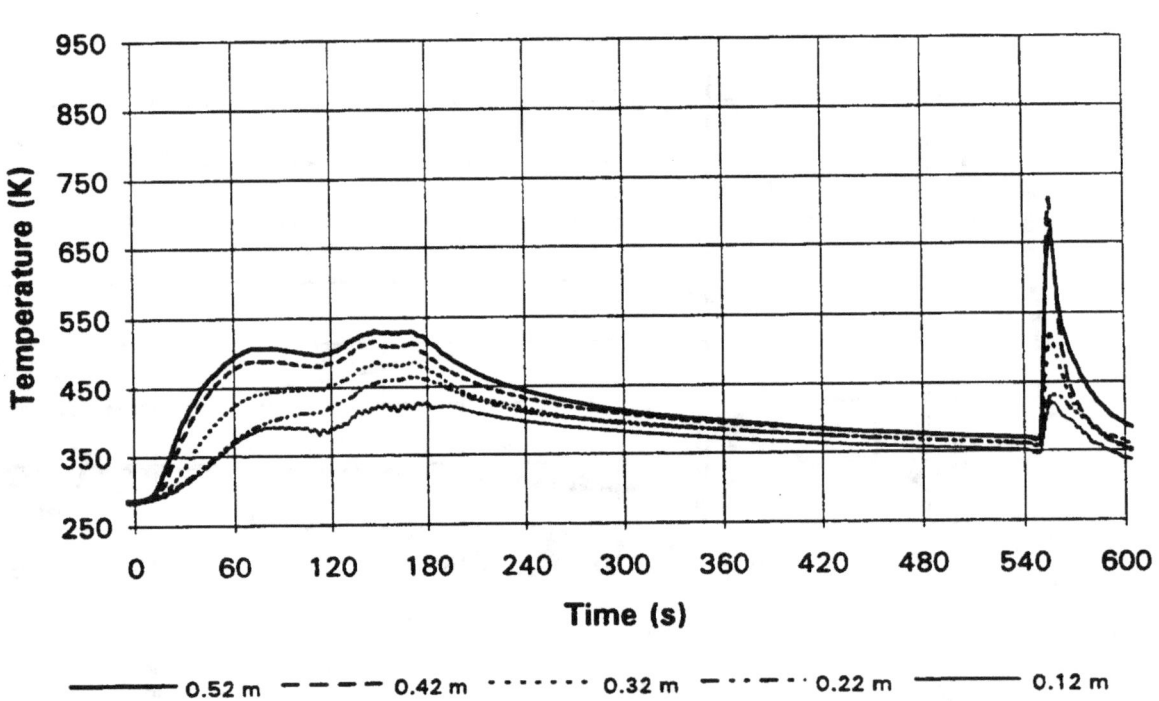

Layer Temperature History - TC Tree Data

Run: P3EXP33

Two Zone Approximation - Temperature & Interface Histories

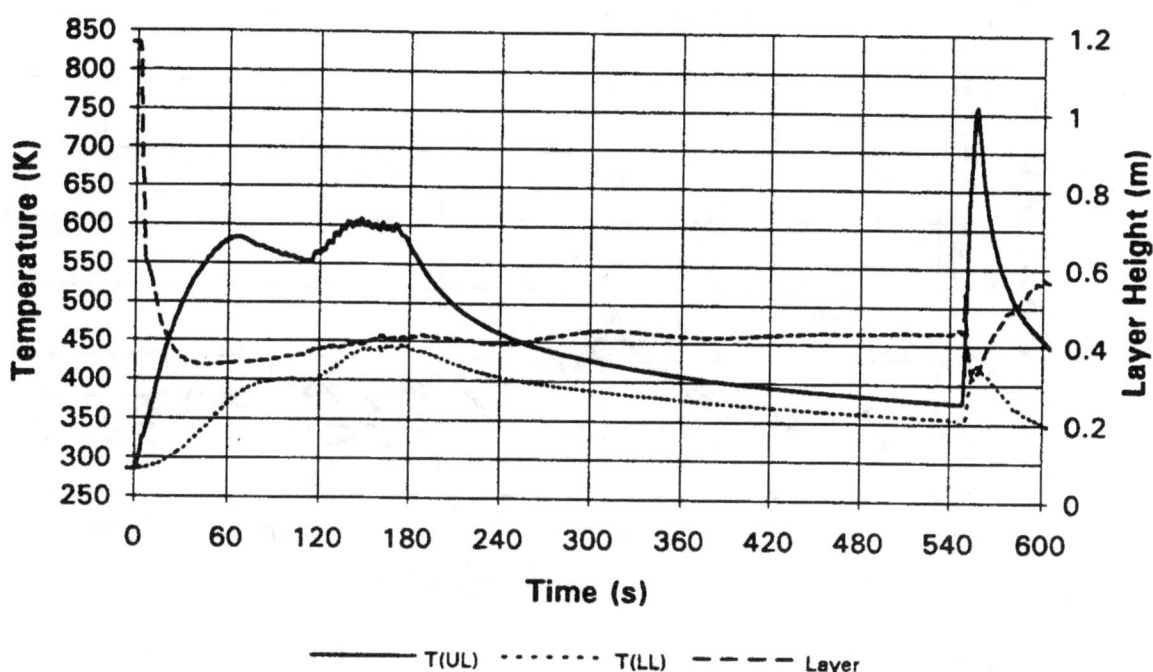

— T(UL) ······ T(LL) – – – Layer

Mass Fraction History - Compartment Gases

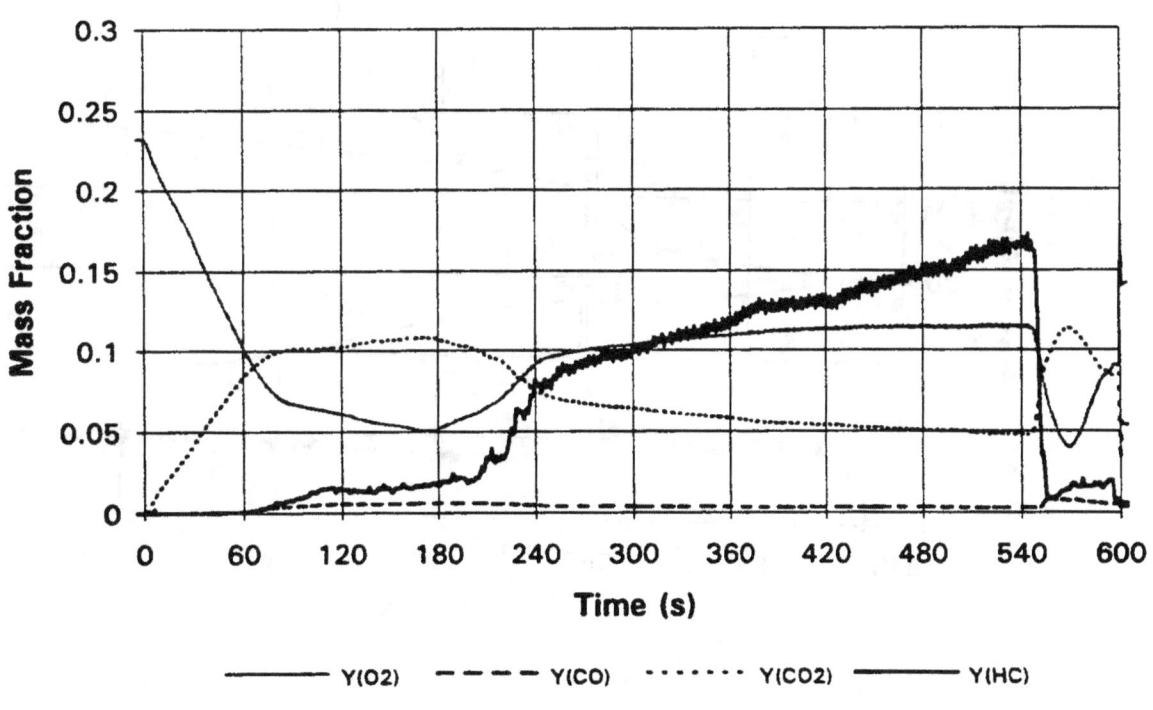

— Y(O2)   – – – Y(CO)   ······ Y(CO2)   ▬▬ Y(HC)

Run: P3EXP33

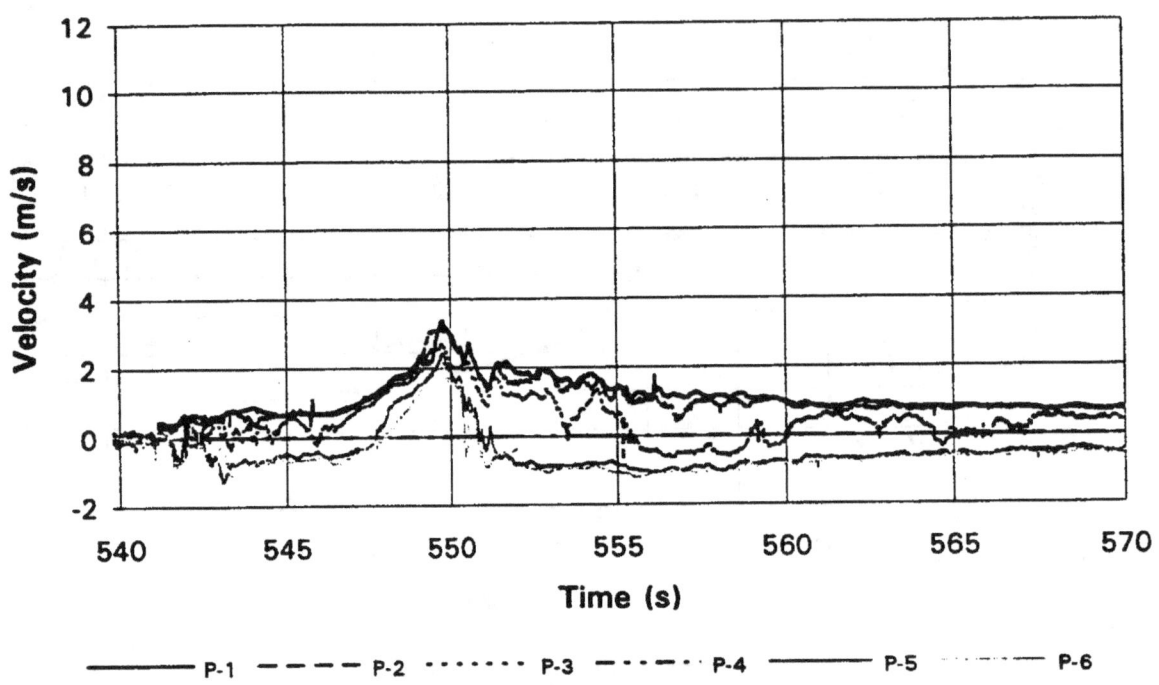

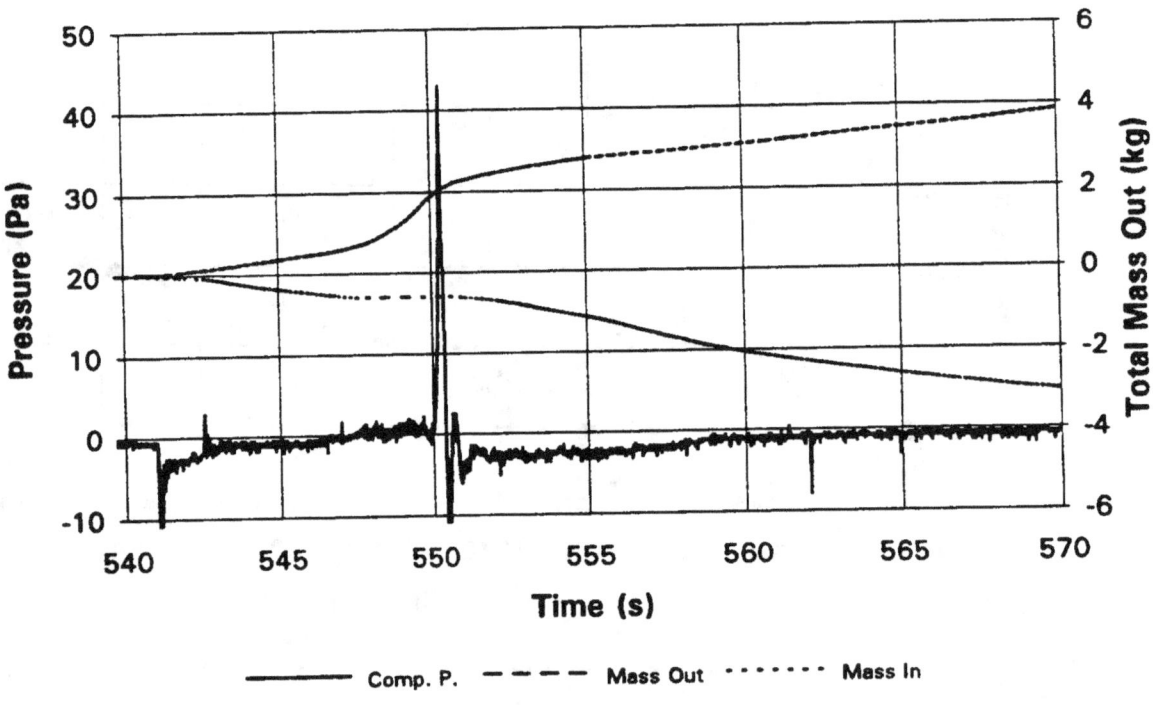

Run: P3EXP33

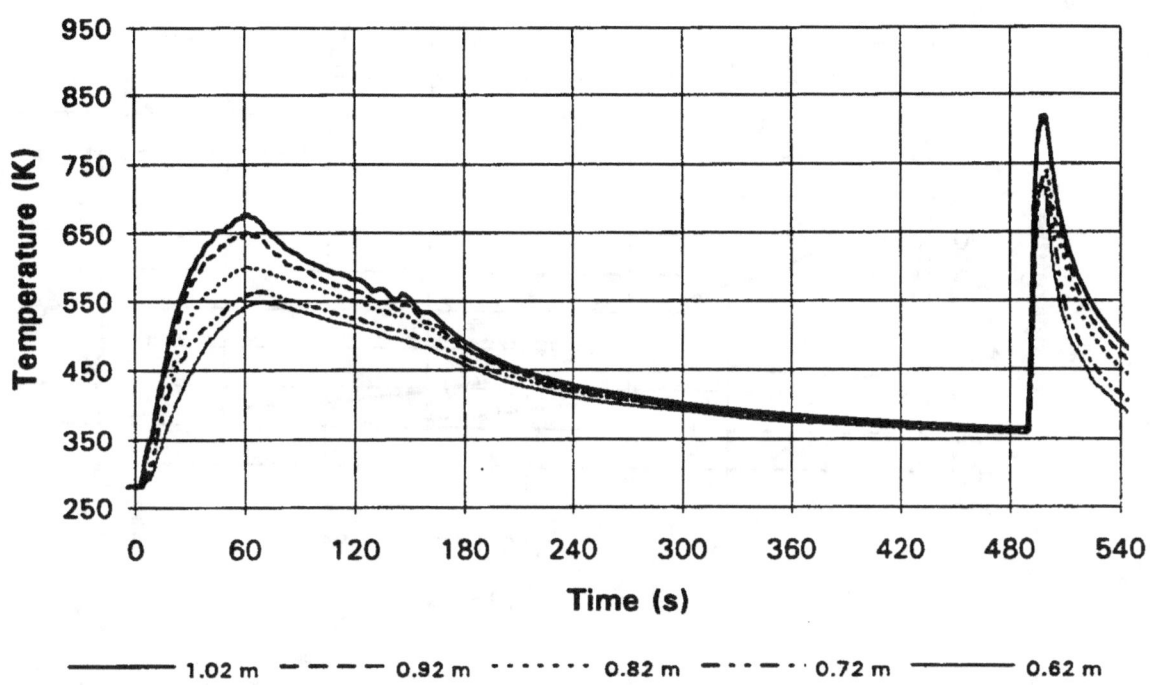

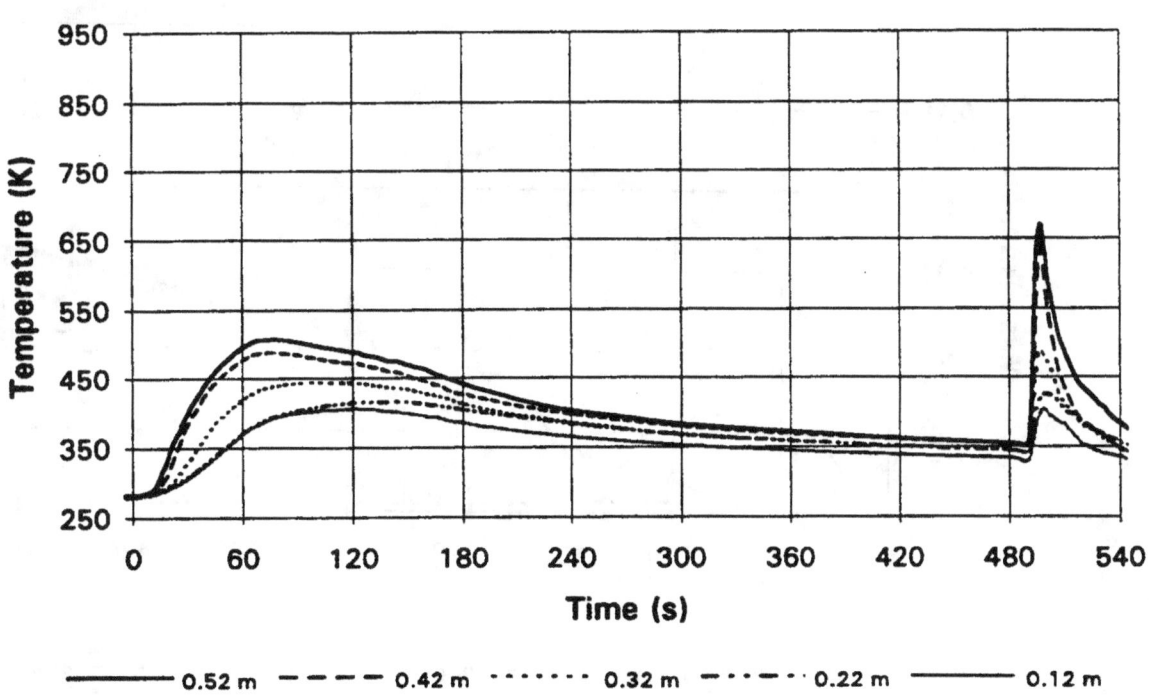

Run: P3EXP34

Two Zone Approximation - Temperature & Interface Histories

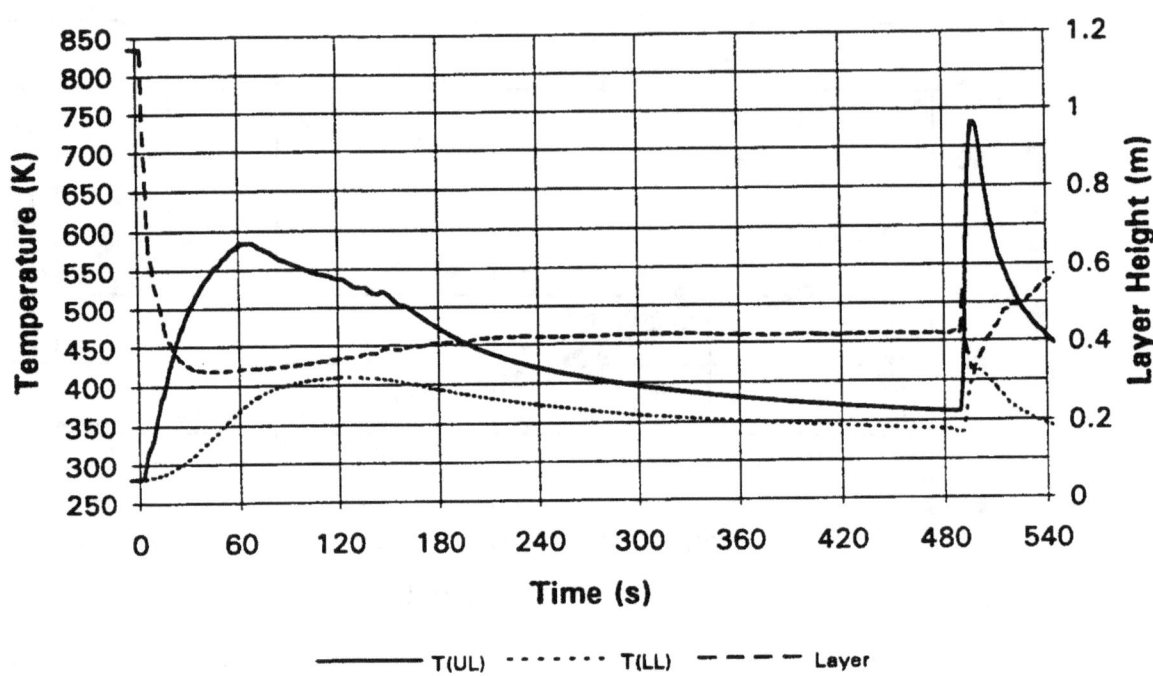

Mass Fraction History - Compartment Gases

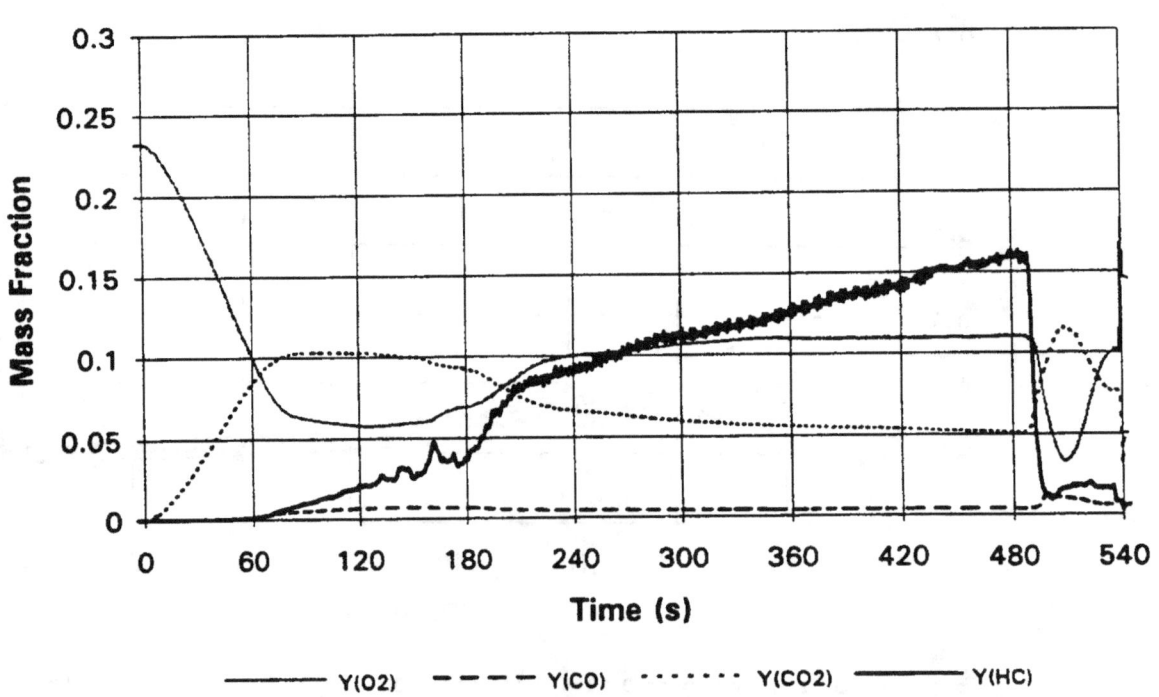

Run: P3EXP34

Bidirectional Probe Velocity History

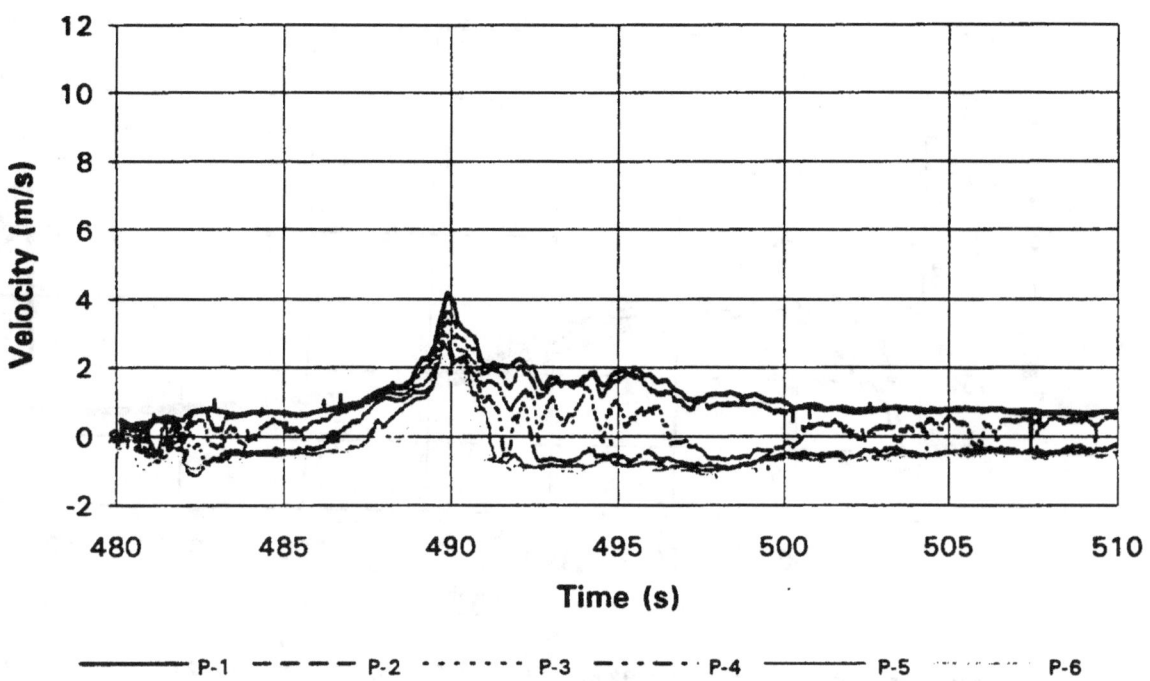

Pressure & Total Mass Out Histories

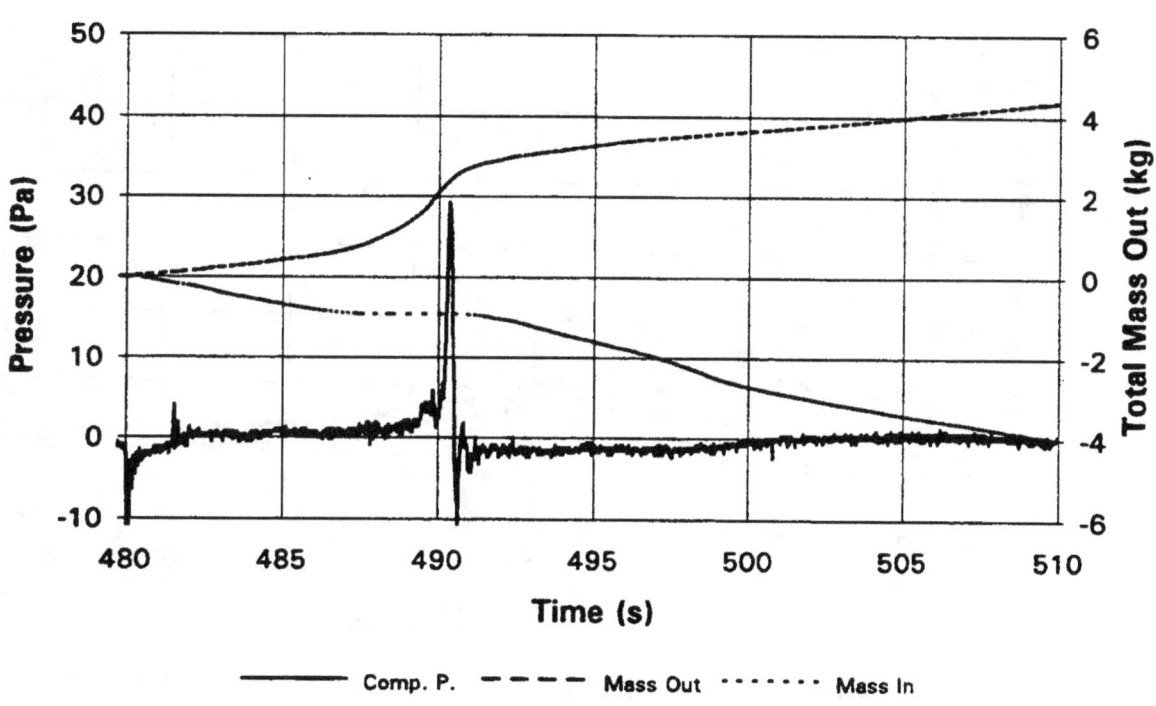

Run: P3EXP34

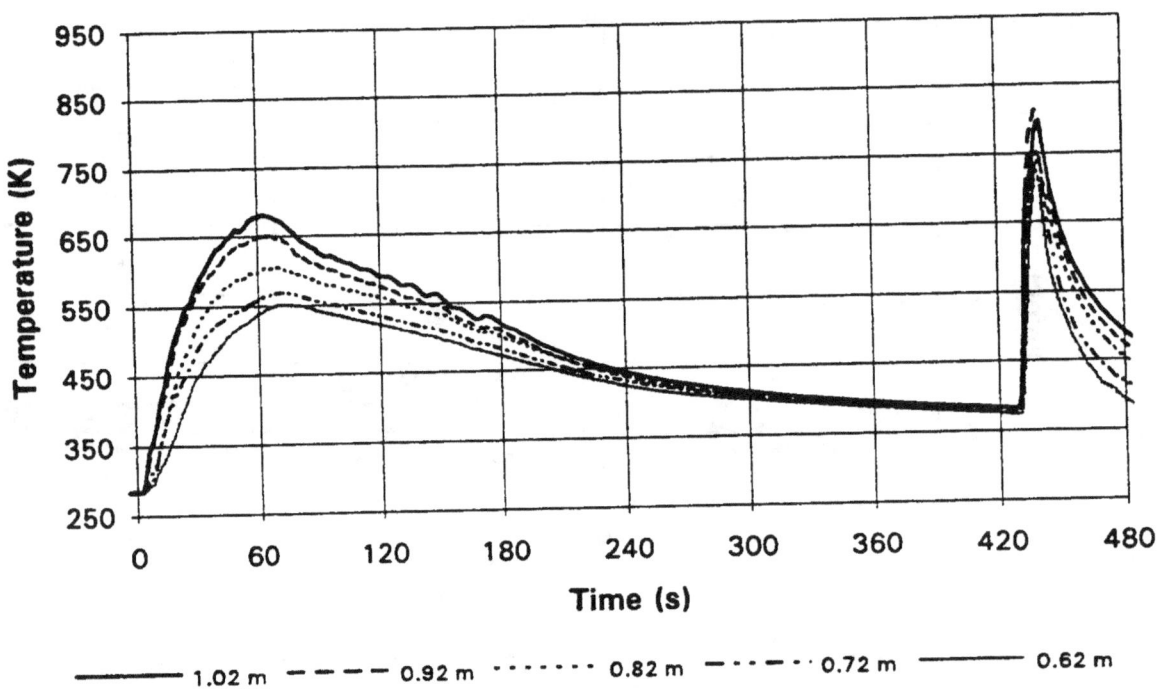

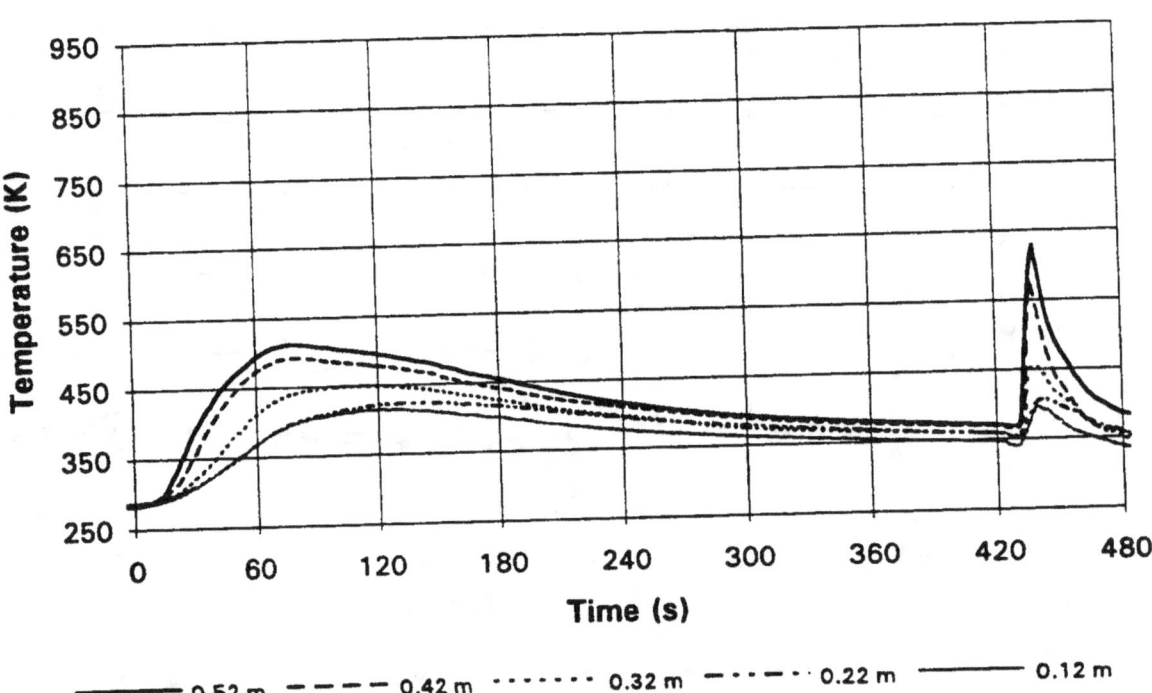

Run: P3EXP35

Two Zone Approximation - Temperature & Interface Histories

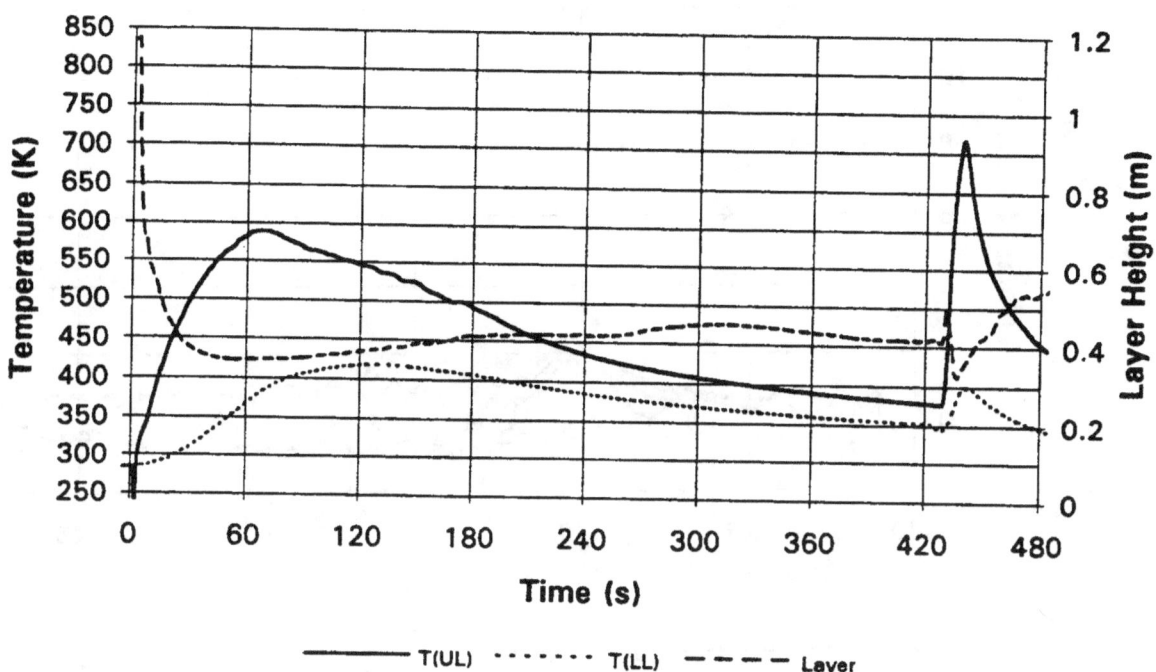

Mass Fraction History - Compartment Gases

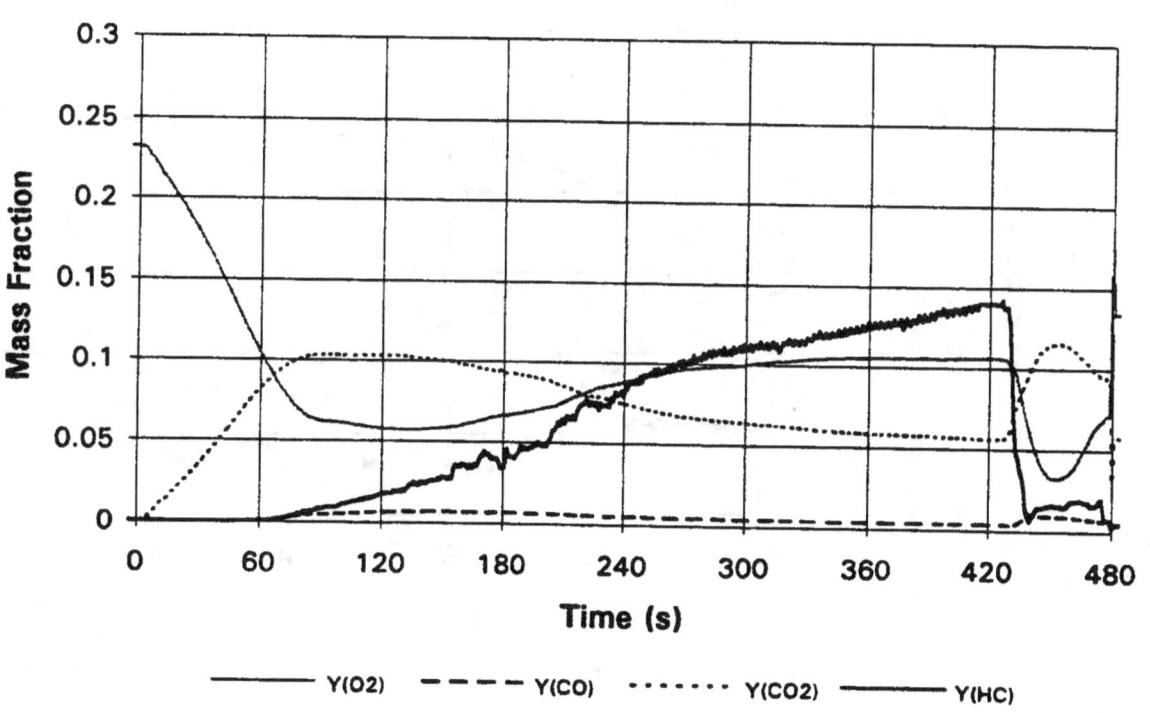

Run: P3EXP35

Bidirectional Probe Velocity History

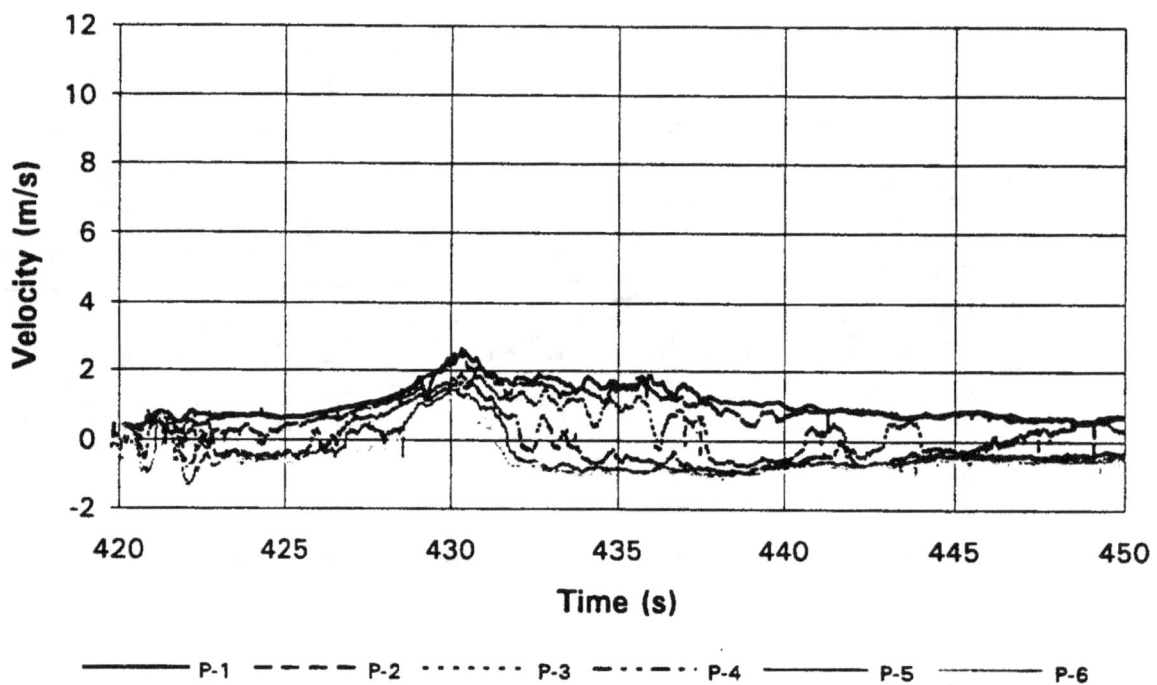

Pressure & Total Mass Out Histories

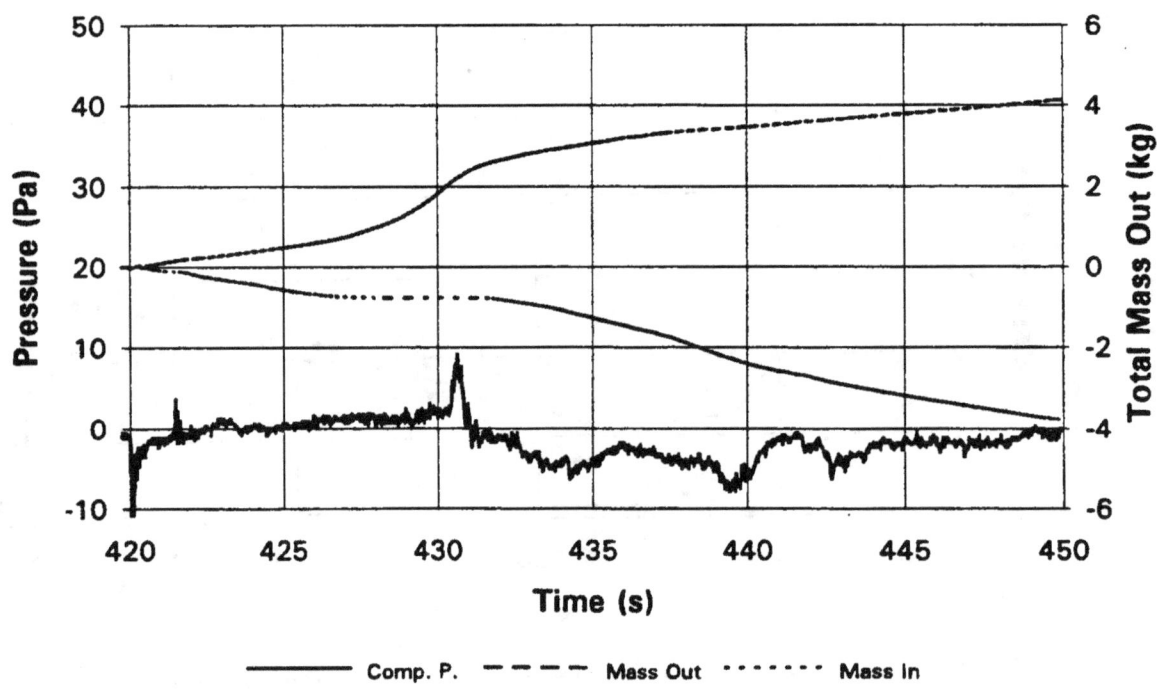

Run: P3EXP35

Layer Temperature History - TC Tree Data

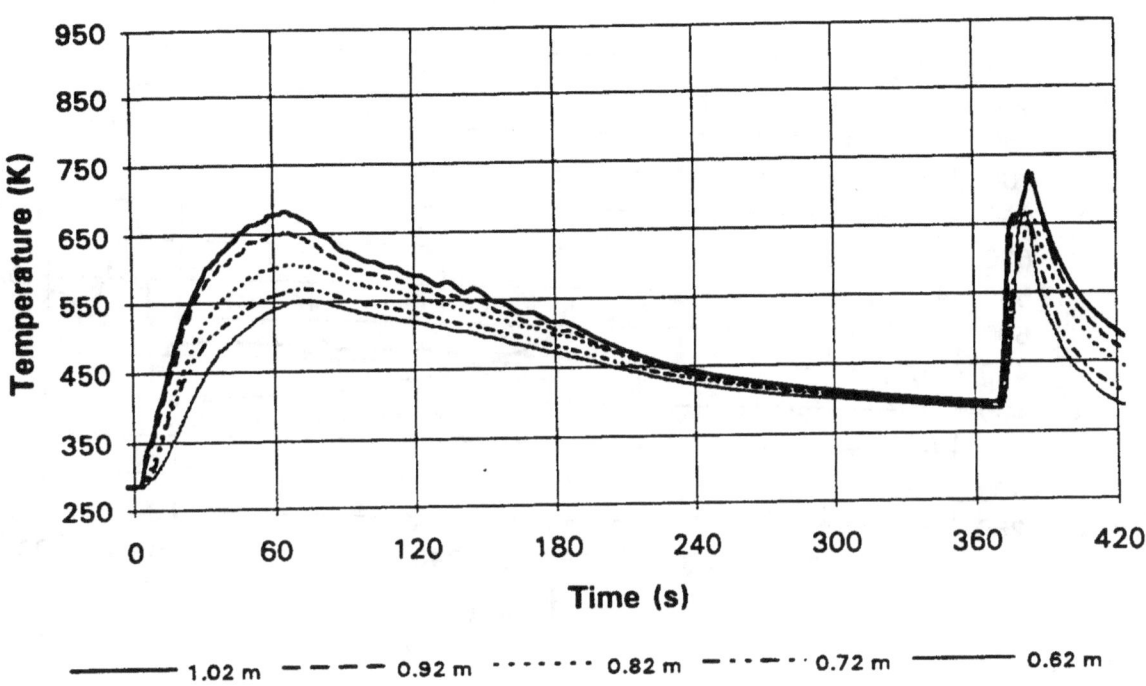

Layer Temperature History - TC Tree Data

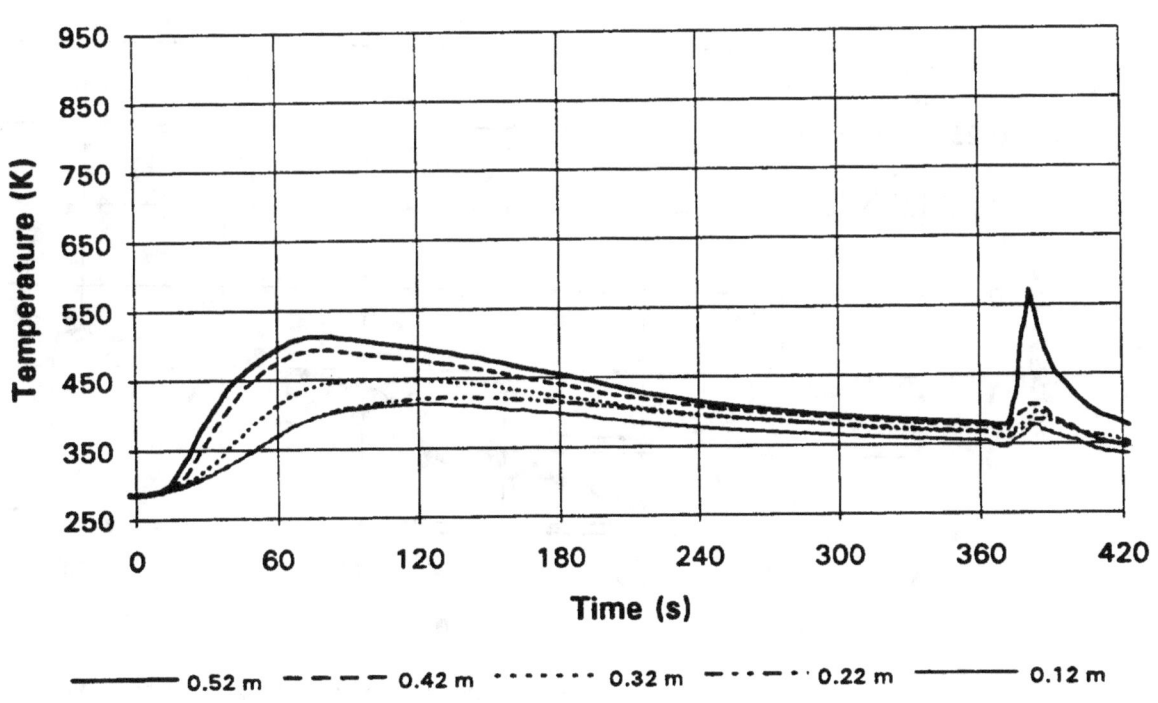

Run: P3EXP36

Two Zone Approximation - Temperature & Interface Histories

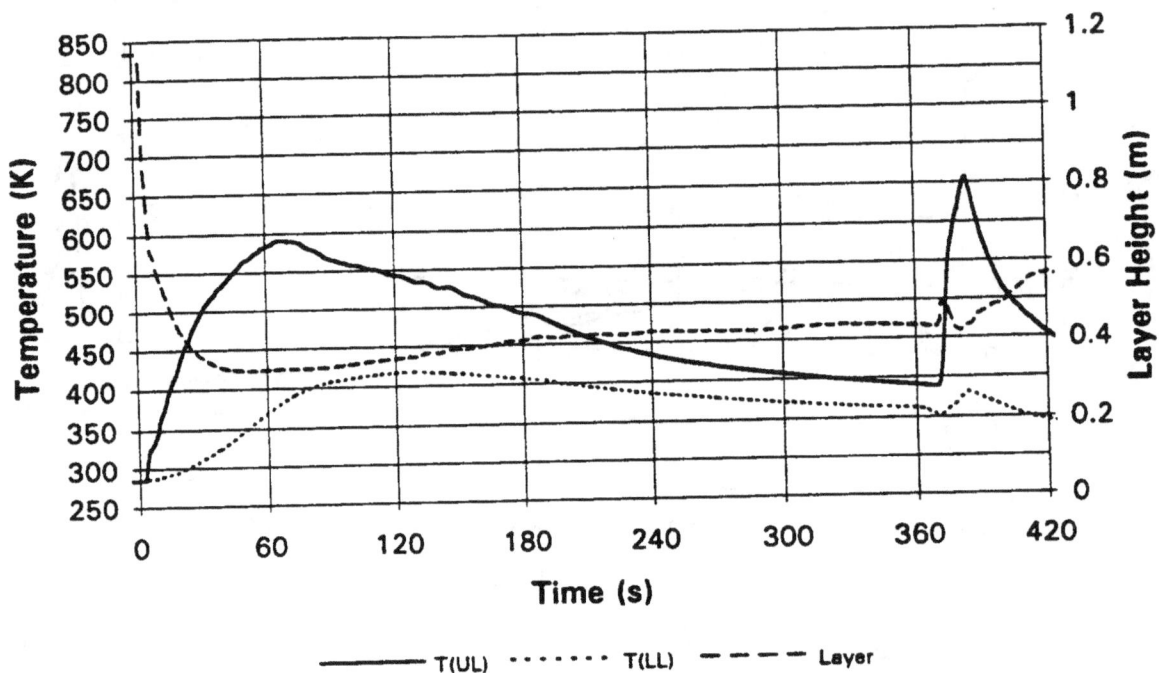

Mass Fraction History - Compartment Gases

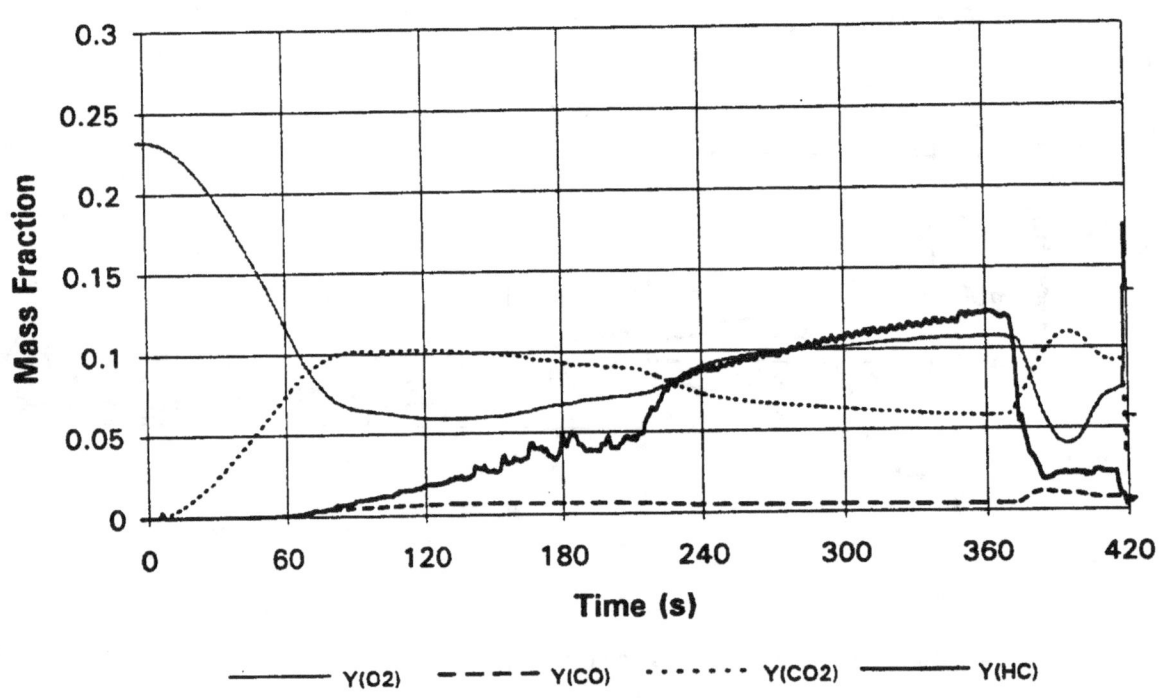

Run: P3EXP36

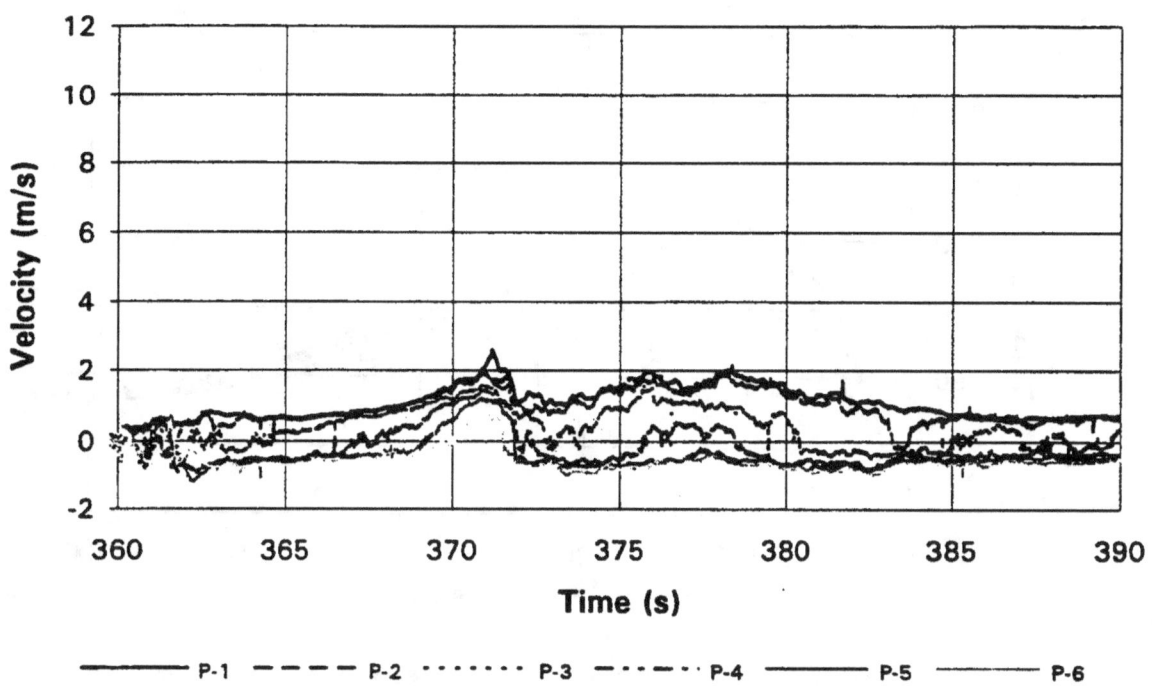

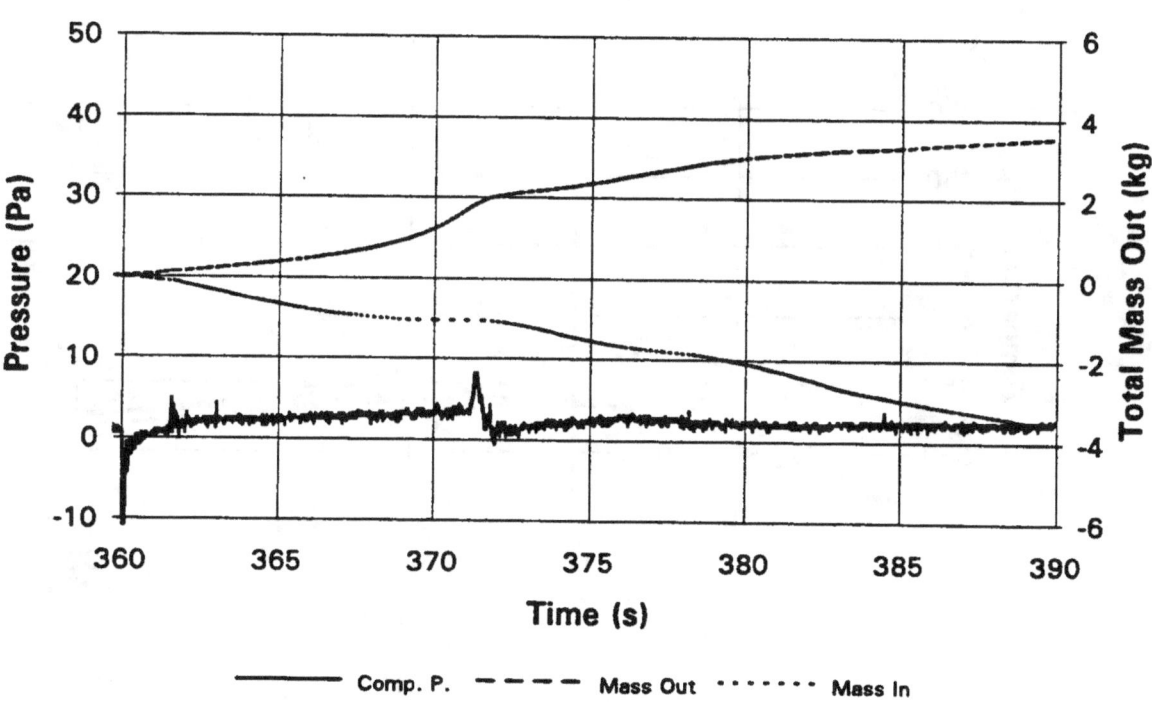

Run: P3EXP36

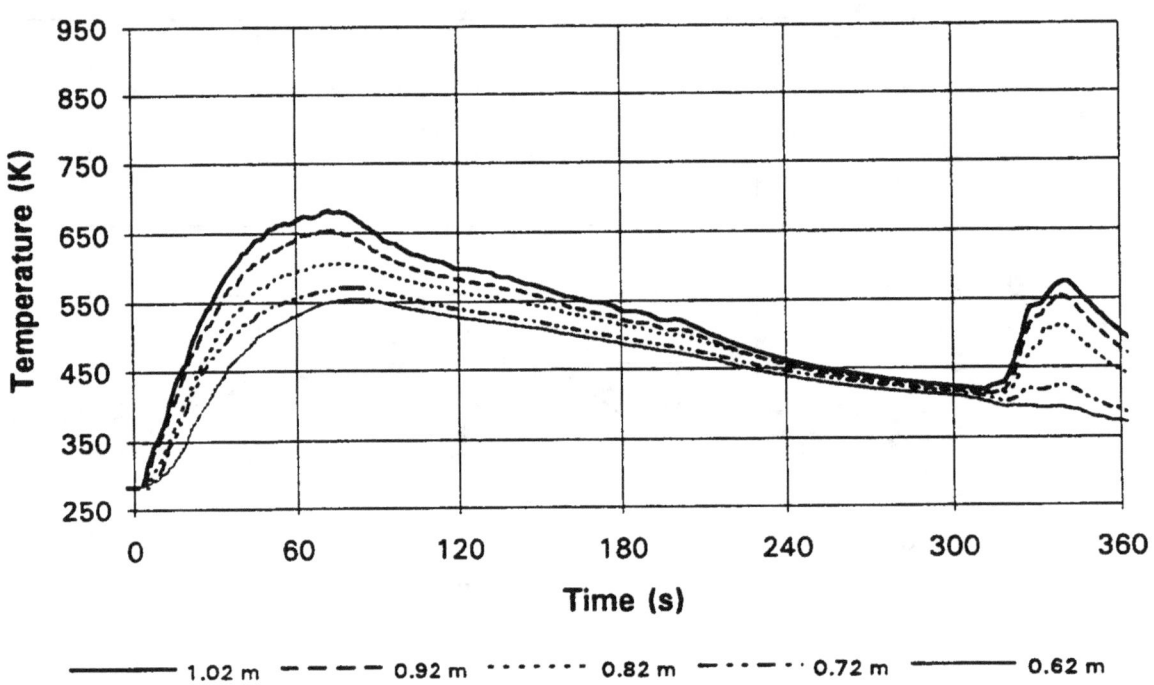

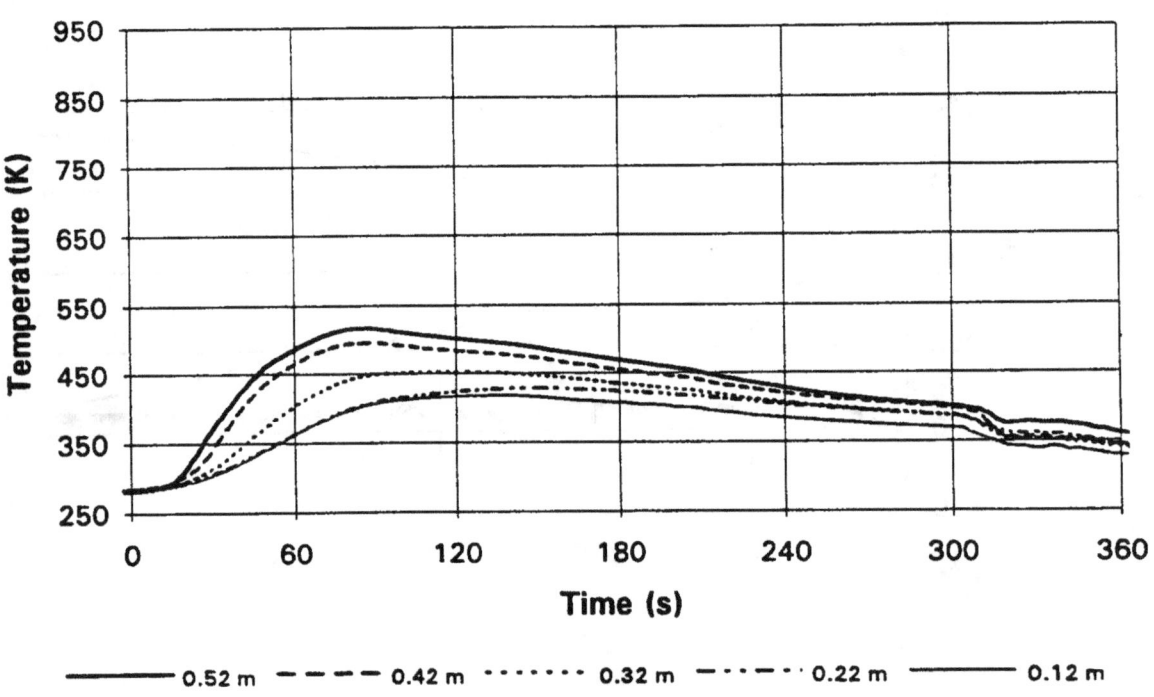

Run: P3EXP38

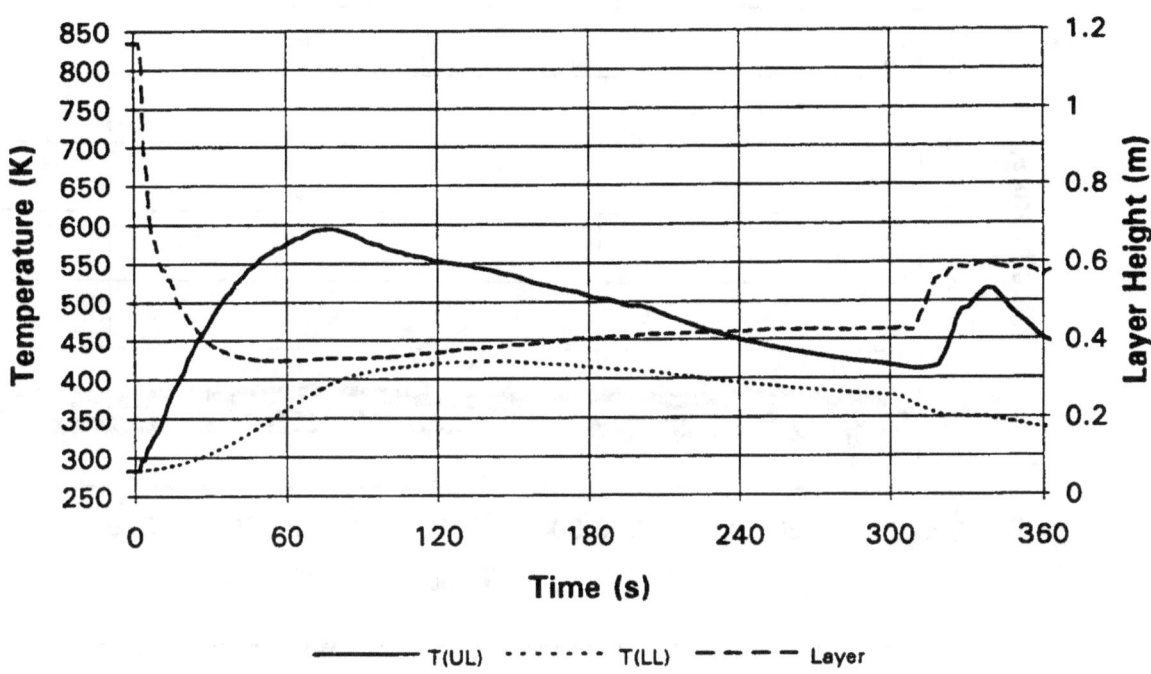

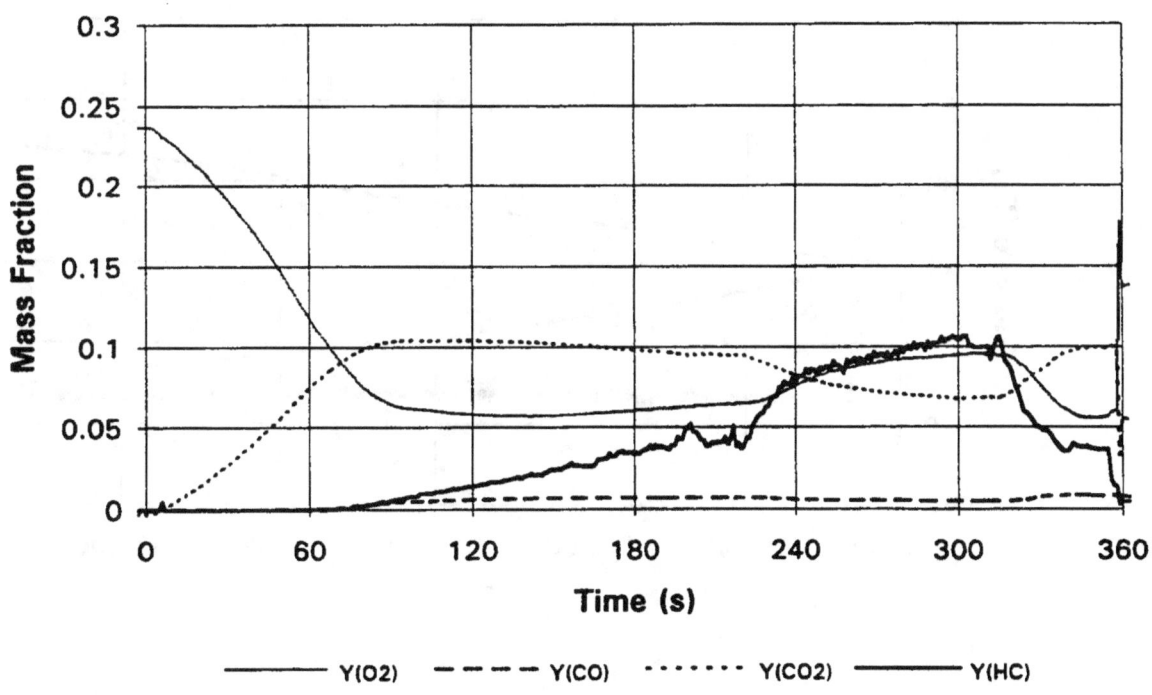

Run: P3EXP38

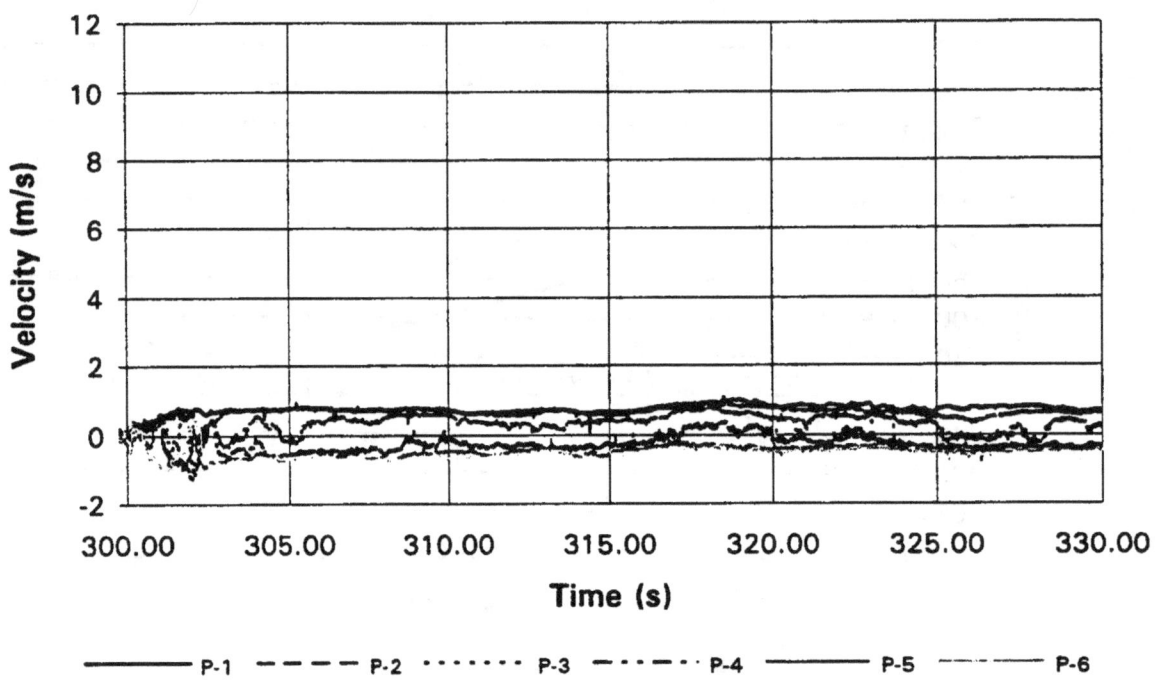

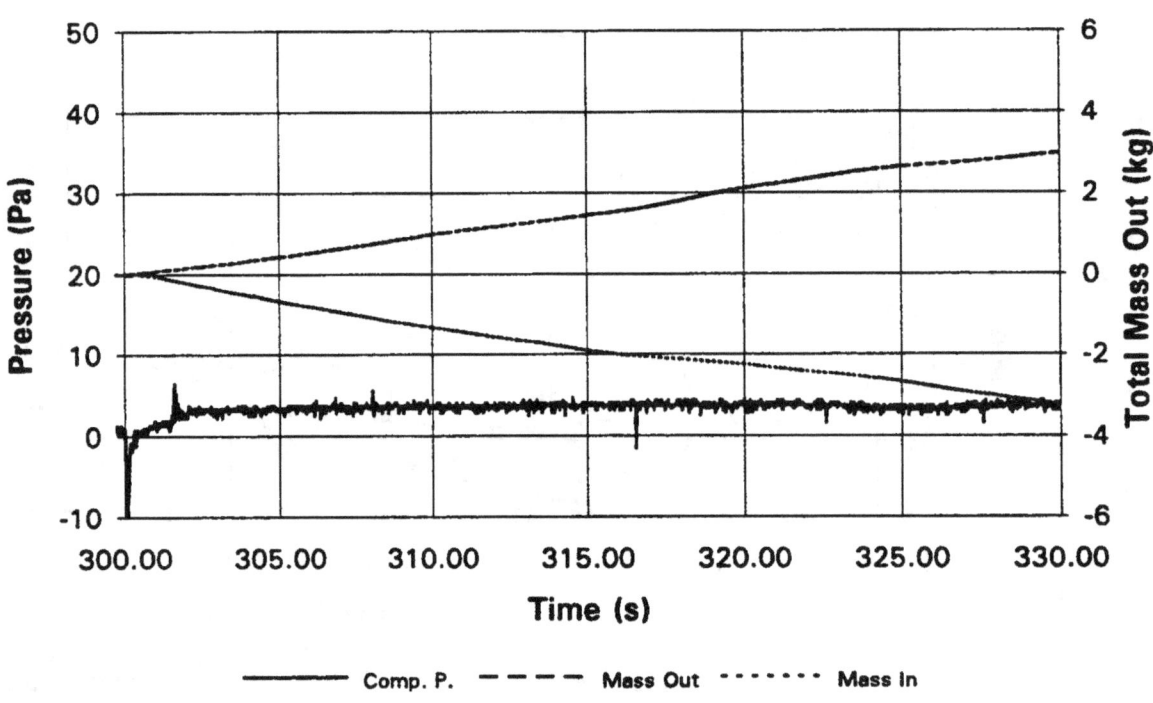

Run: P3EXP38

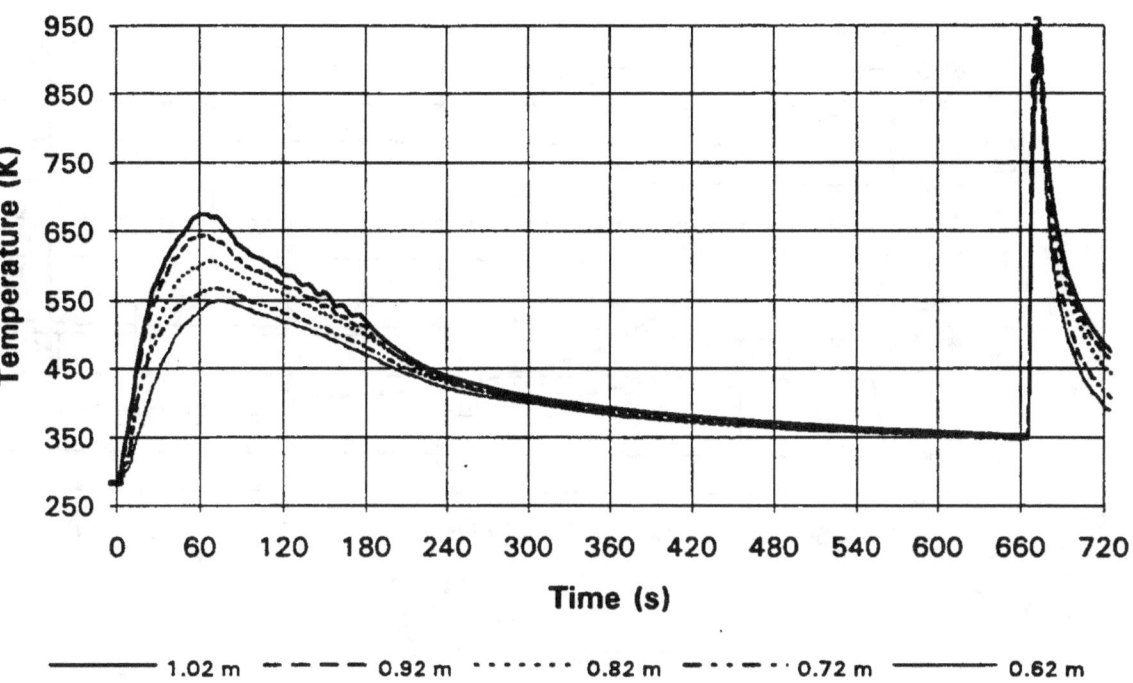

Layer Temperature History - TC Tree Data

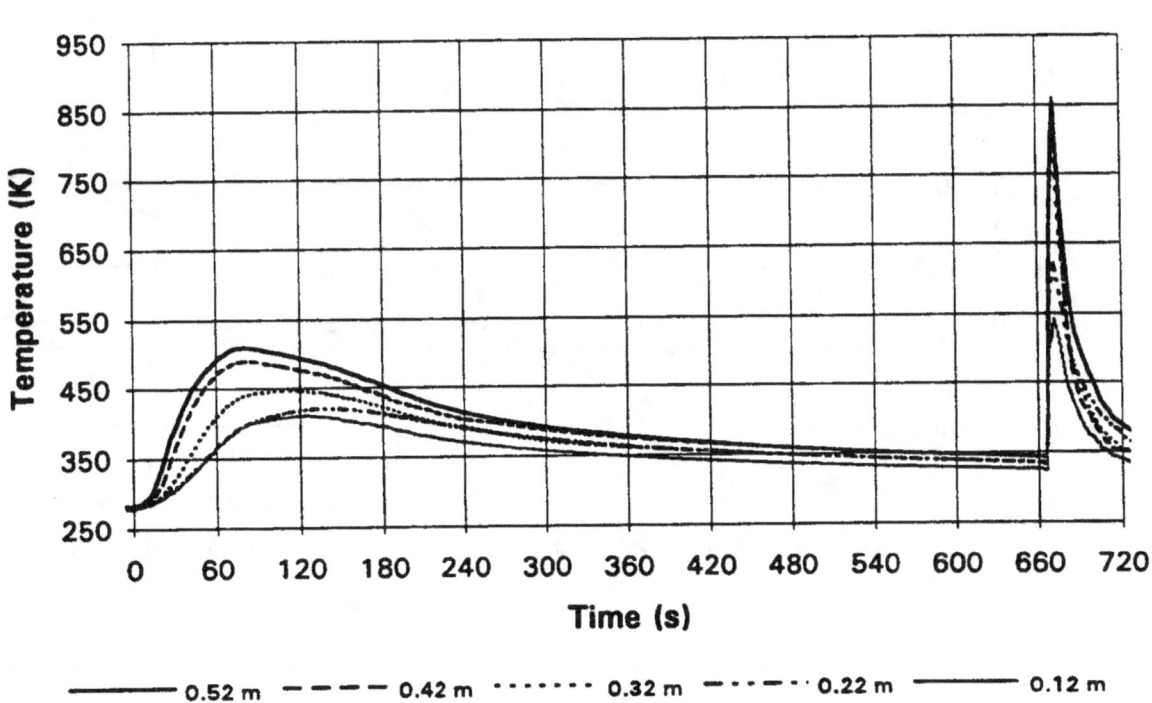

Layer Temperature History - TC Tree Data

Run: P3EXP39

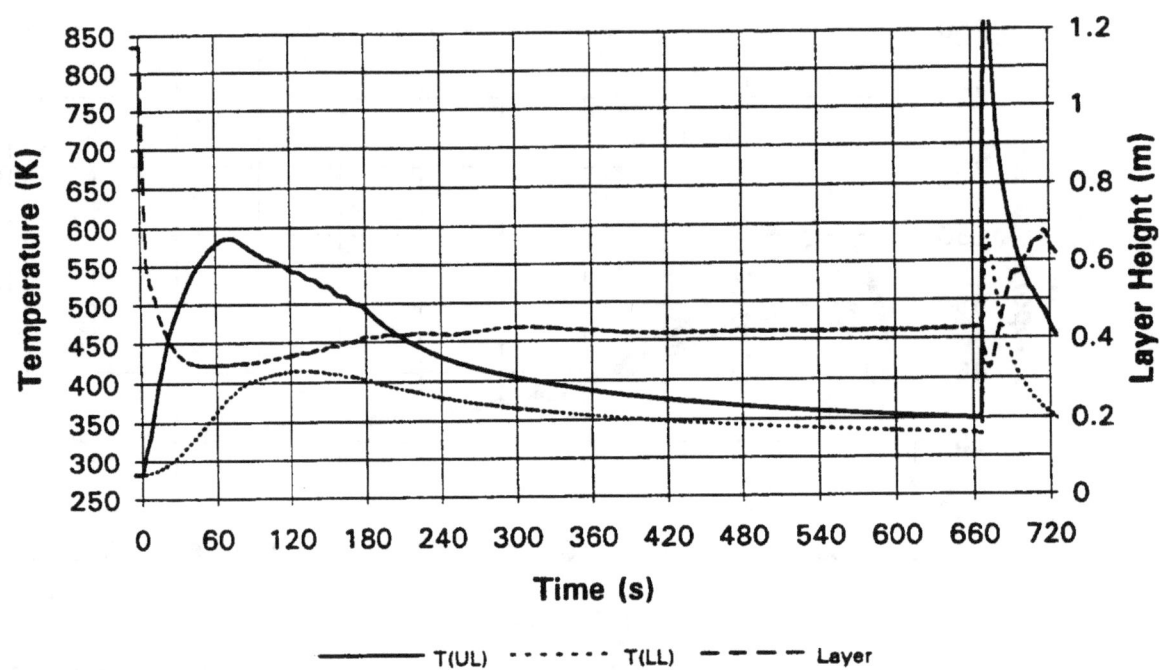

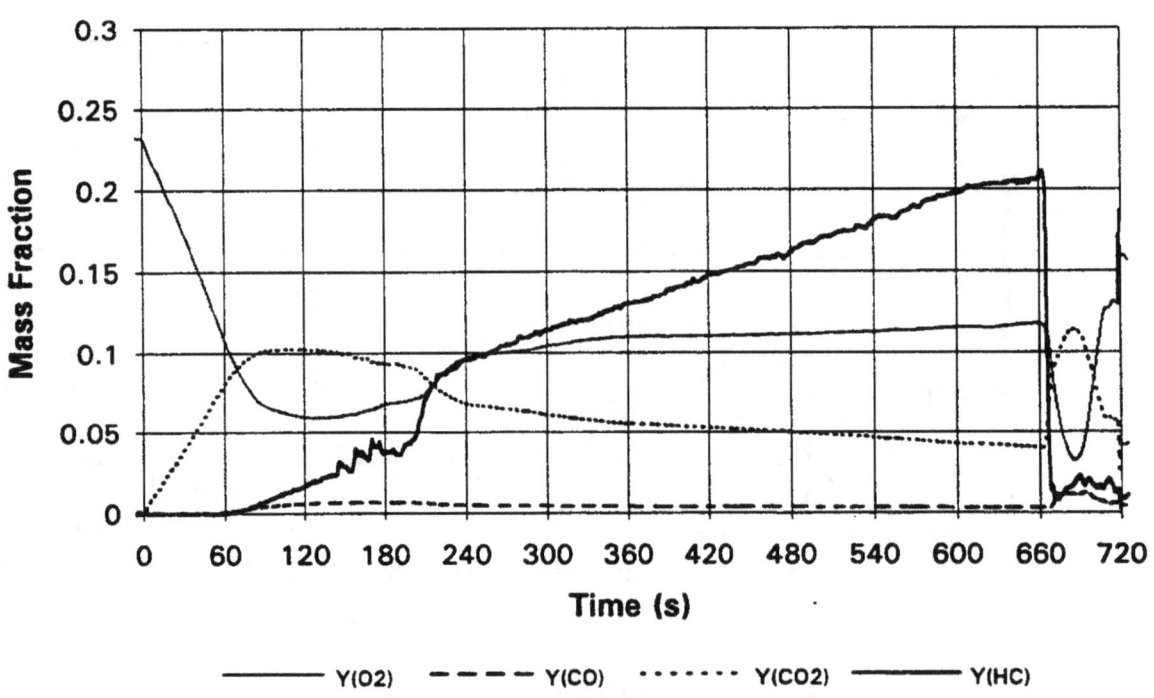

Run: P3EXP39

Bidirectional Probe Velocity History

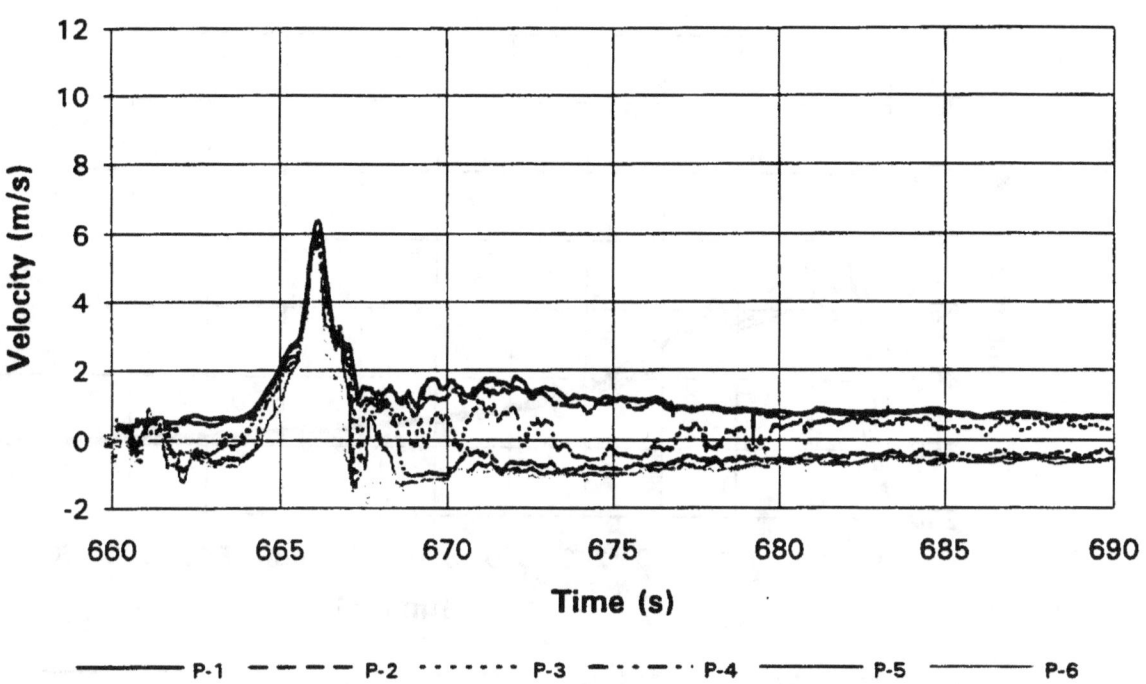

Pressure & Total Mass Out Histories

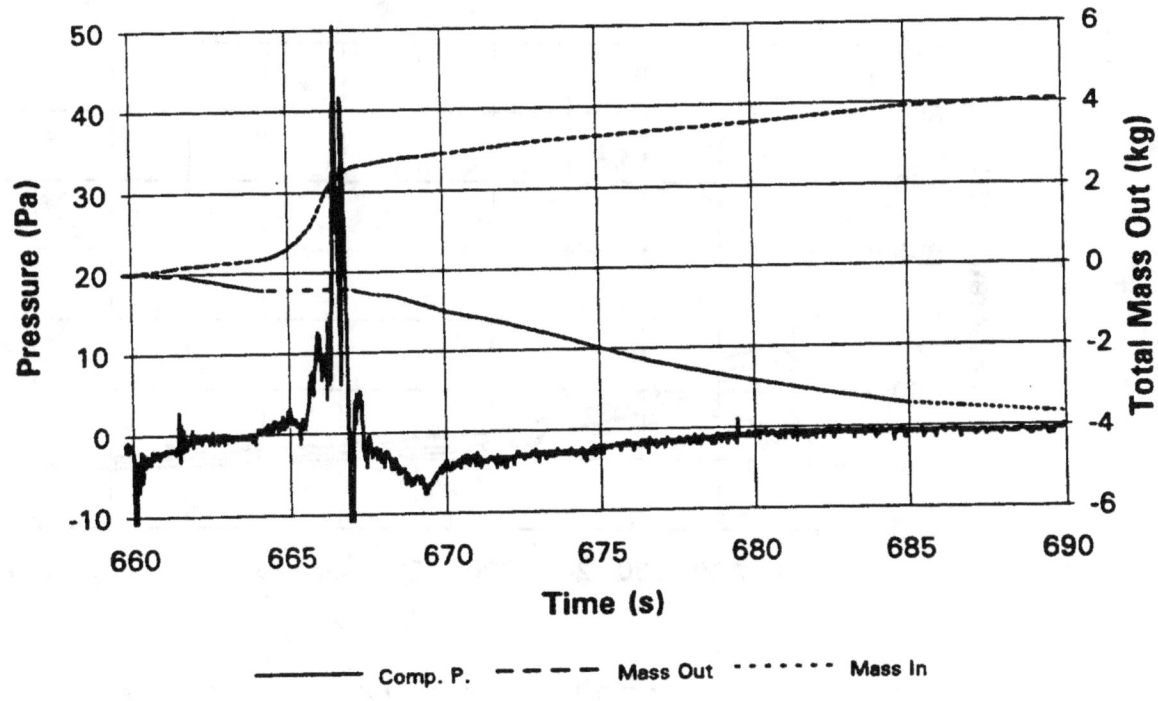

Run: P3EXP39

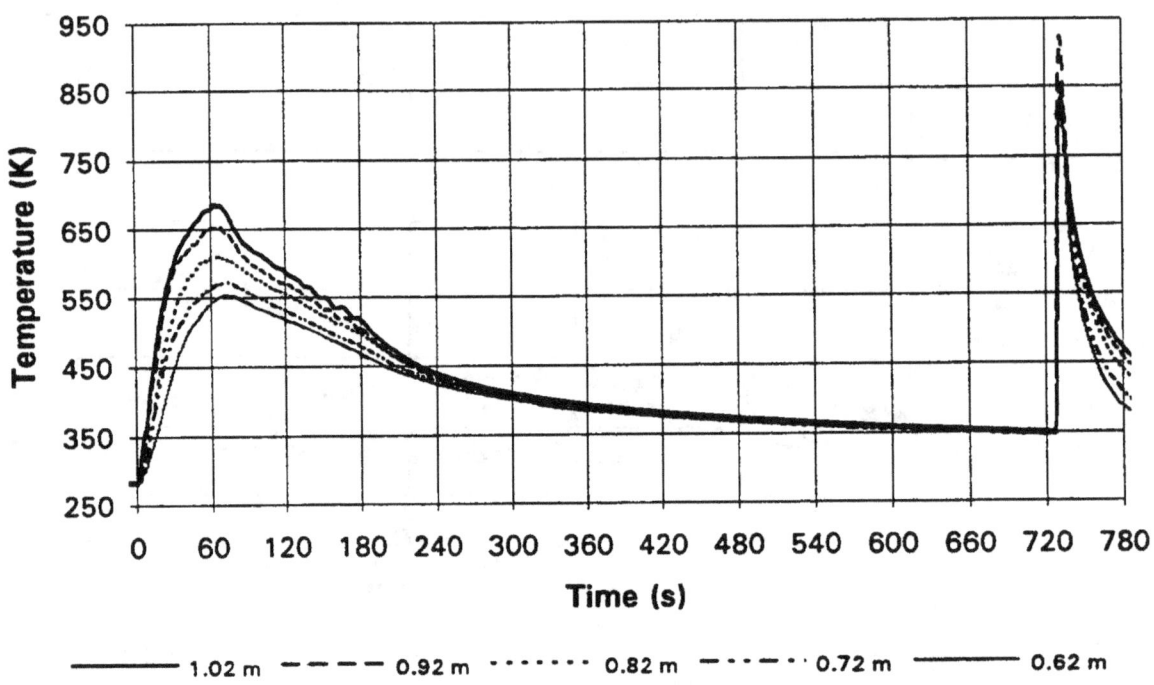

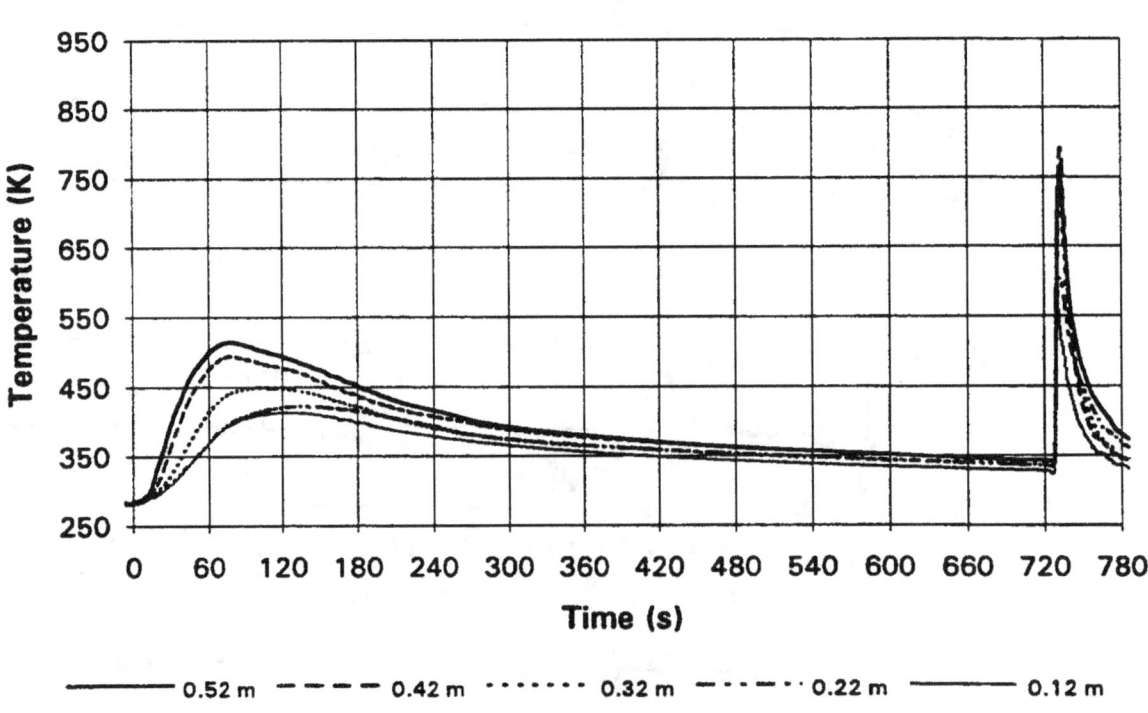

Run: P3EXP40

172

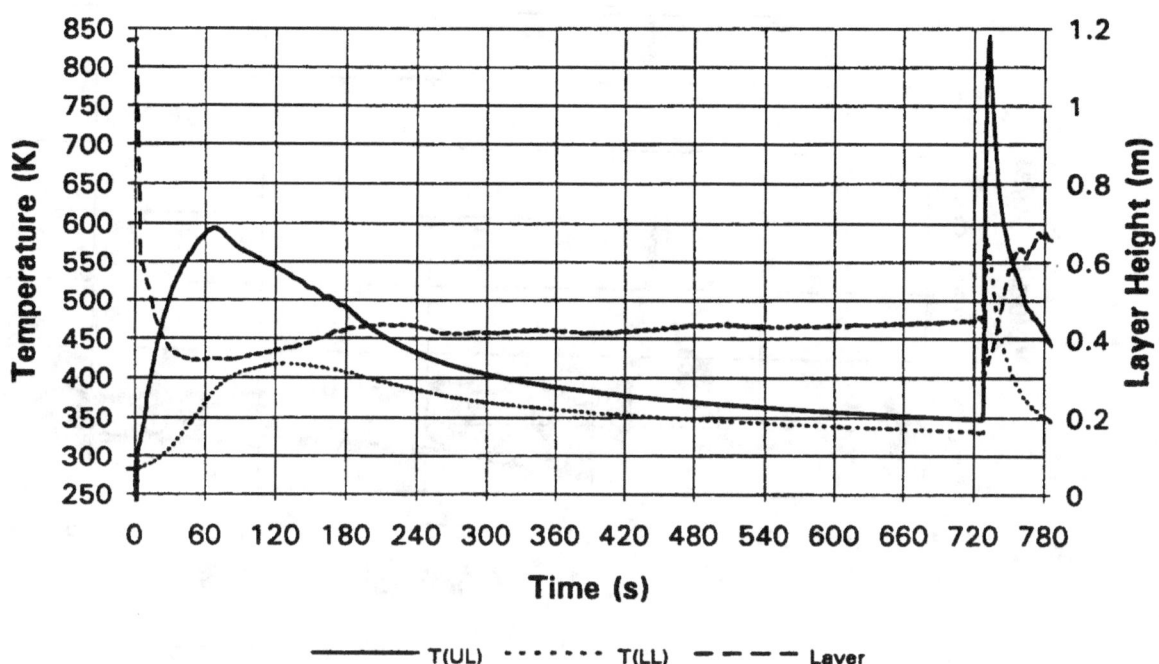

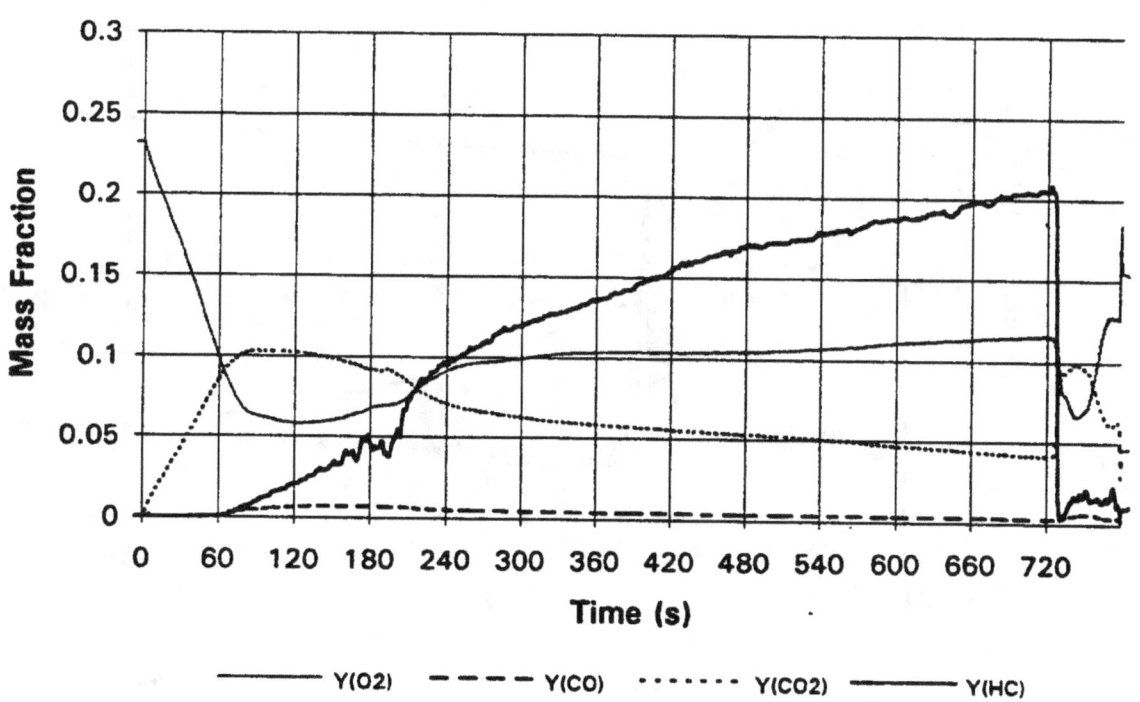

Run: P3EXP40

173

Bidirectional Probe Velocity History

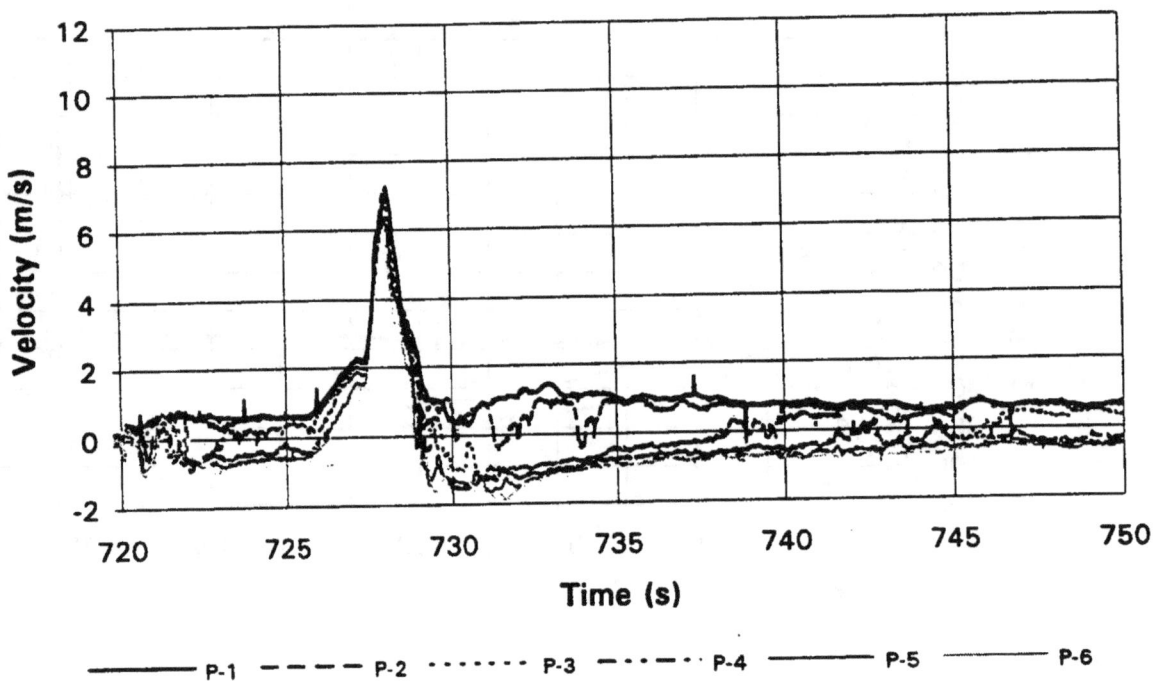

Pressure & Total Mass Out Histories

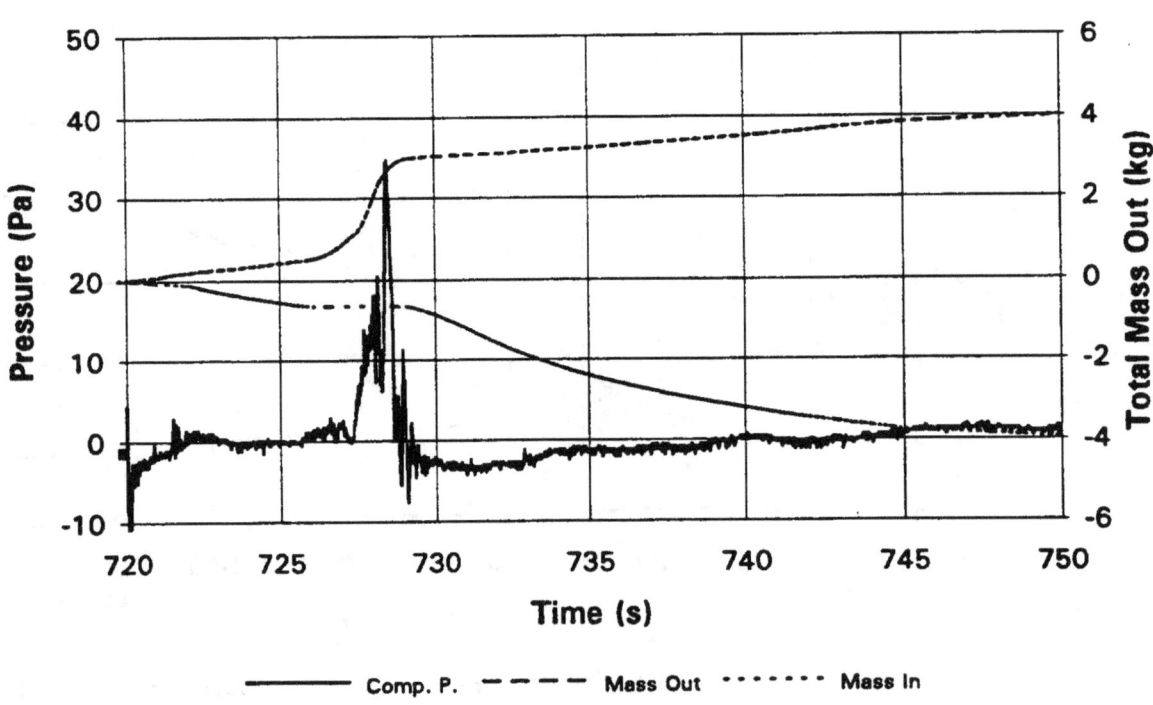

Run: P3EXP40

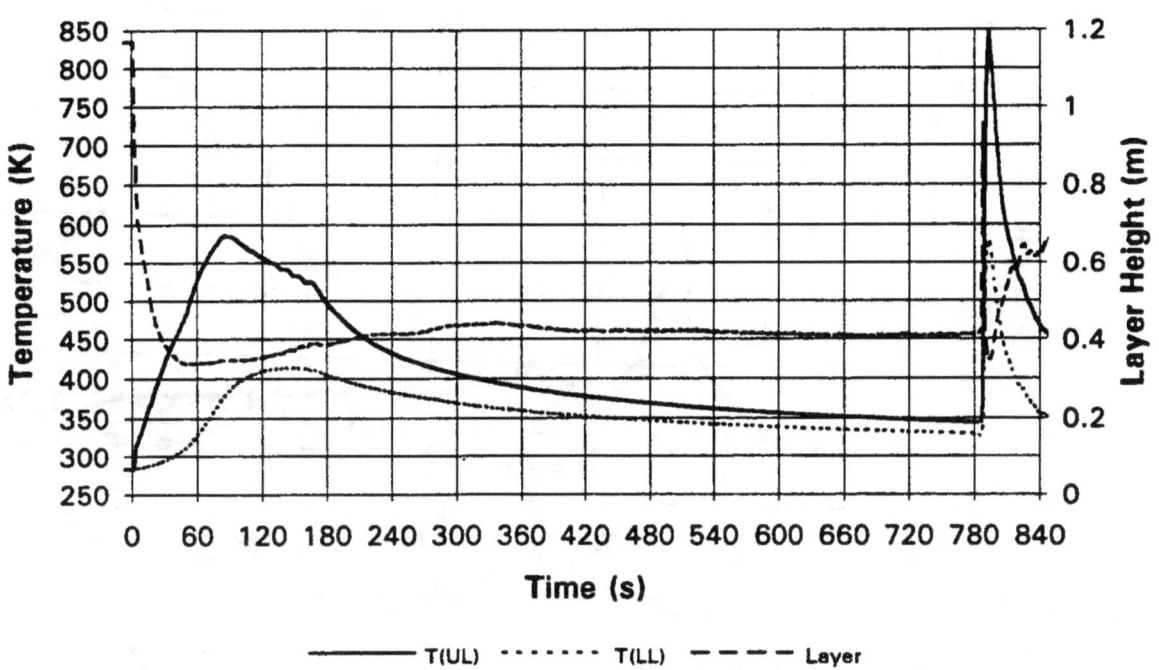

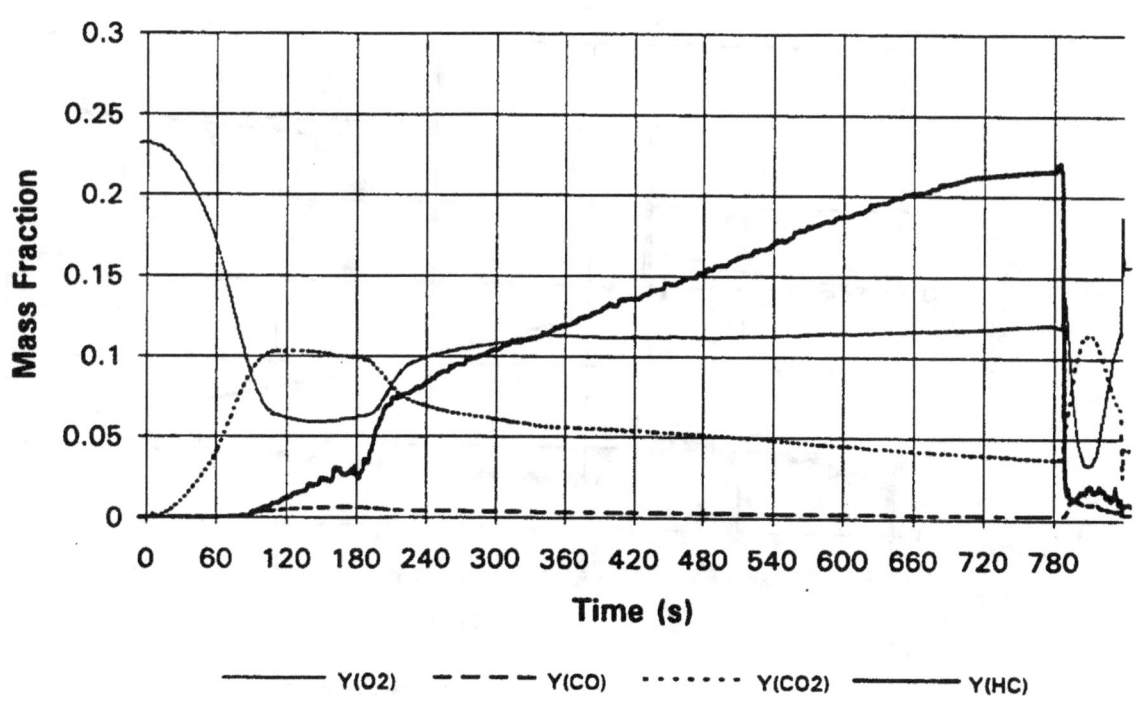

Run: P3EXP41

Bidirectional Probe Velocity History

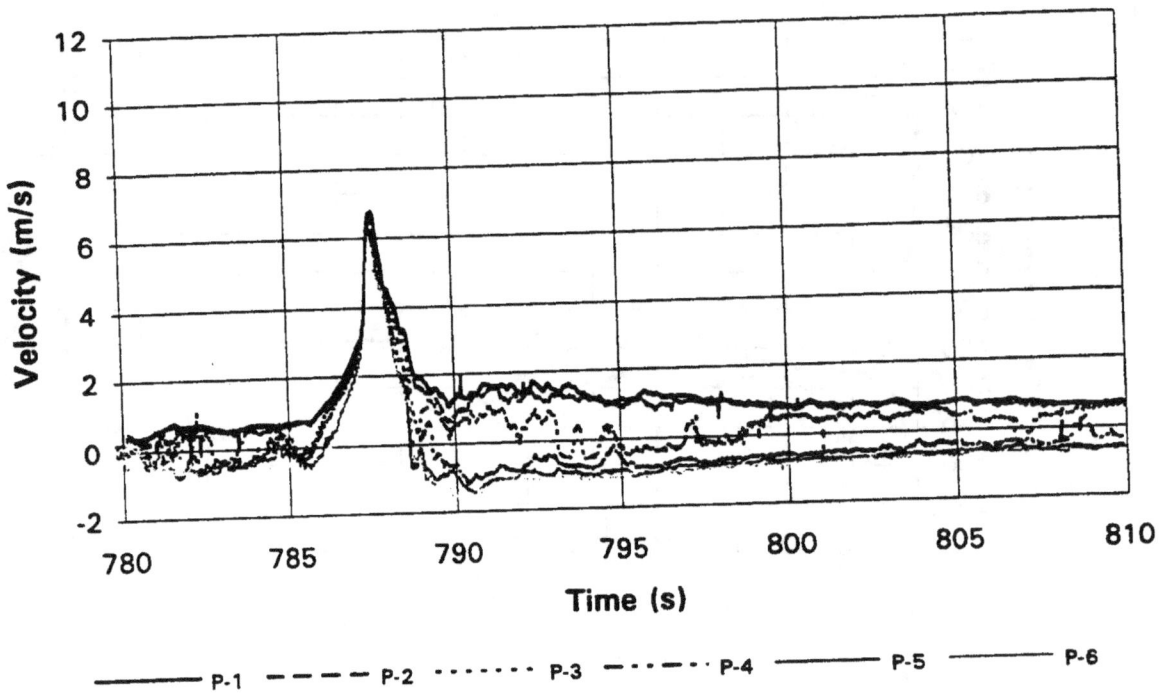

Pressure & Total Mass Out Histories

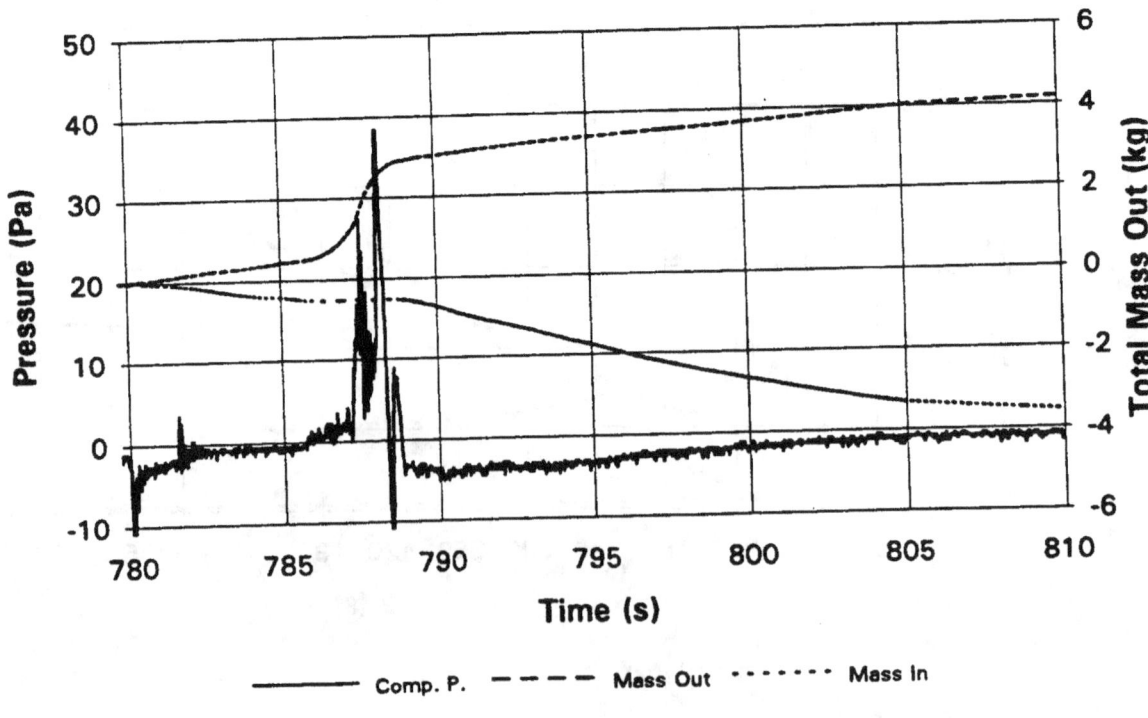

Run: P3EXP41

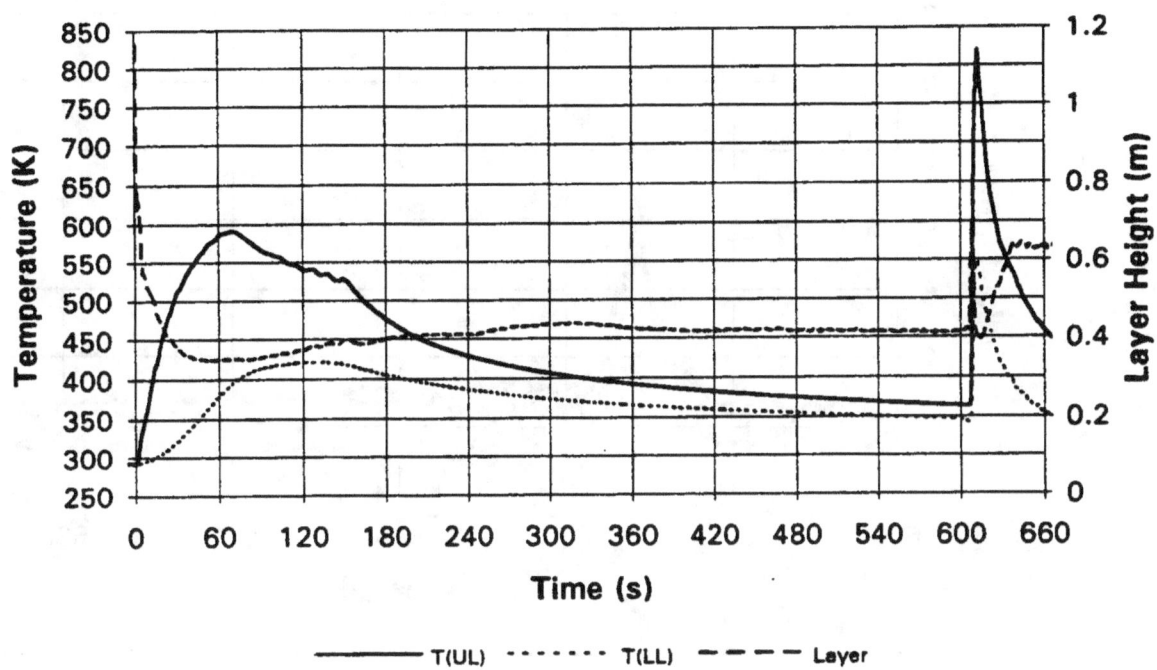

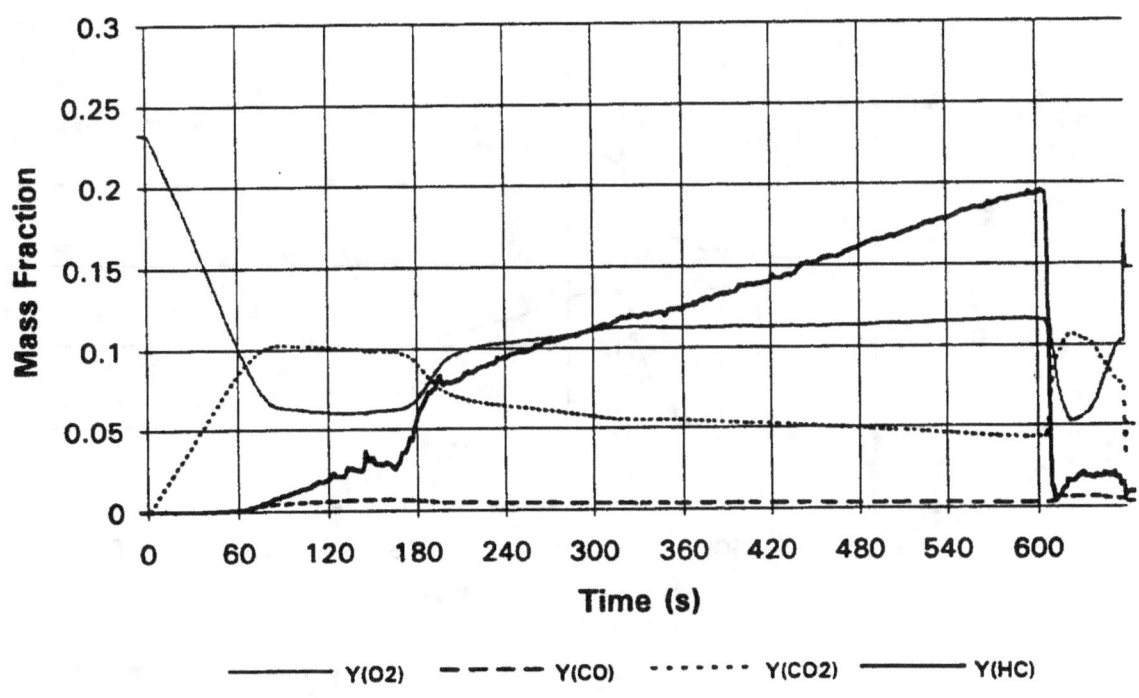

Run: P3EXP42

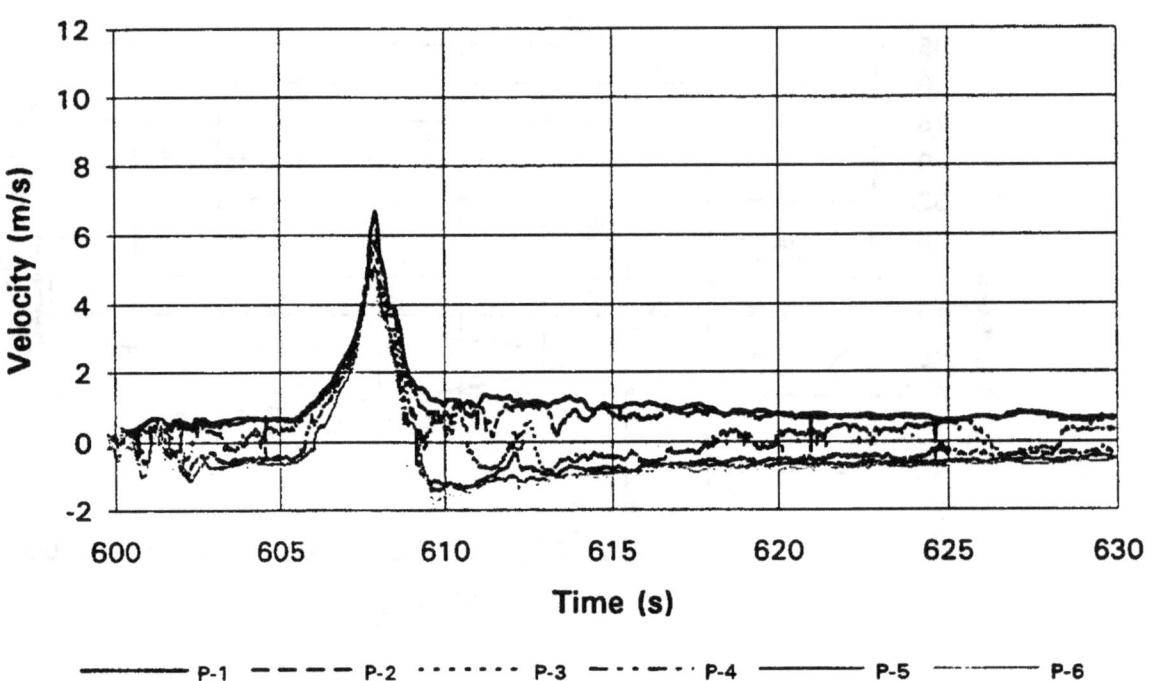

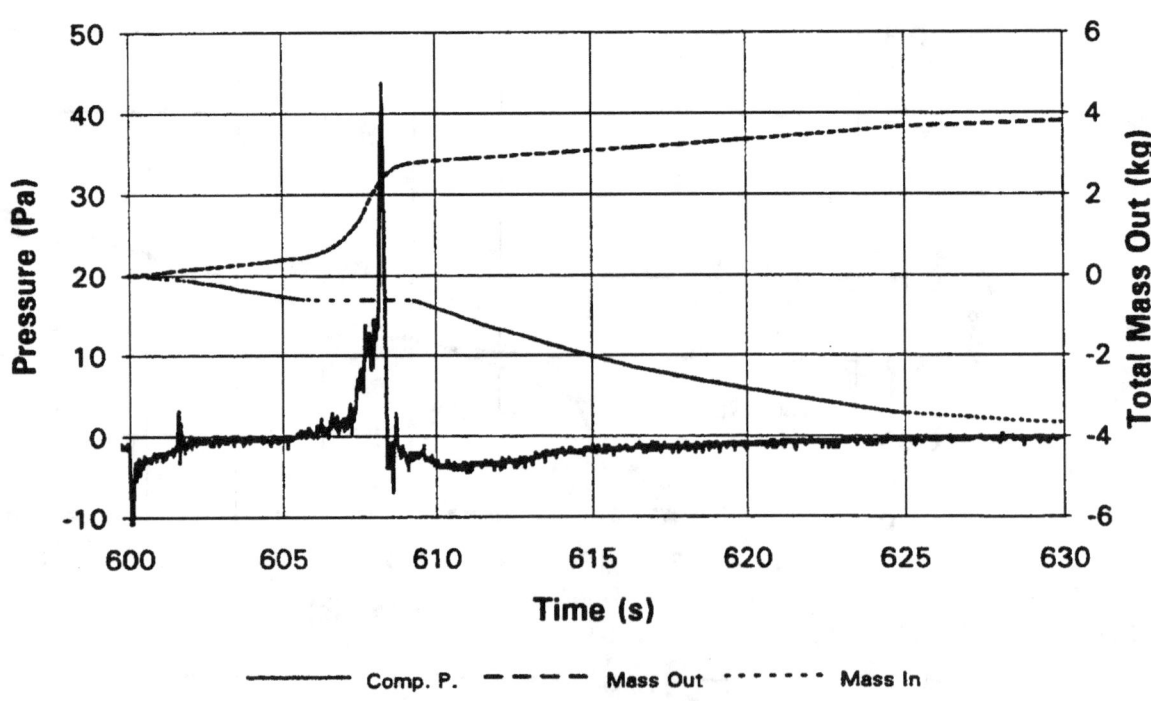

Run: P3EXP42

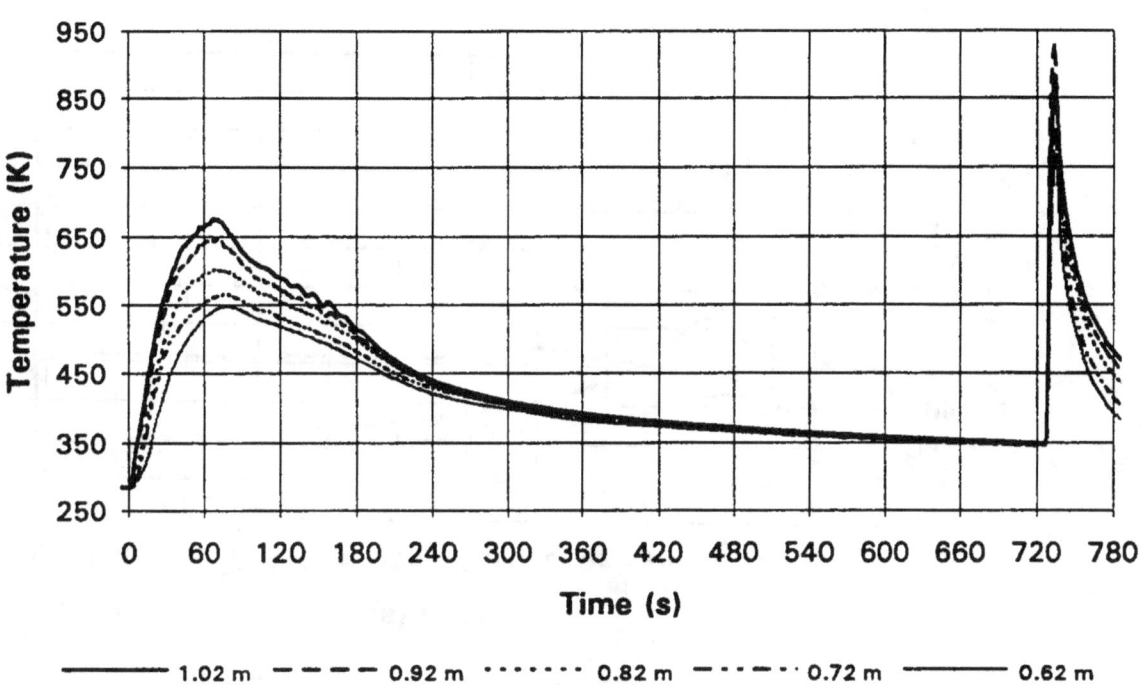

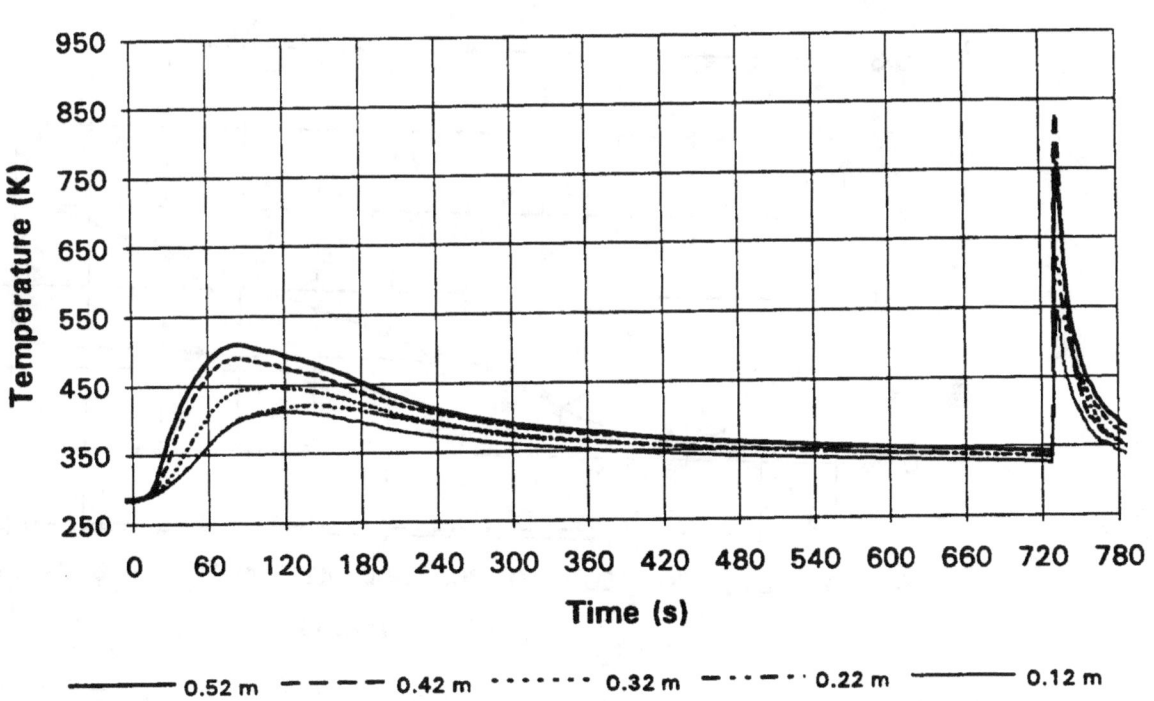

Run: P3EXP43

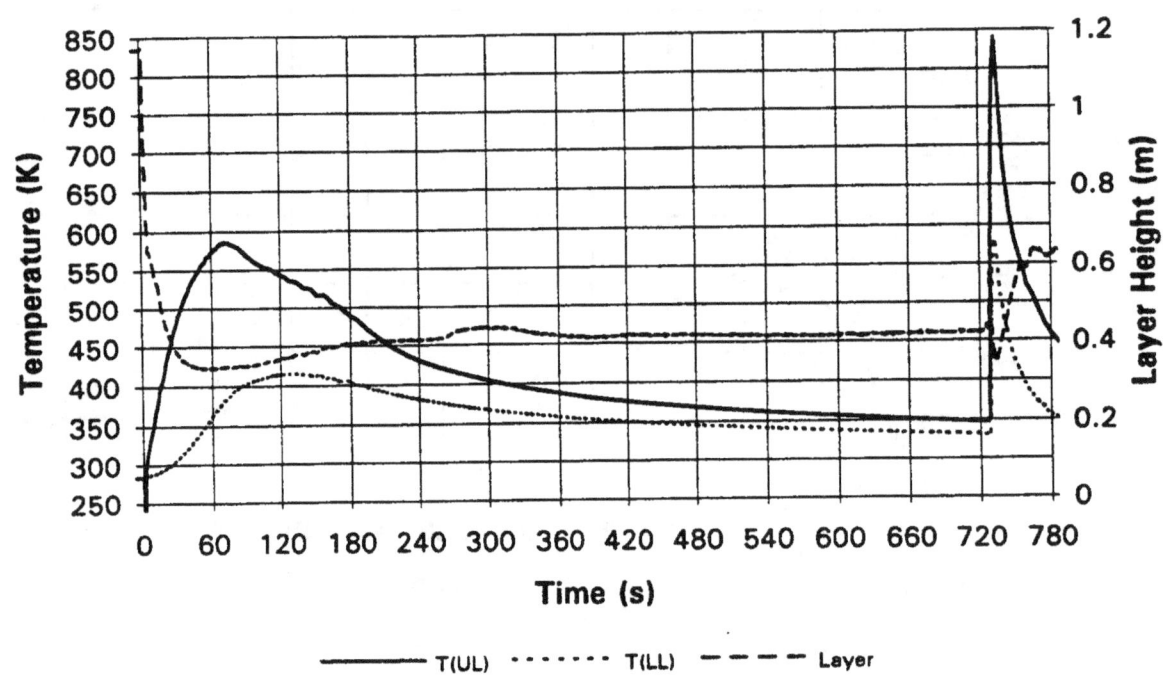

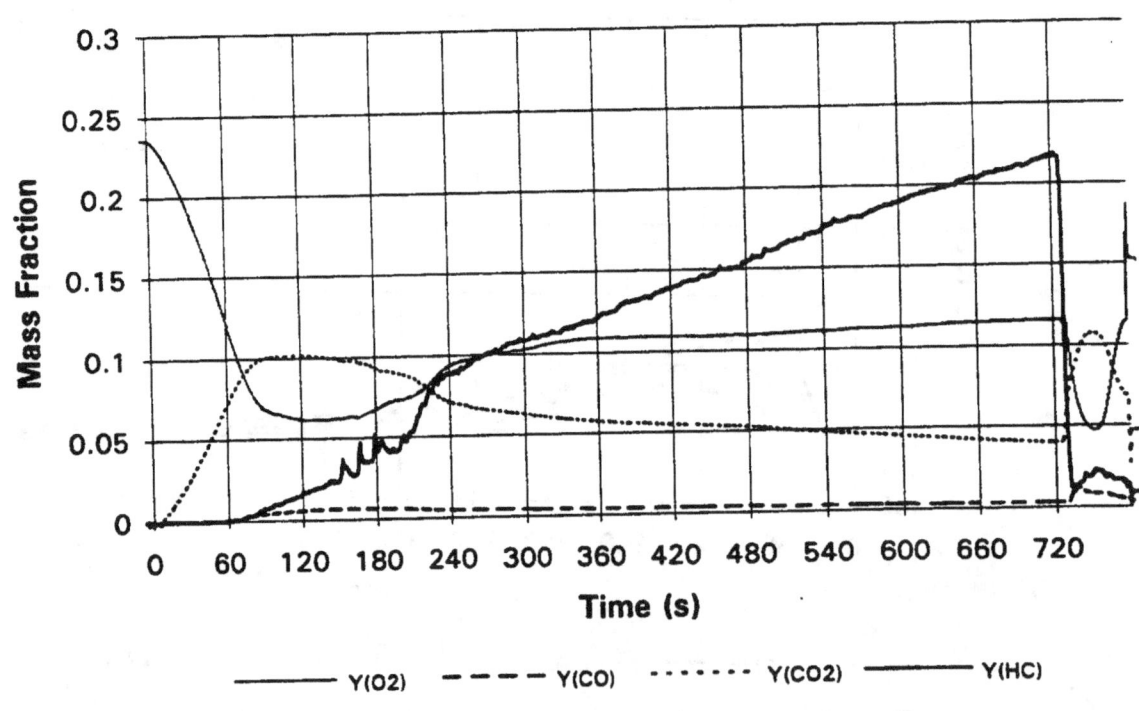

Run: P3EXP43

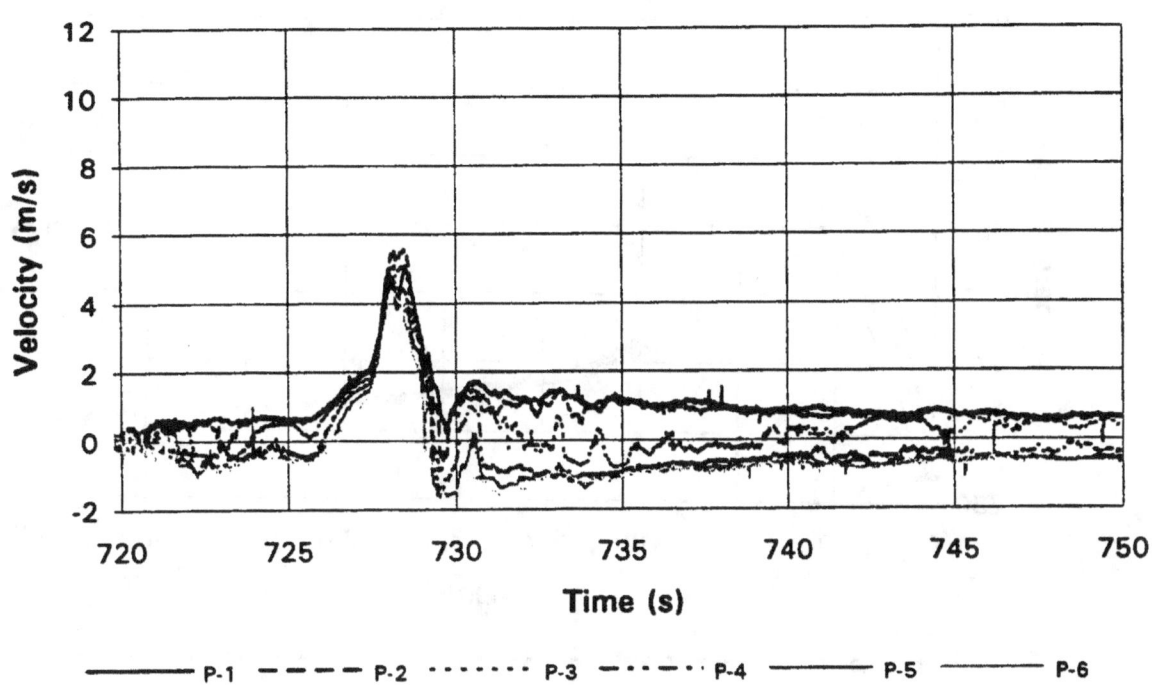

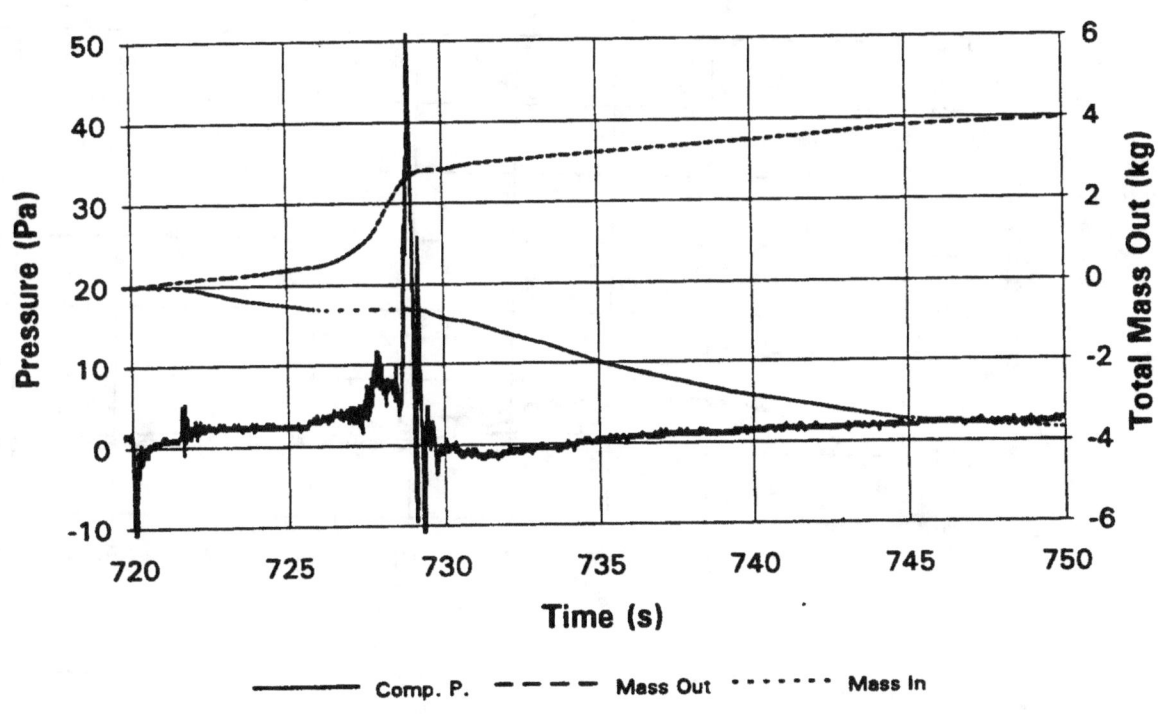

Run: P3EXP43

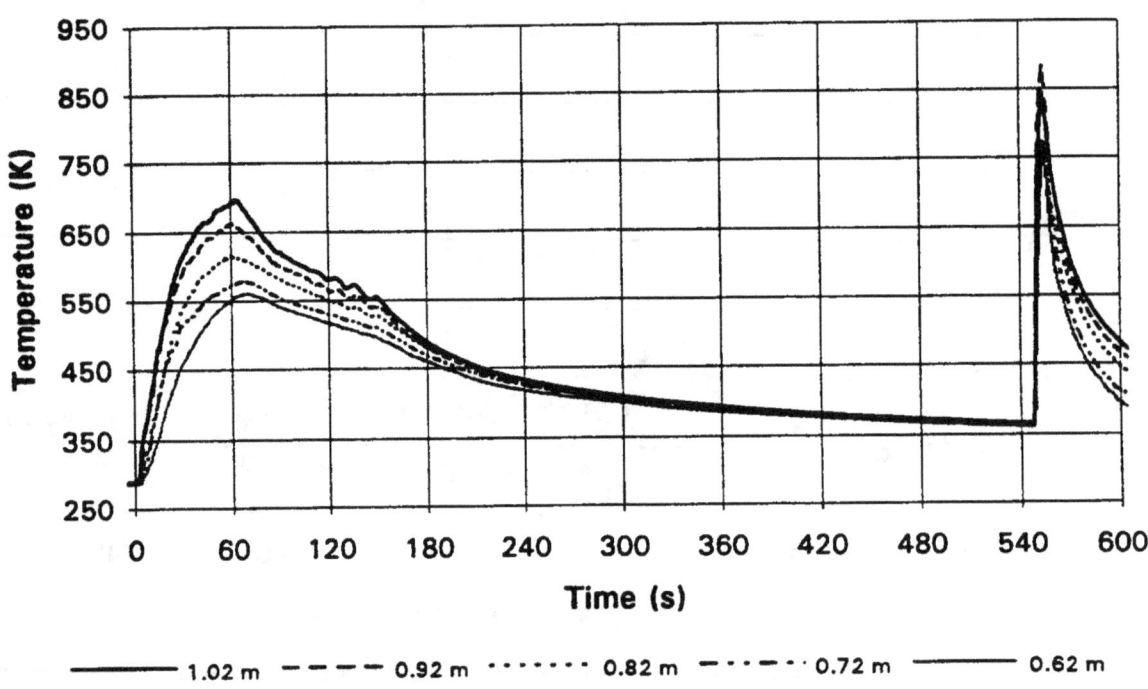

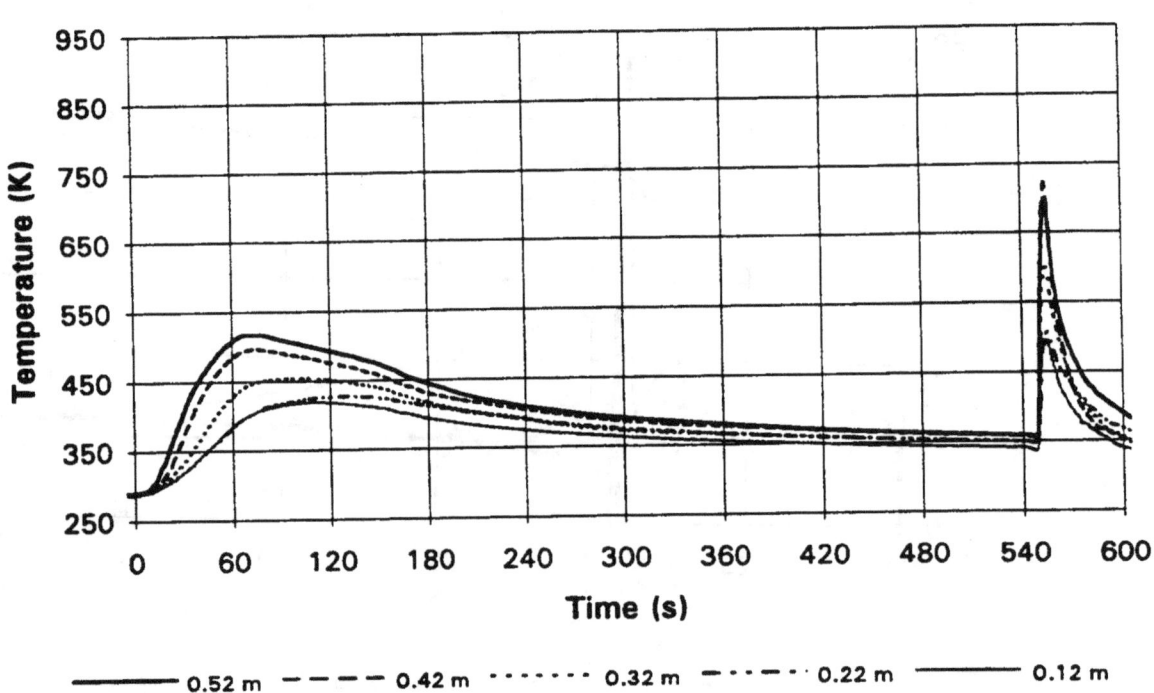

Run: P3EXP45

Two Zone Approximation - Temperature & Interface Histories

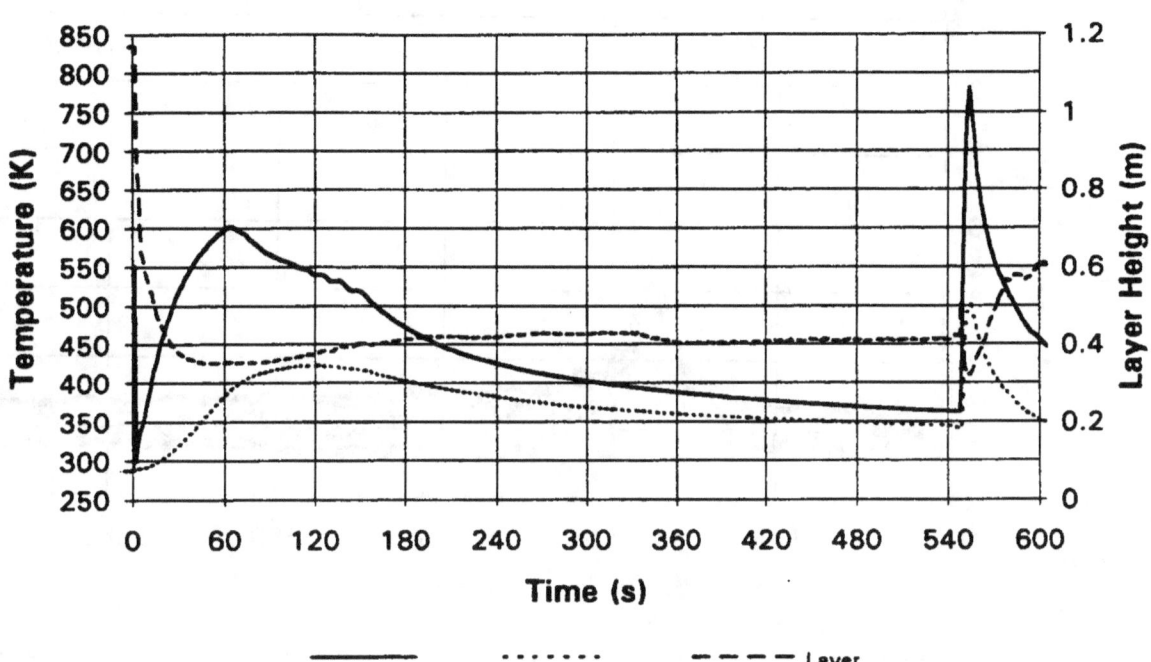

Mass Fraction History - Compartment Gases

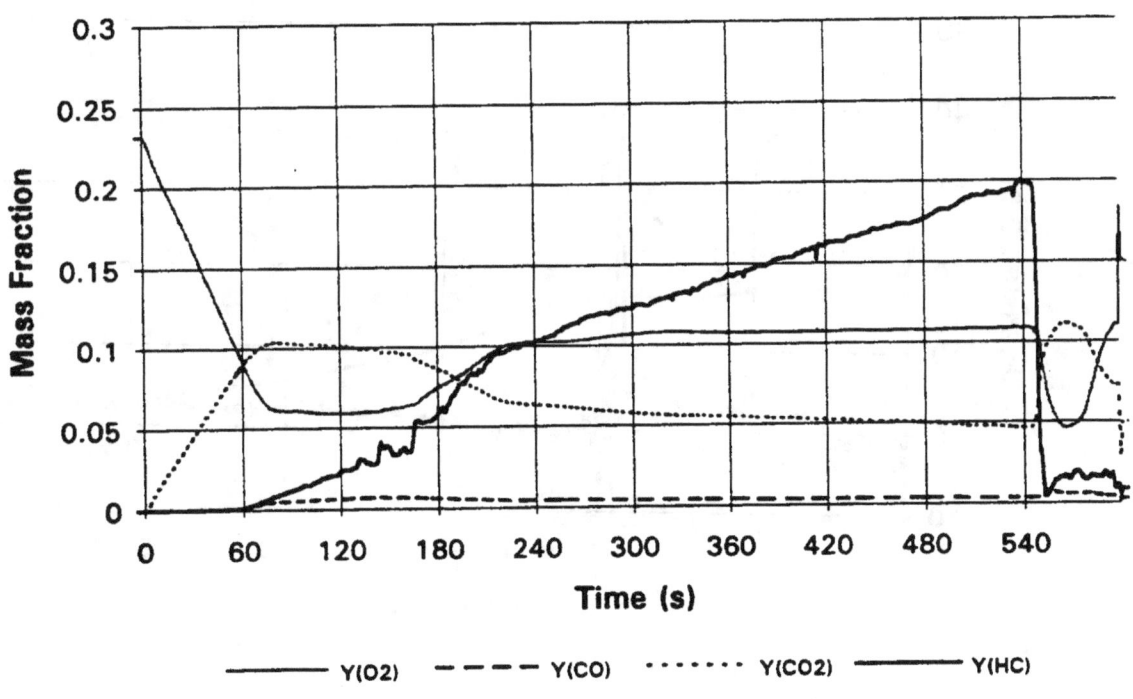

Run: P3EXP45

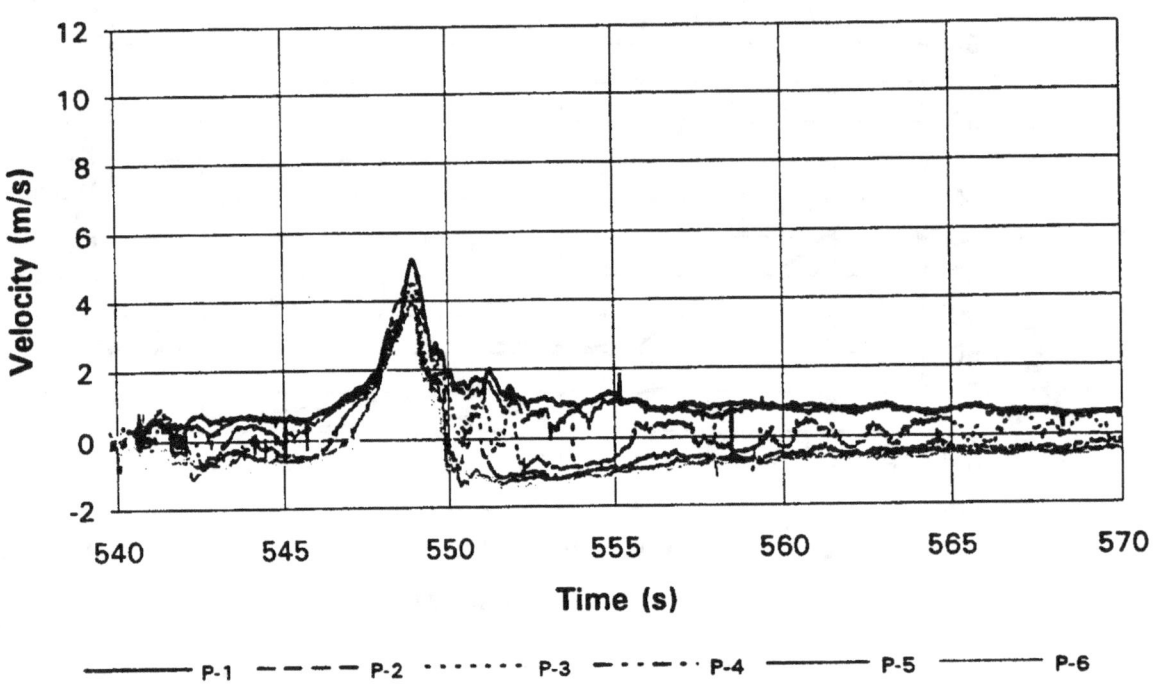

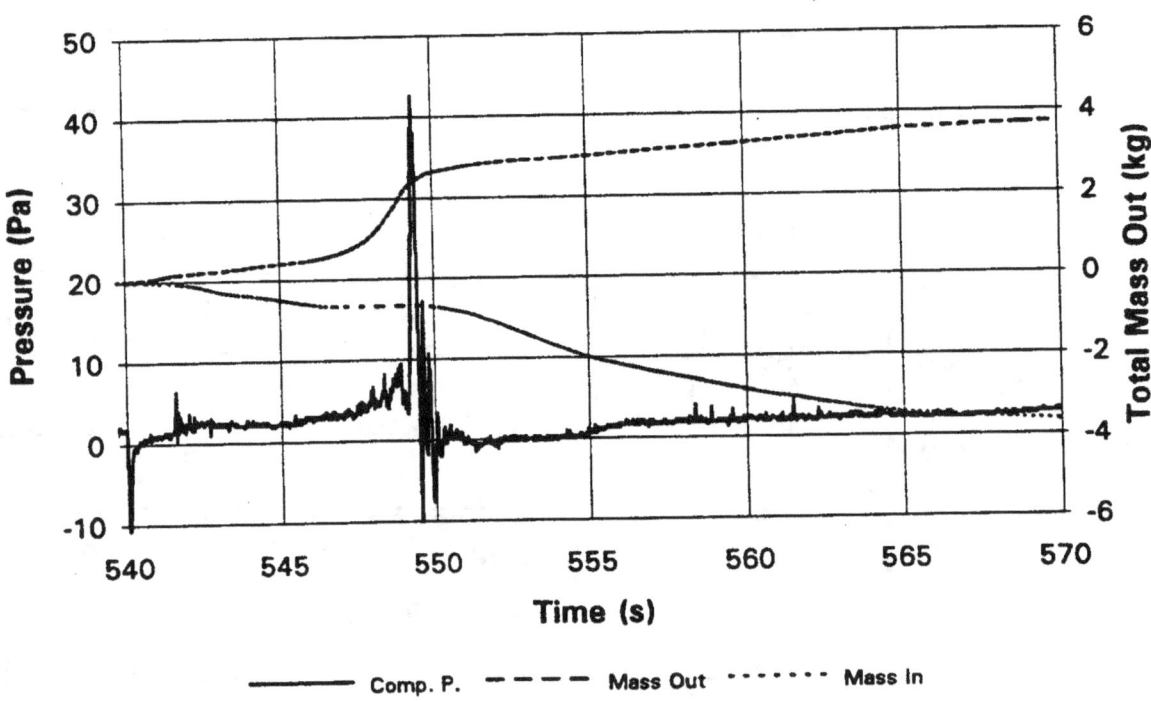

Run: P3EXP45

Two Zone Approximation -Temperature & Interface Histories

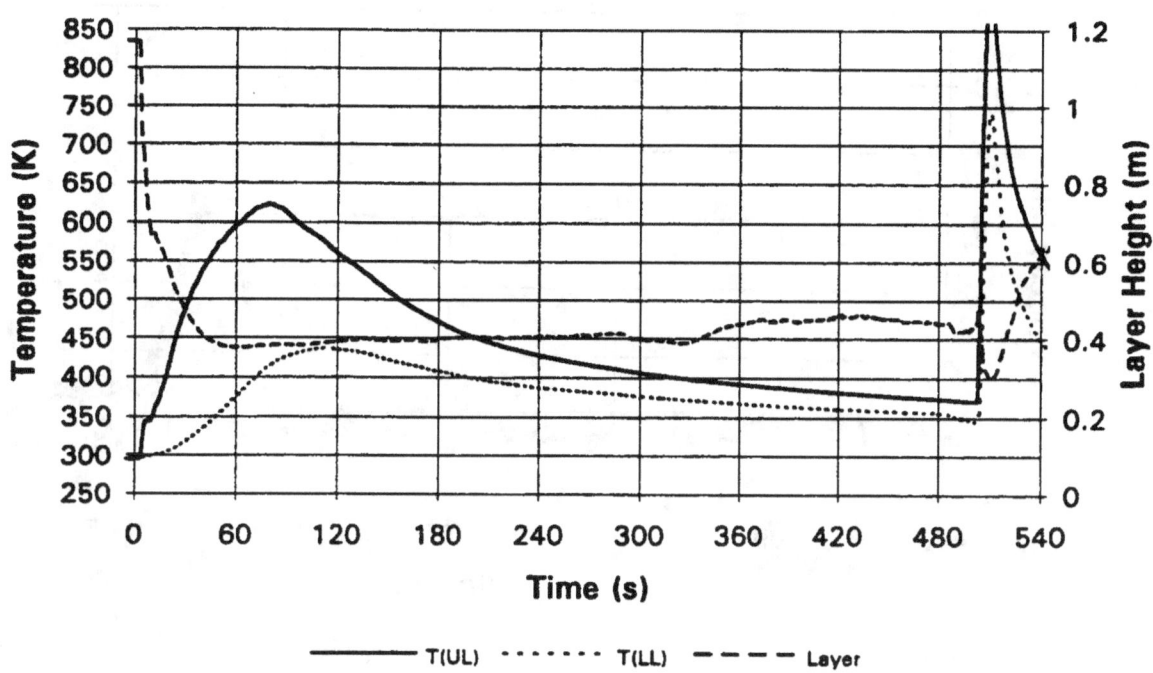

Mass Fraction History - Compartment Gases

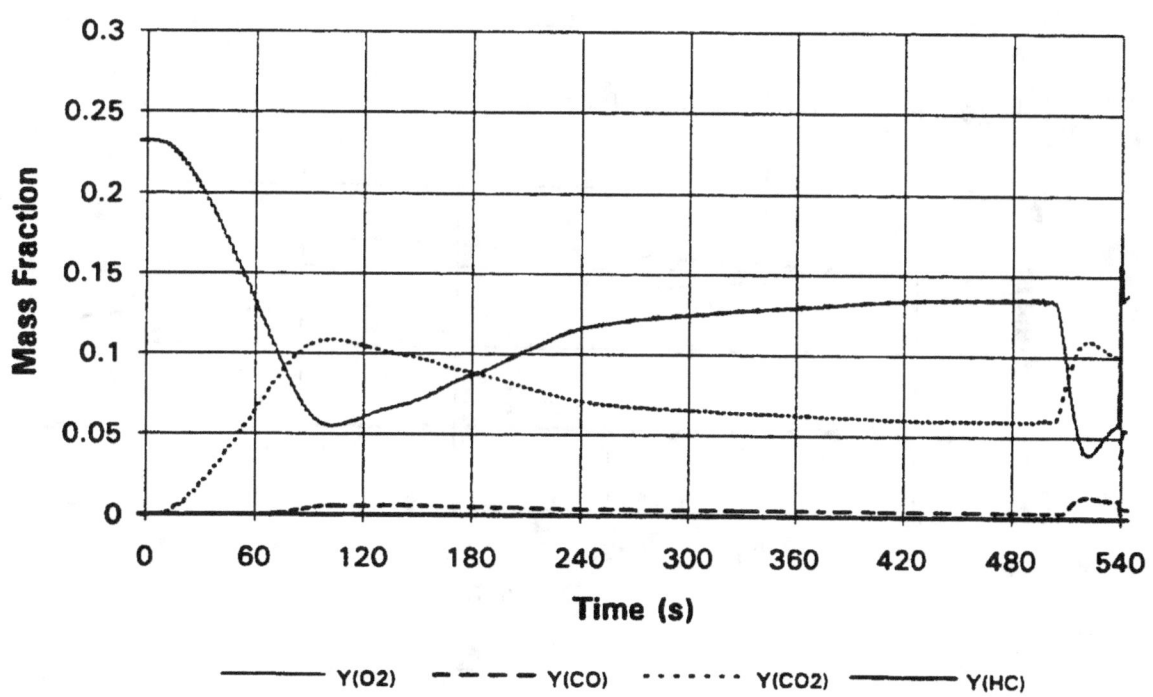

Run: P3EXP76

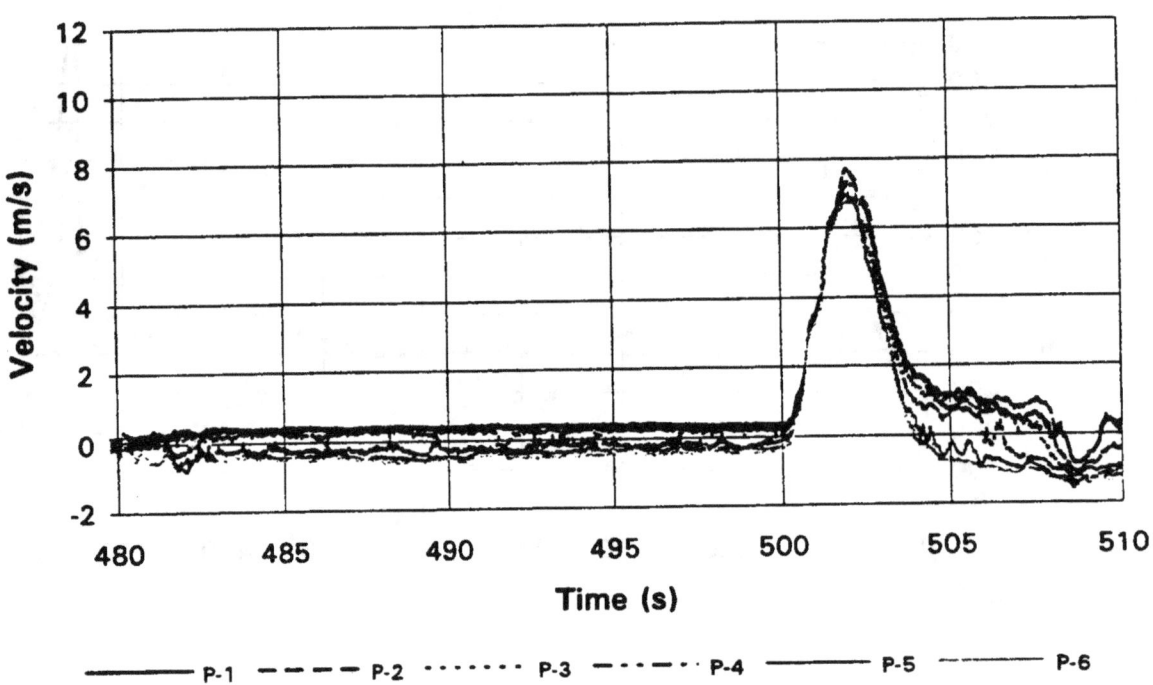

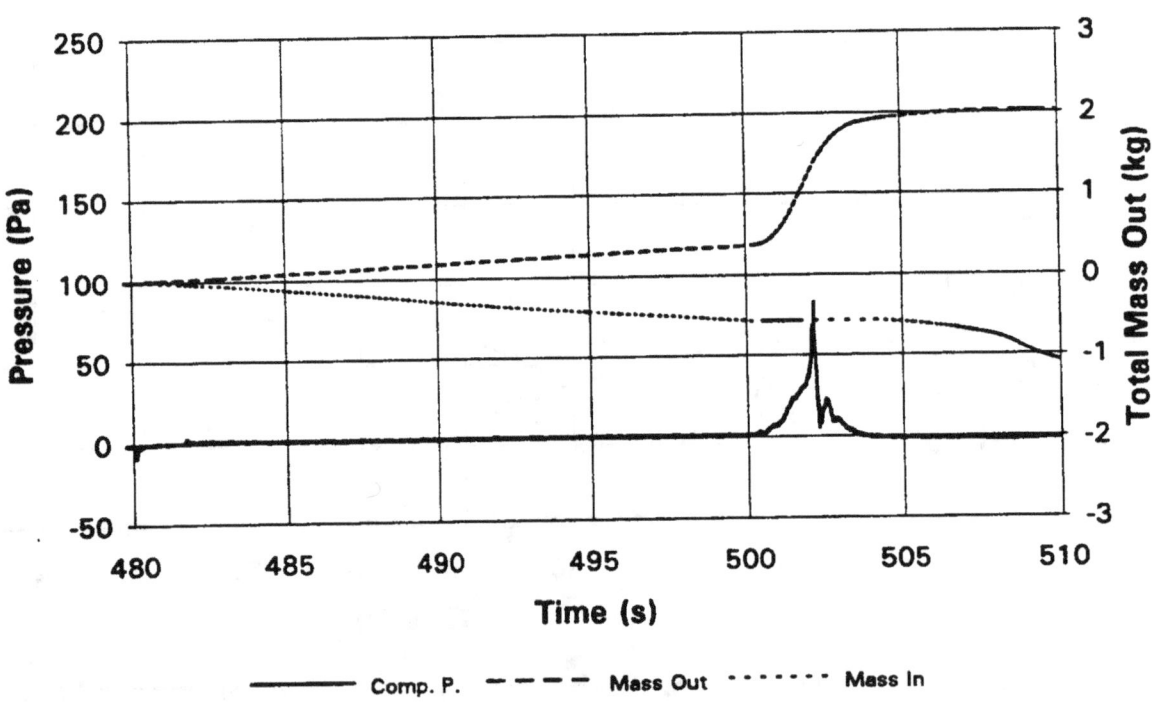

Run: P3EXP76

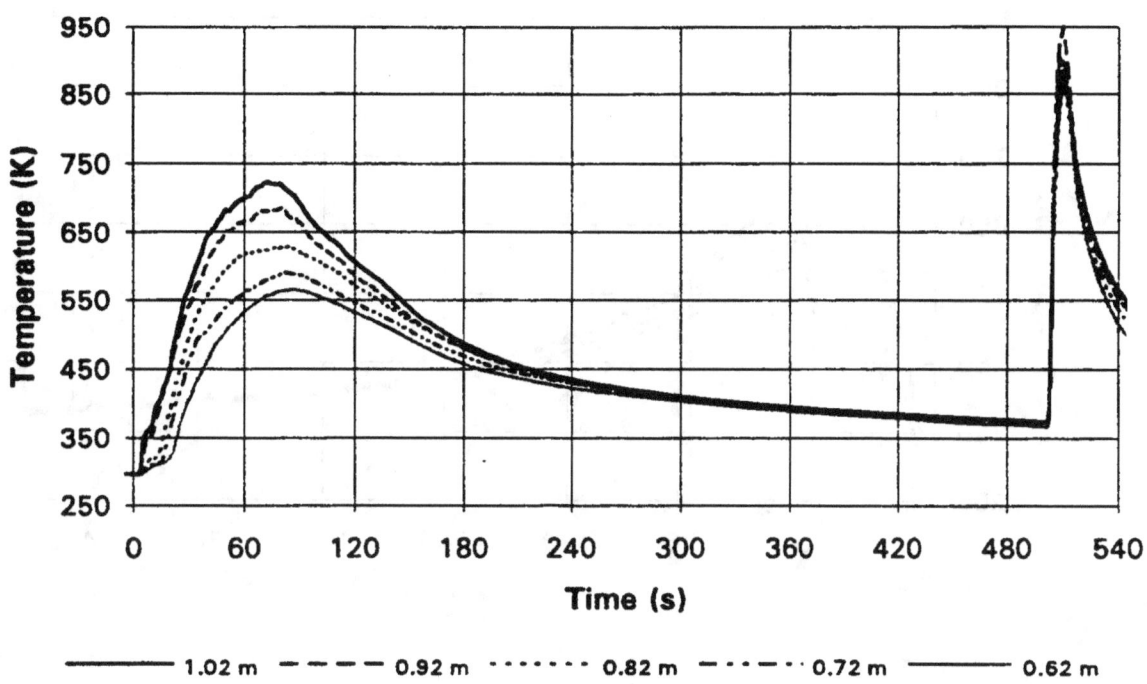

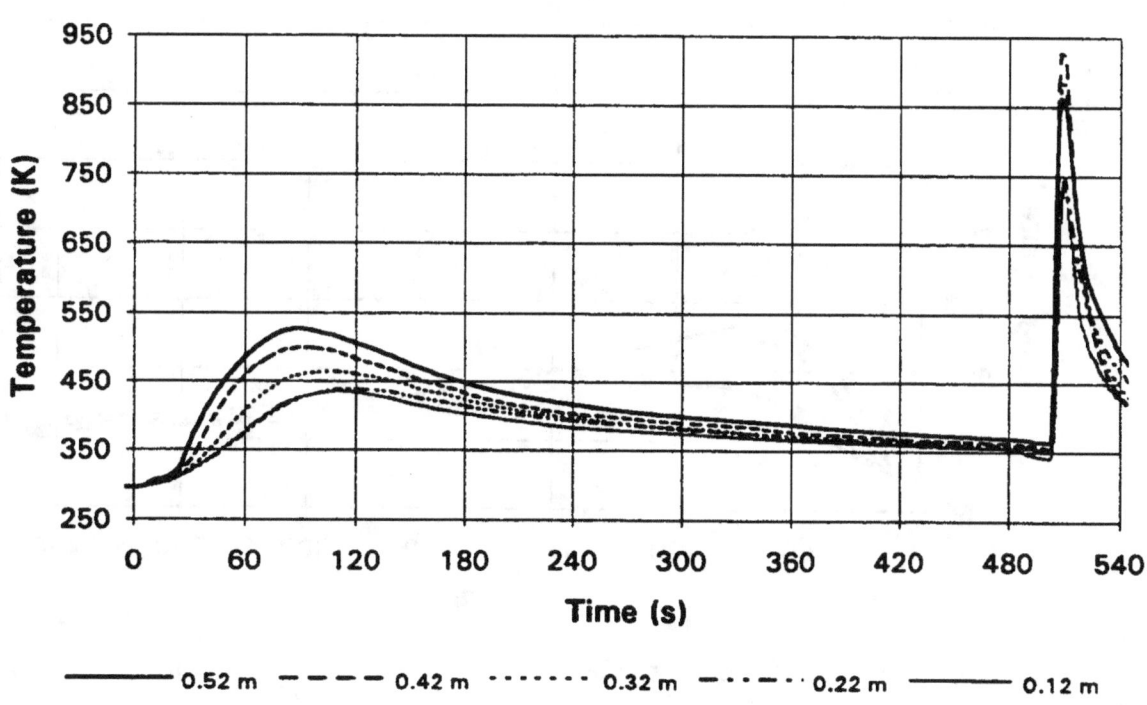

Run: P3EXP76

Layer Temperature History - TC Tree Data

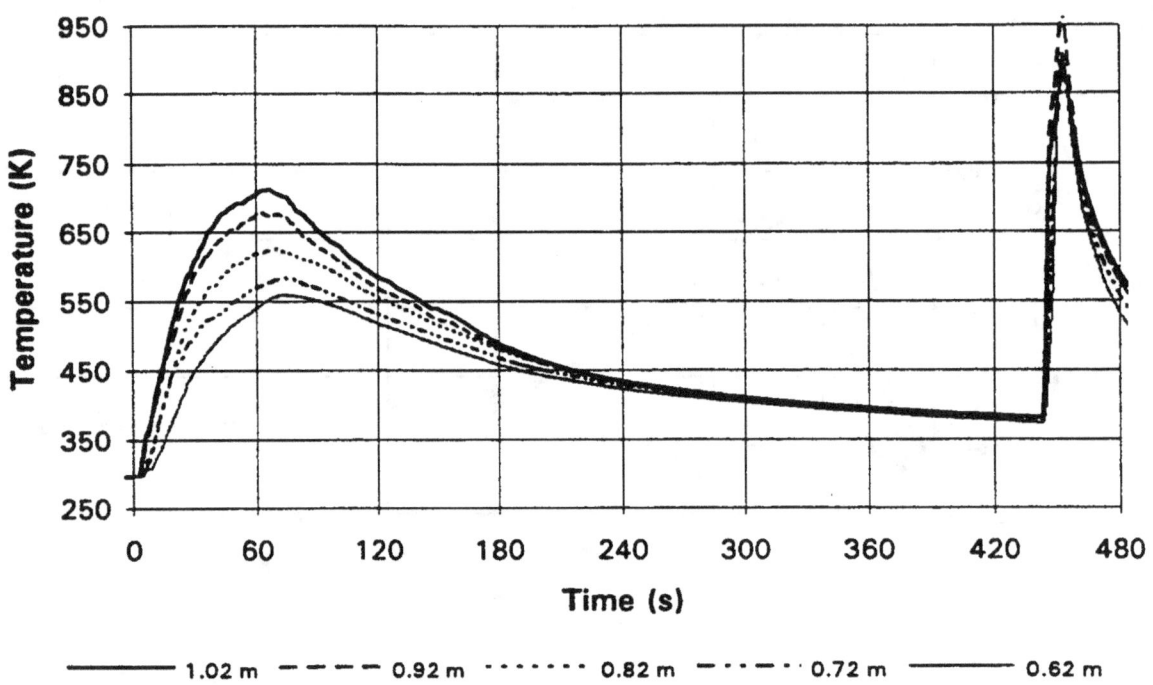

Layer Temperature History - TC Tree Data

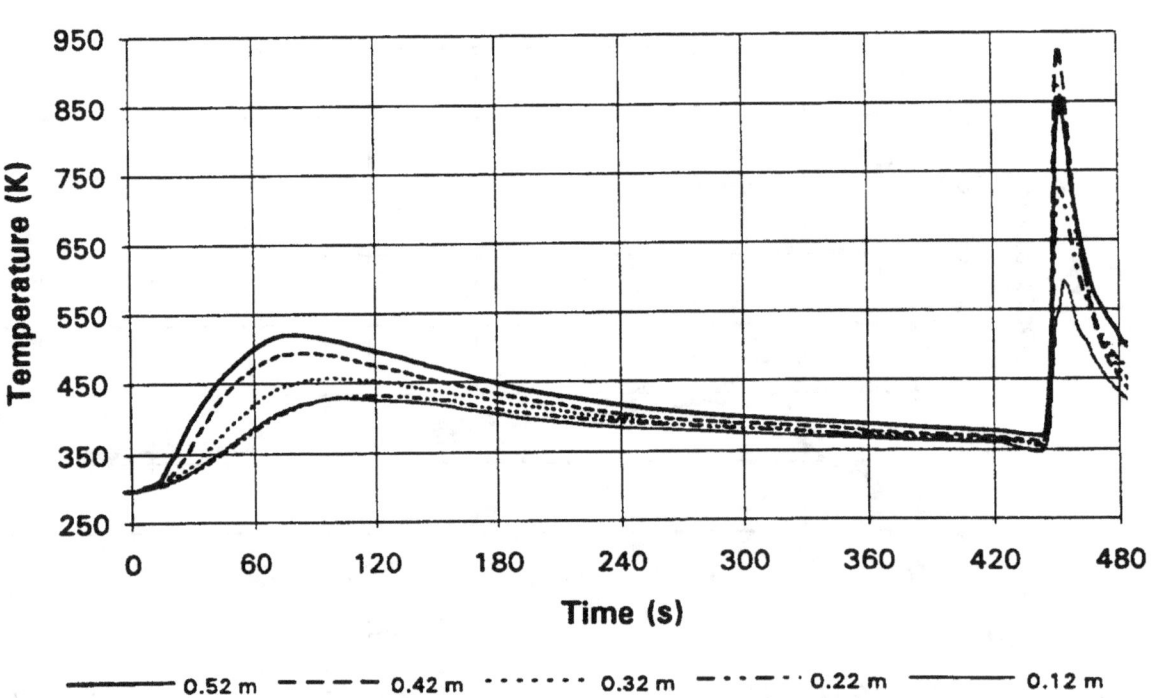

Run: P3EXP77

Two Zone Approximation - Temperature & Interface Histories

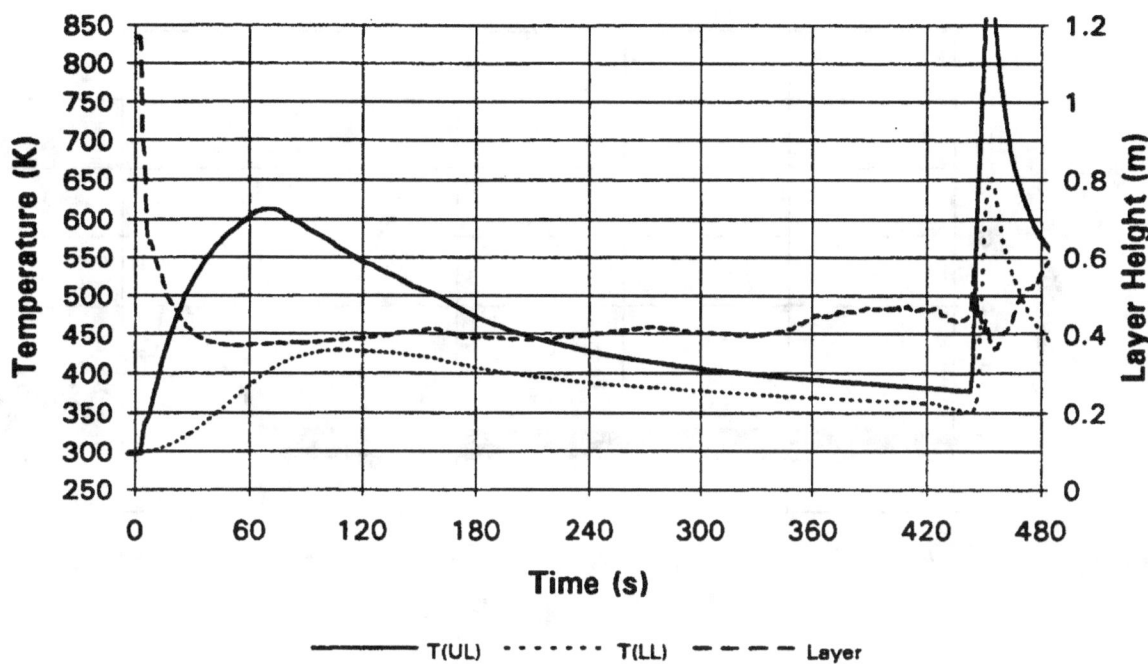

Mass Fraction History - Compartment Gases

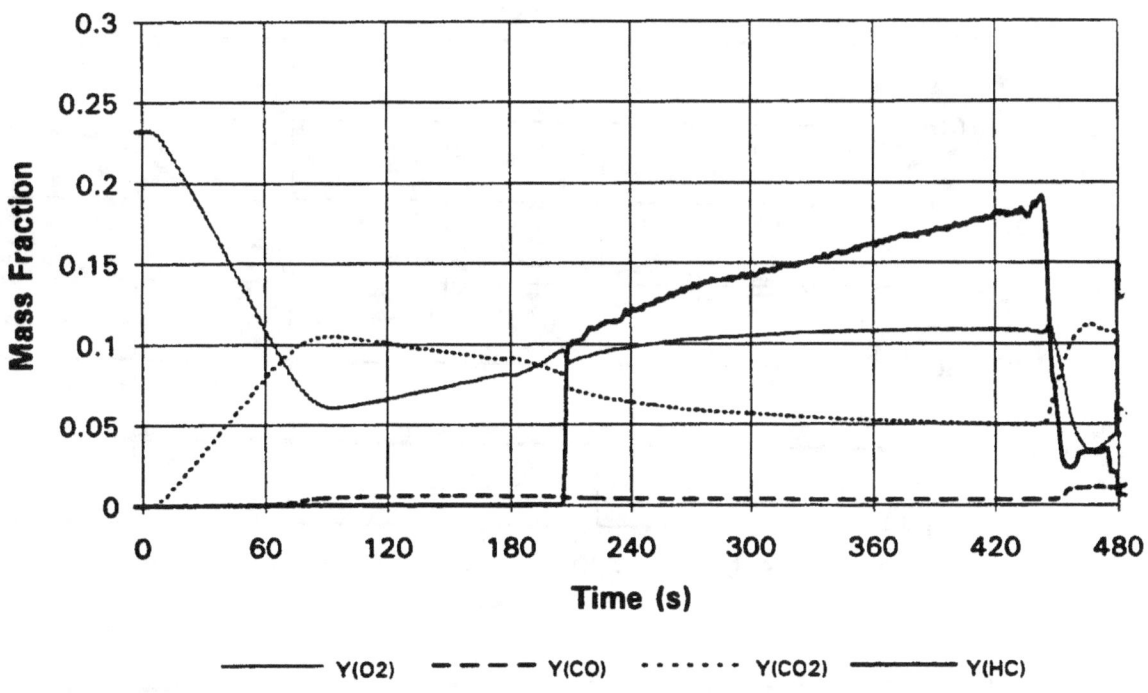

Run: P3EXP77

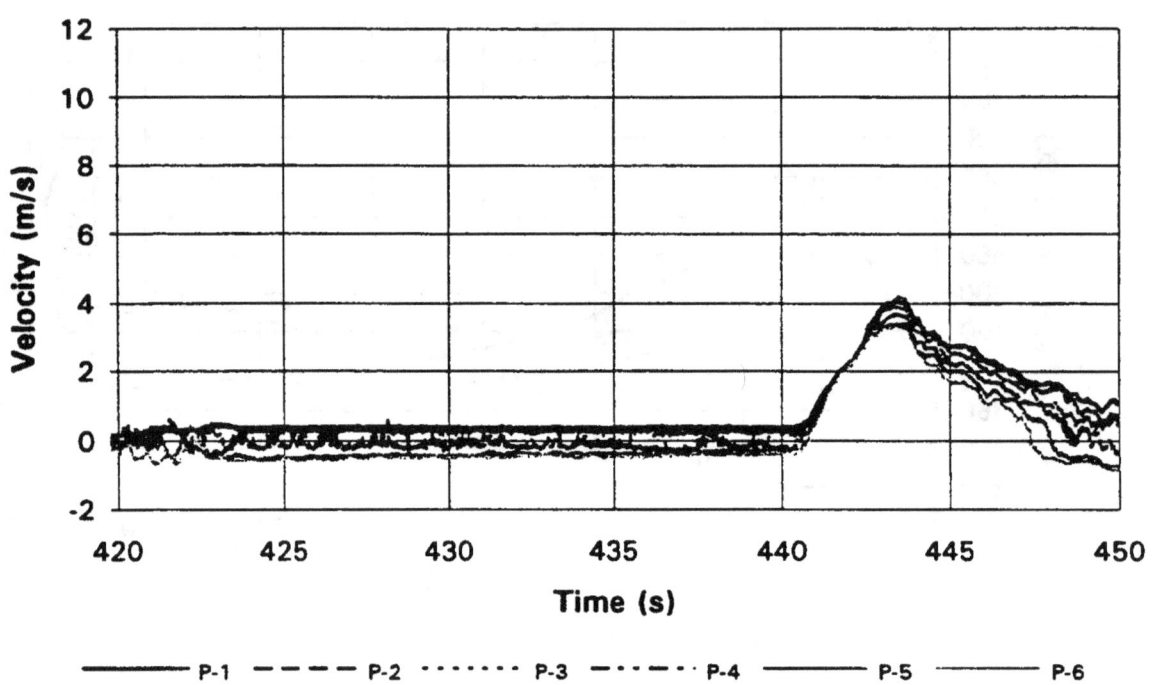

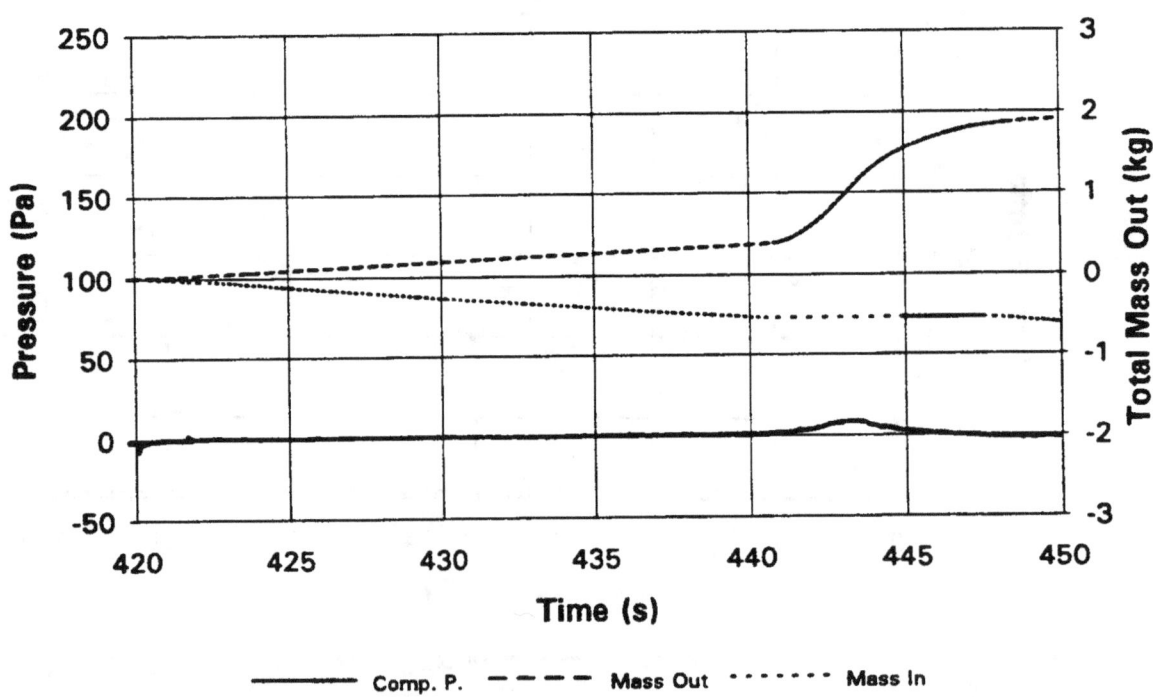

Run: P3EXP77

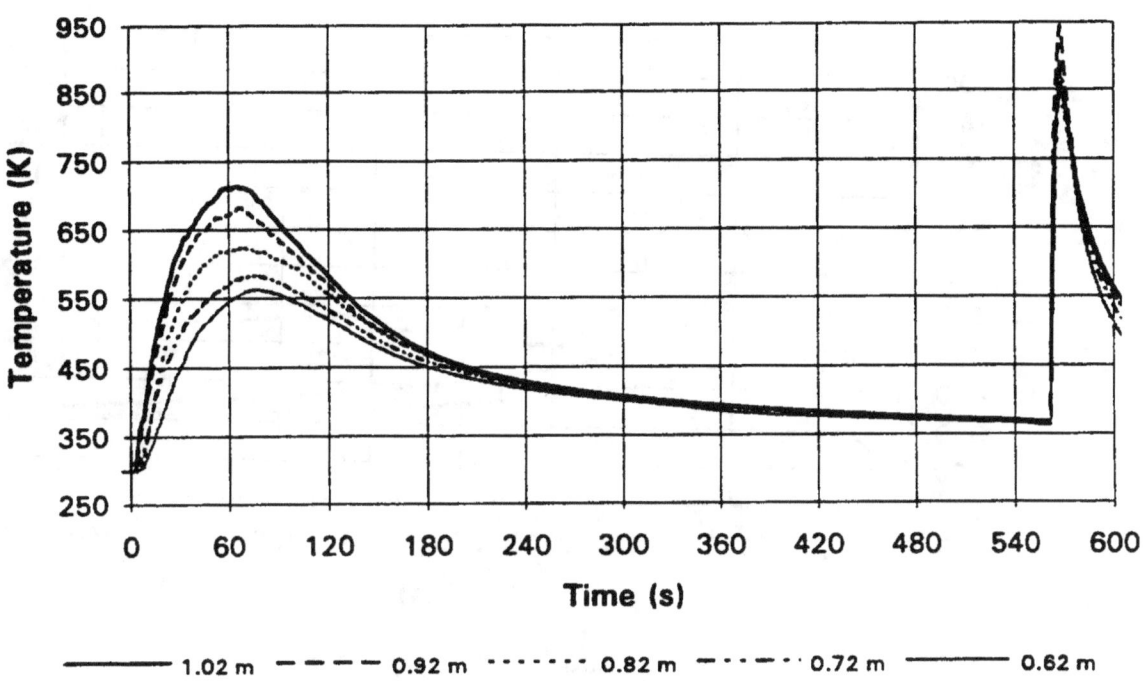

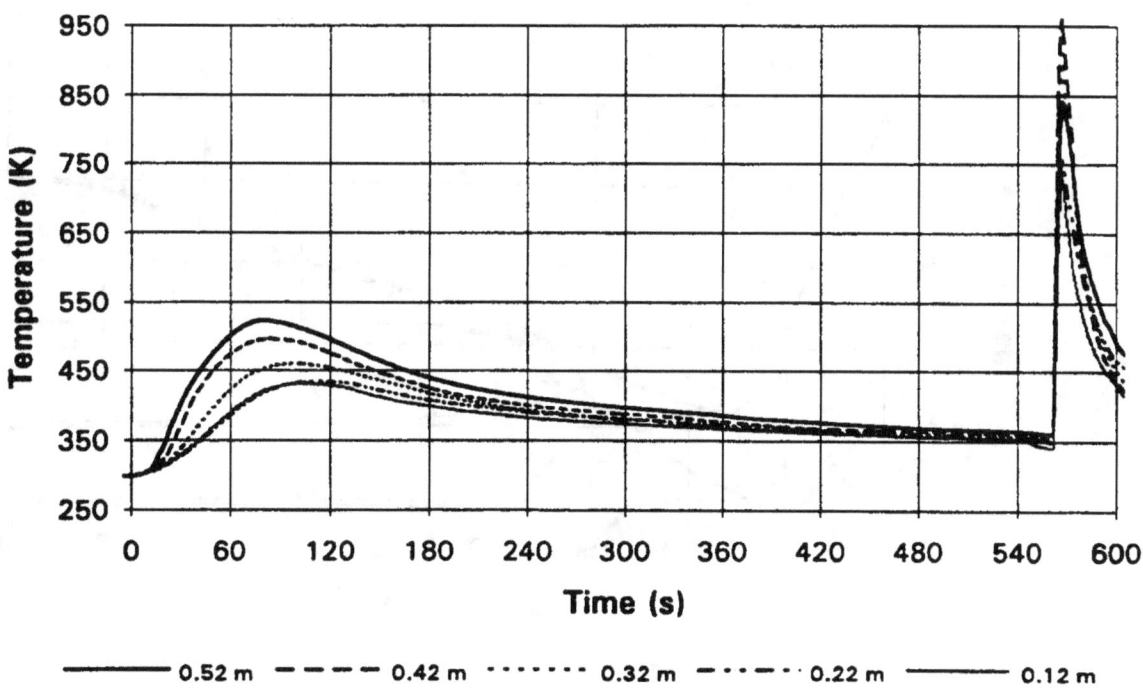

Run: P3EXP78

Two Zone Approximation - Temperature & Interface Histories

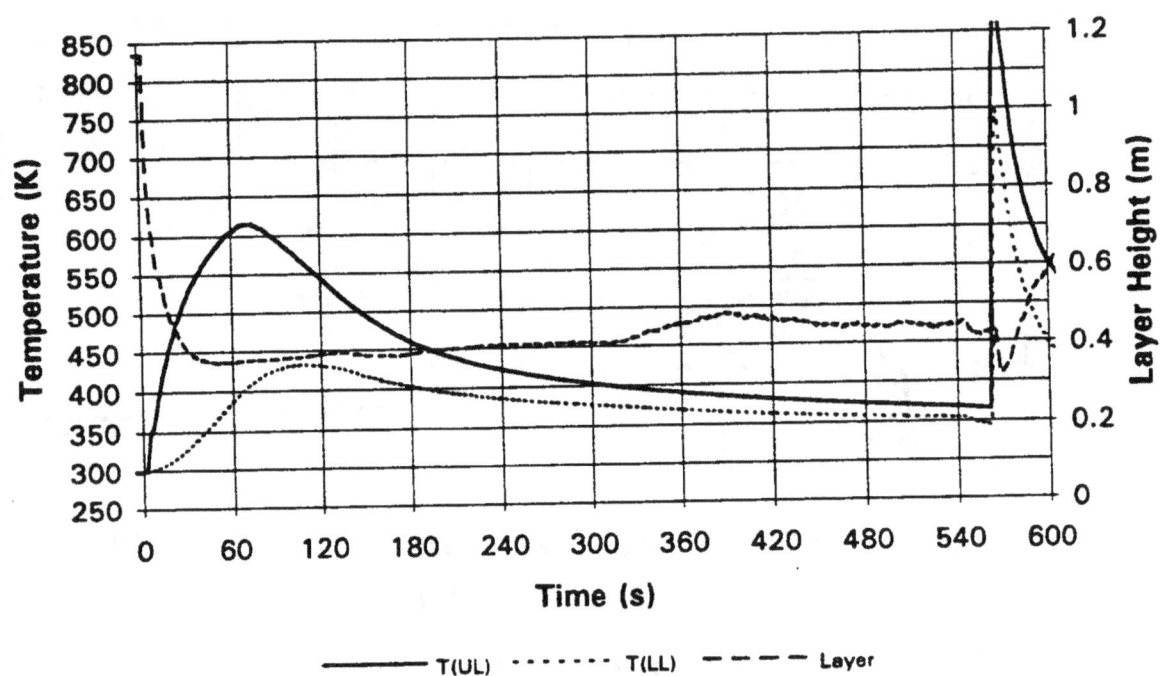

Mass Fraction History - Compartment Gases

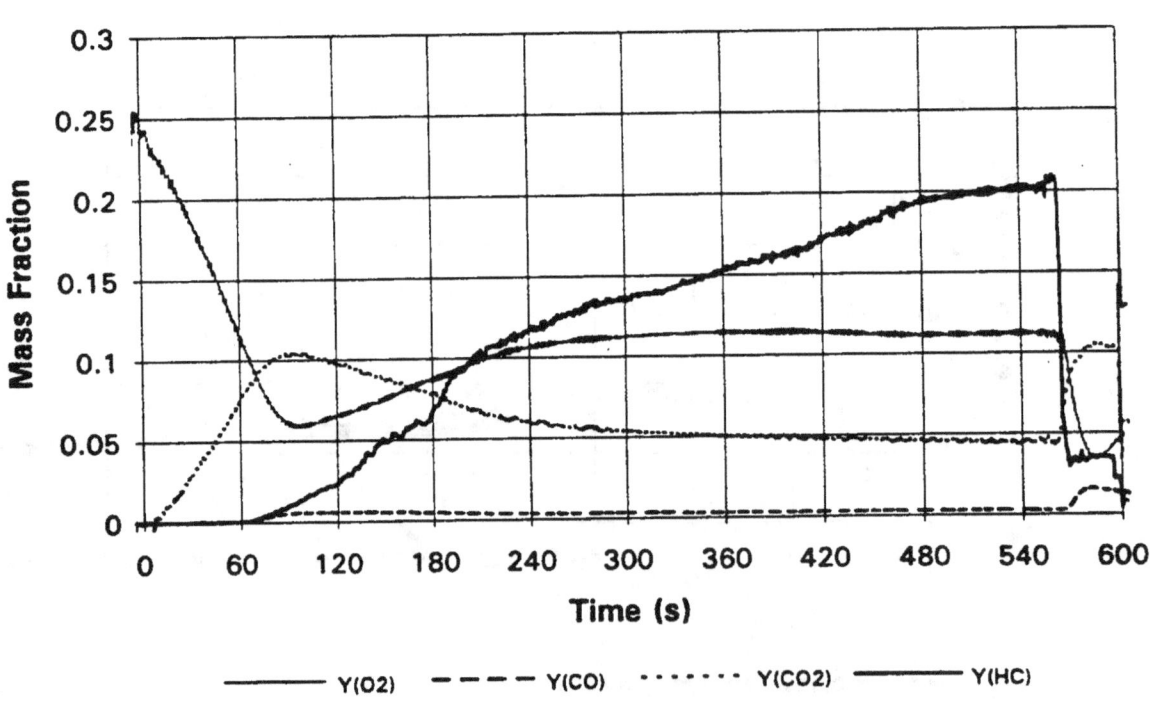

Run: P3EXP78

Bidirectional Probe Velocity History

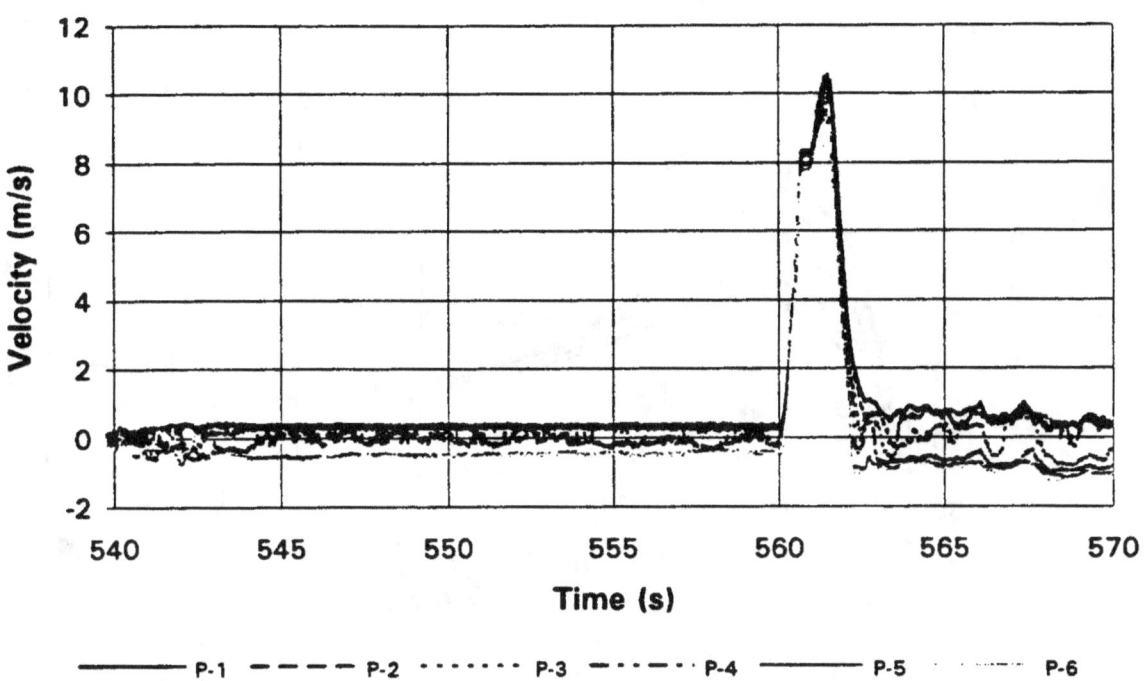

Pressure & Total Mass Out Histories

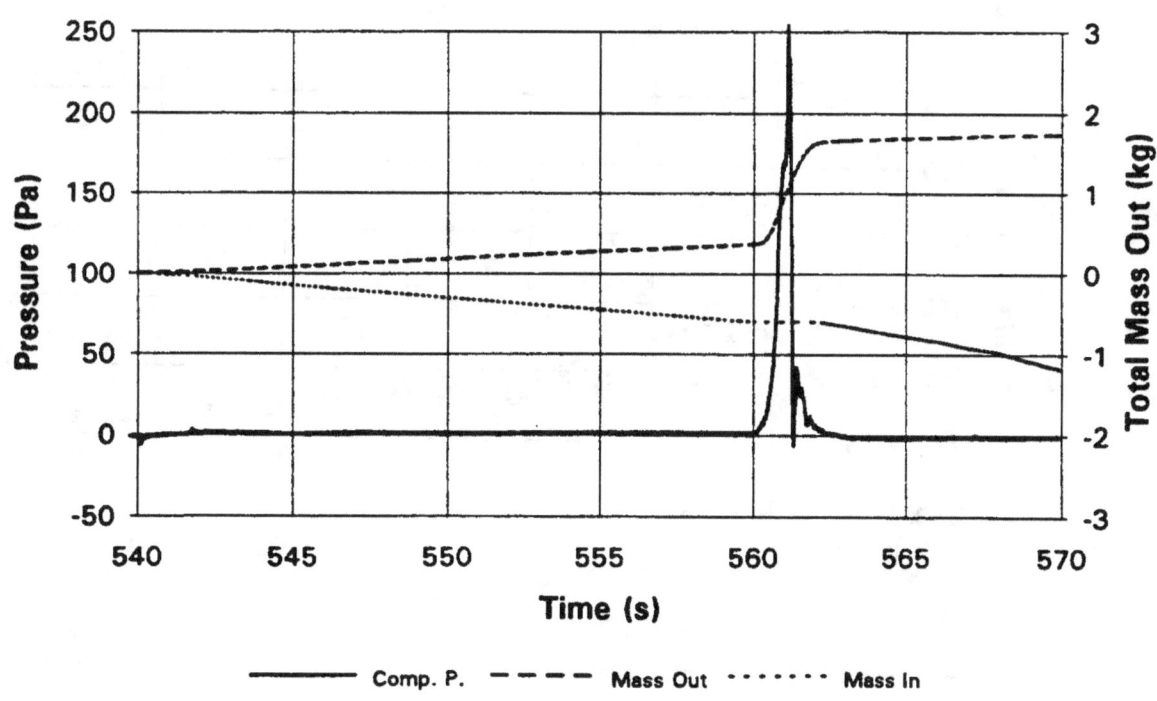

Run: P3EXP78

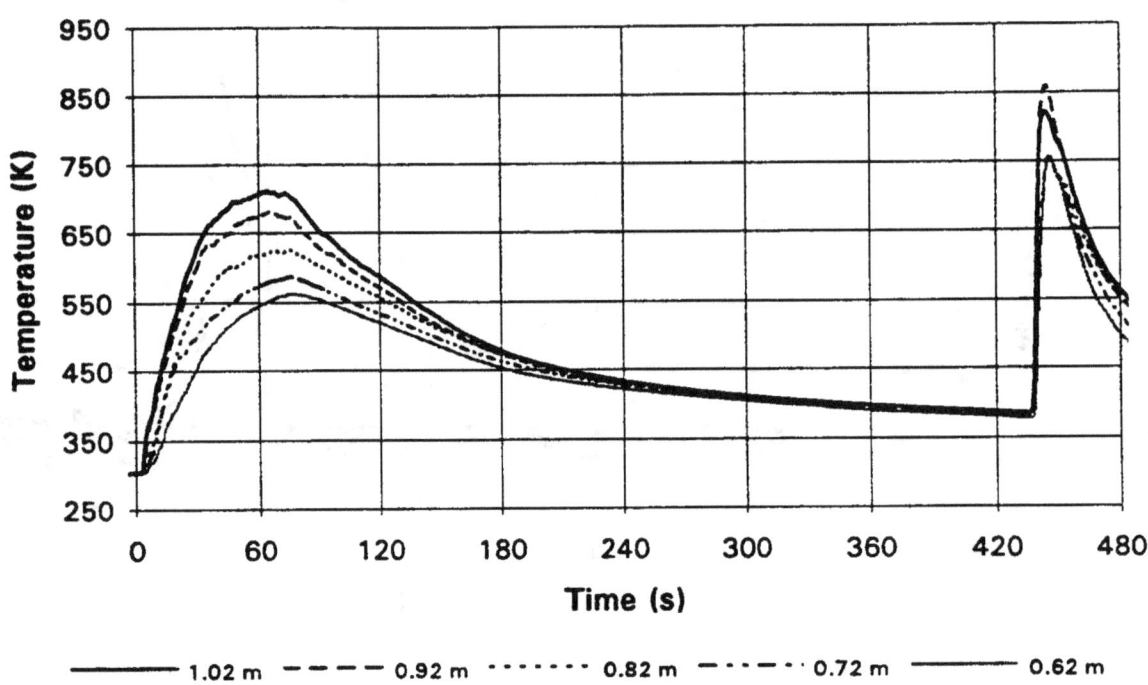

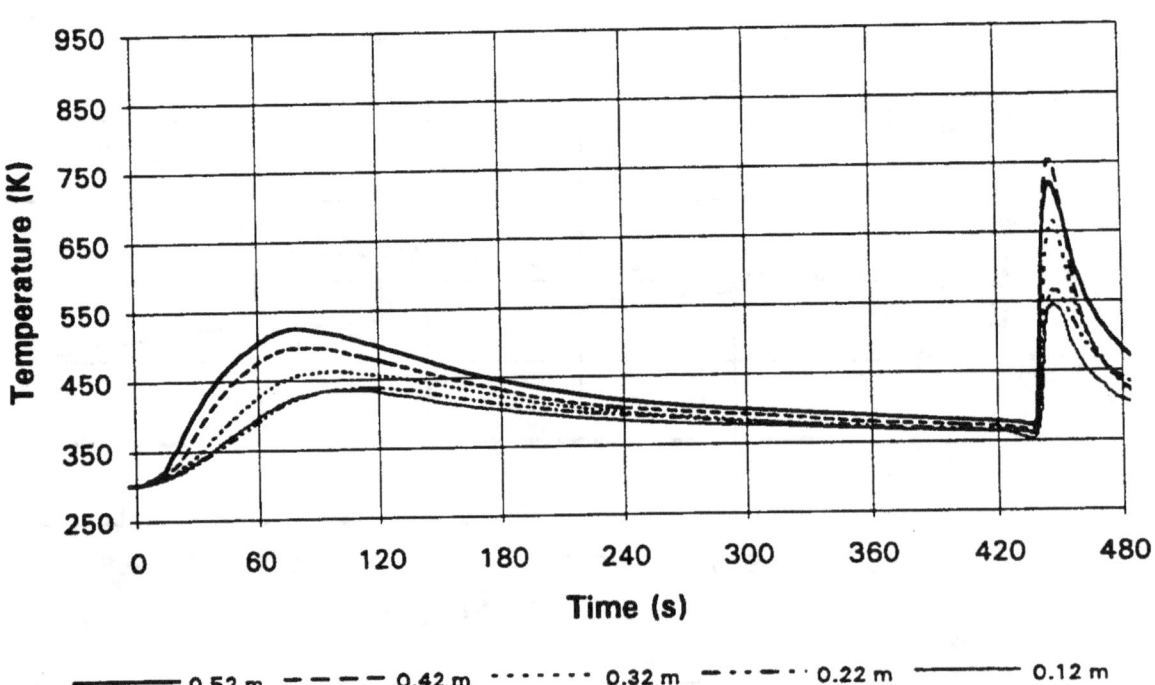

Run: P3EXP79

Two Zone Approximation - Temperature & Interface Histories

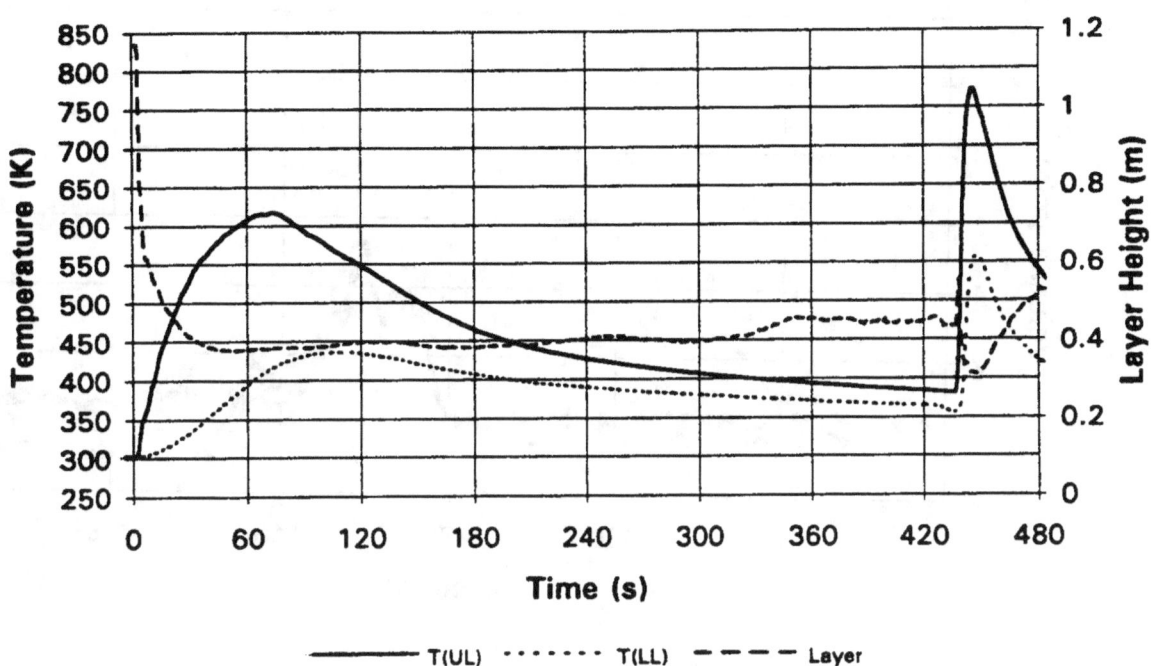

Mass Fraction History - Compartment Gases

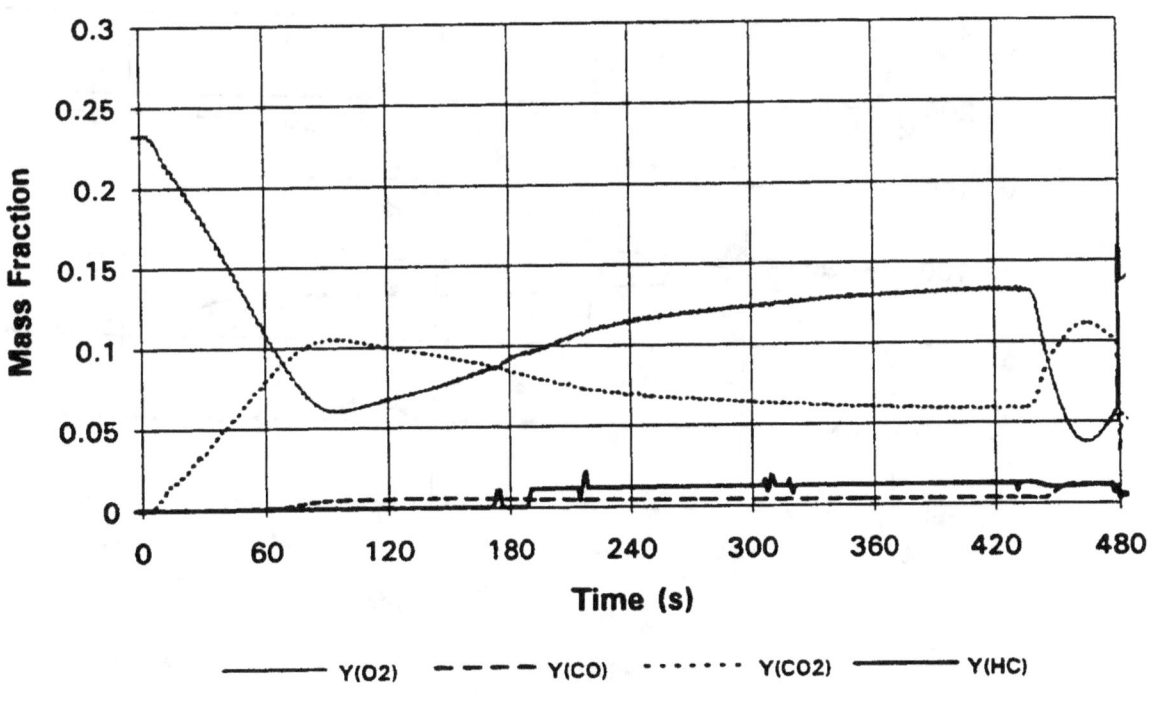

Run: P3EXP79

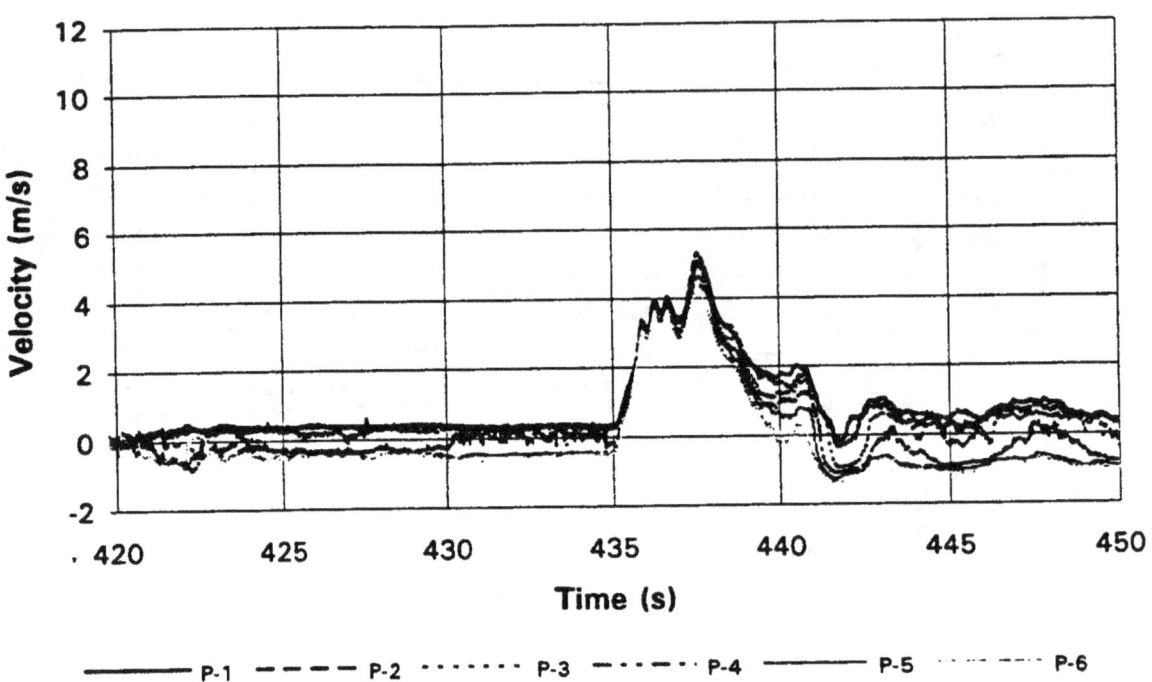

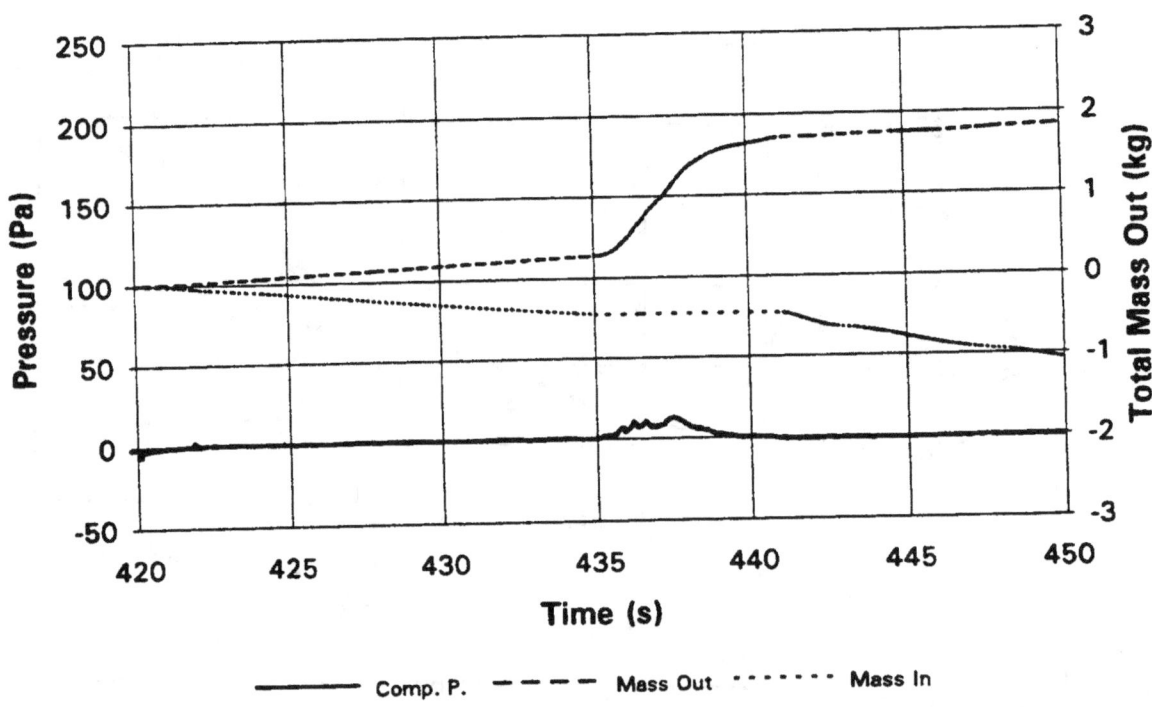

Run: P3EXP79

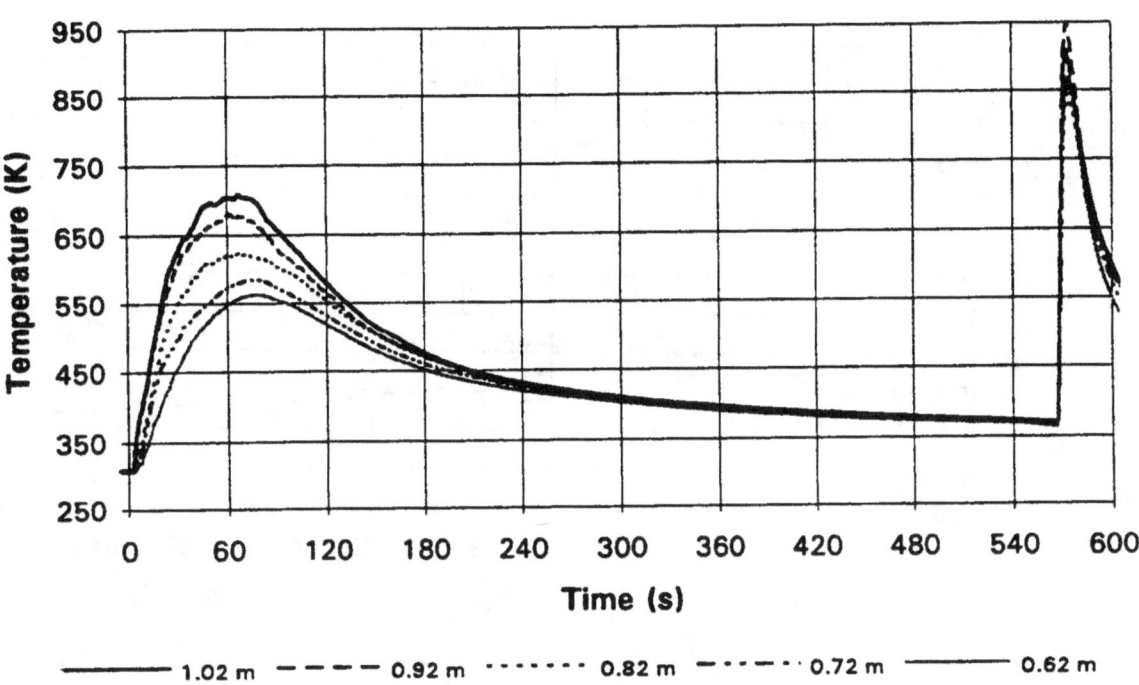

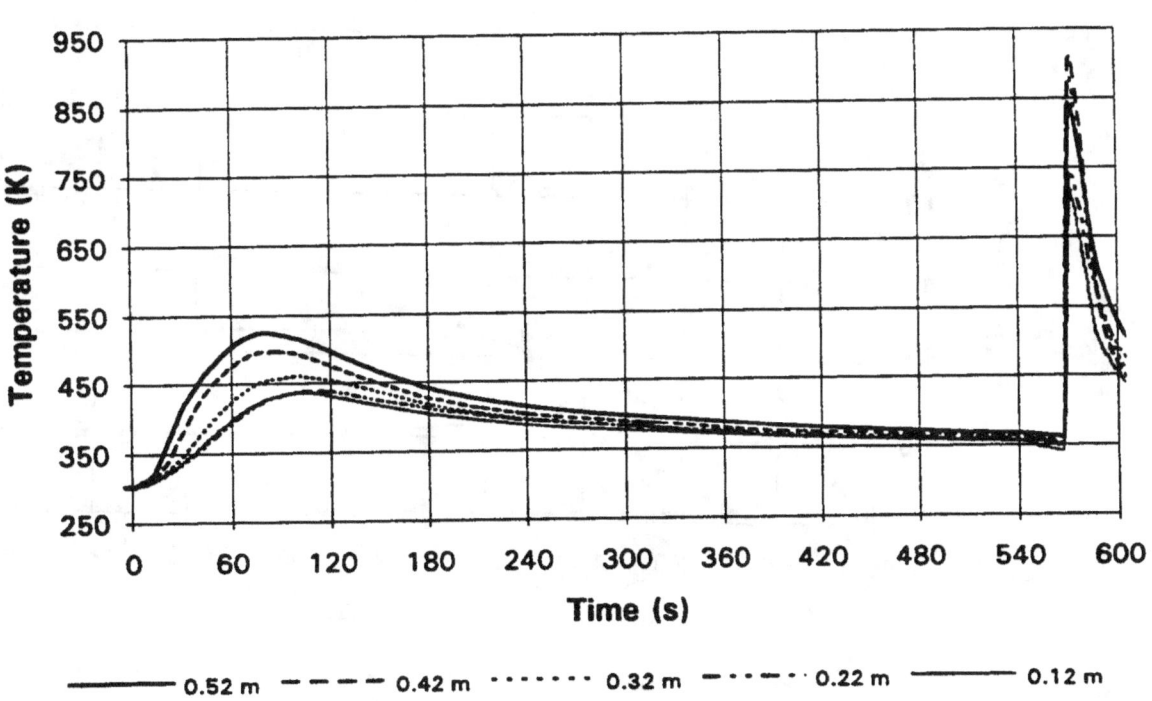

Run: P3EXP80

Two Zone Approximation - Temperature & Interface Histories

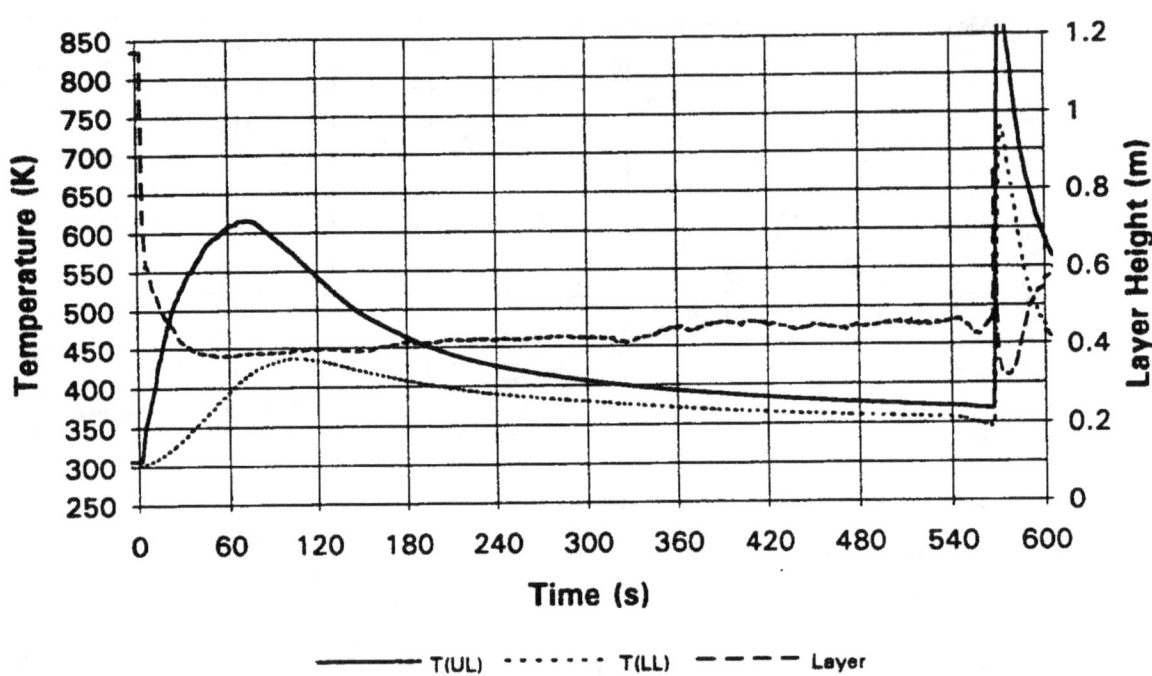

—— T(UL) ······ T(LL) ---- Layer

Mass Fraction History - Compartment Gases

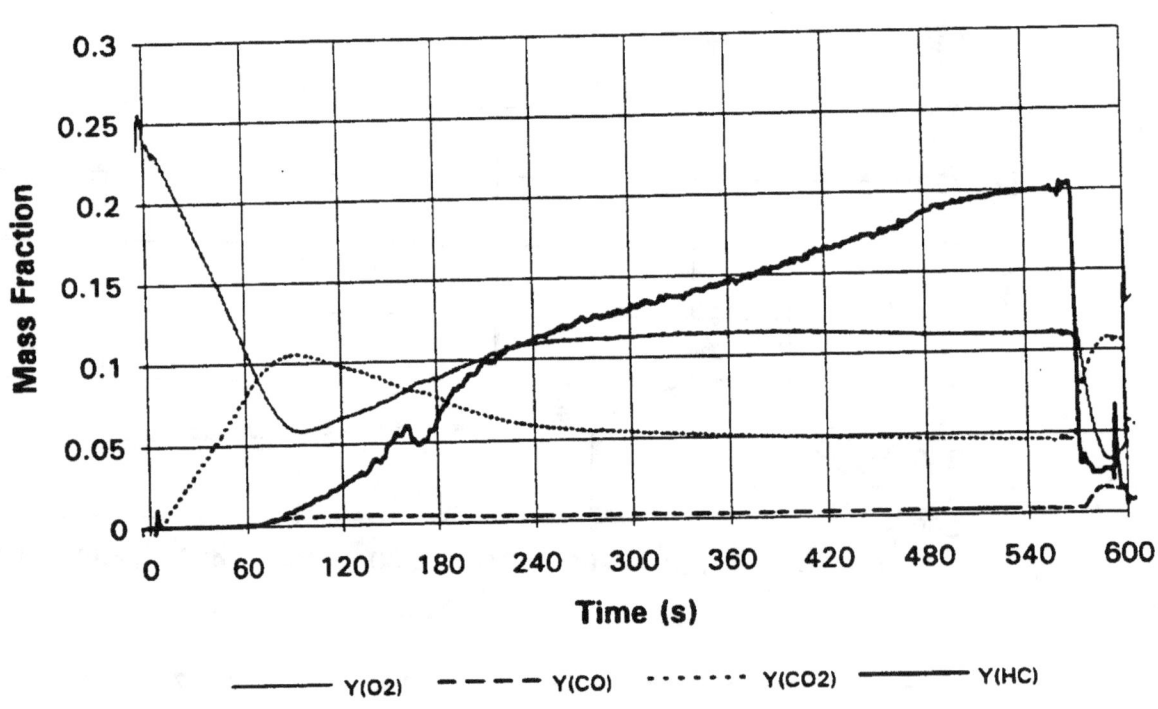

—— Y(O2) ---- Y(CO) ······ Y(CO2) —— Y(HC)

Run: P3EXP80

Bidirectional Probe Velocity History

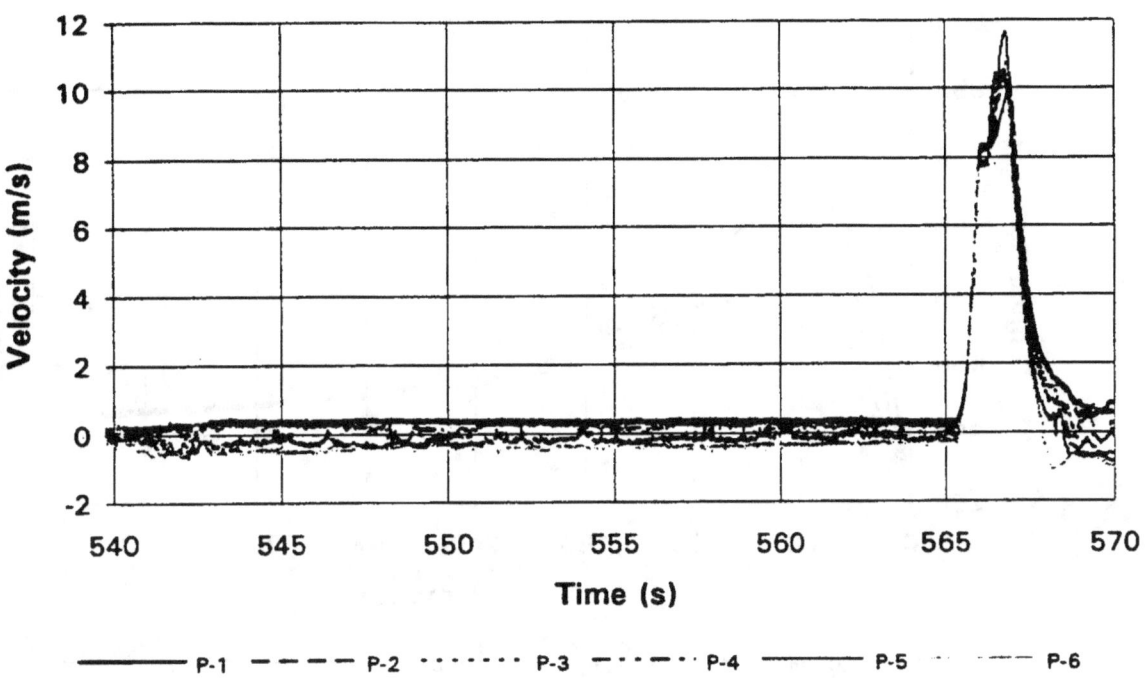

Pressure & Total Mass Out Histories

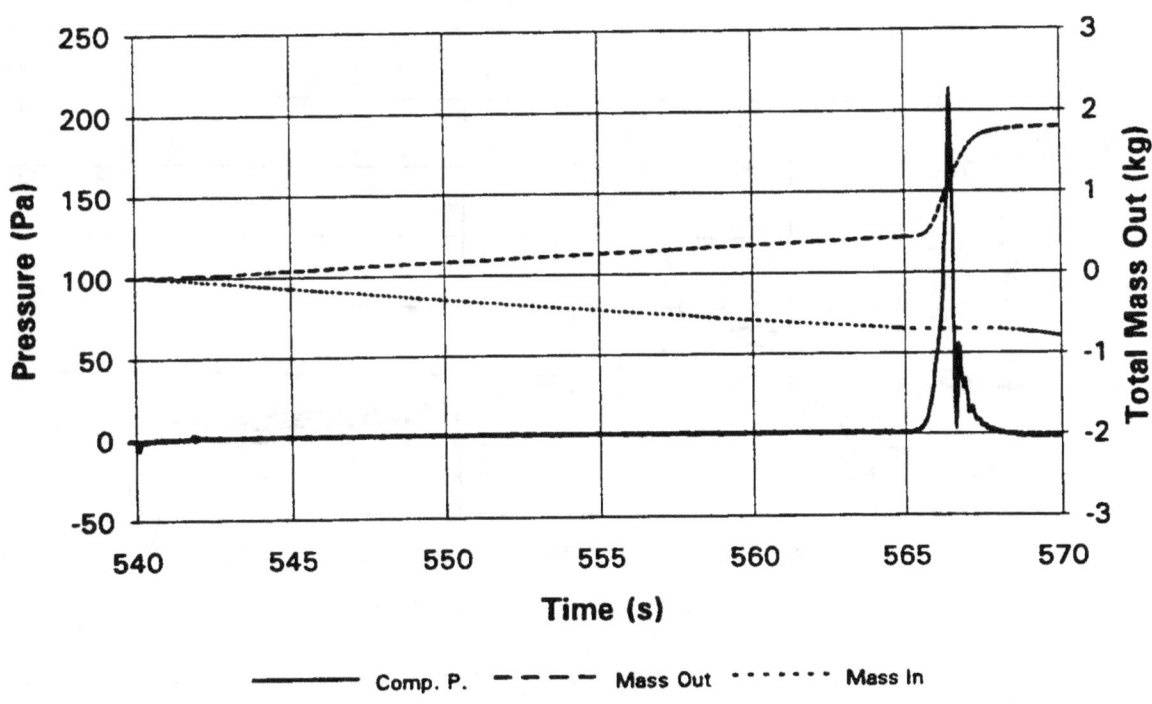

Run: P3EXP80

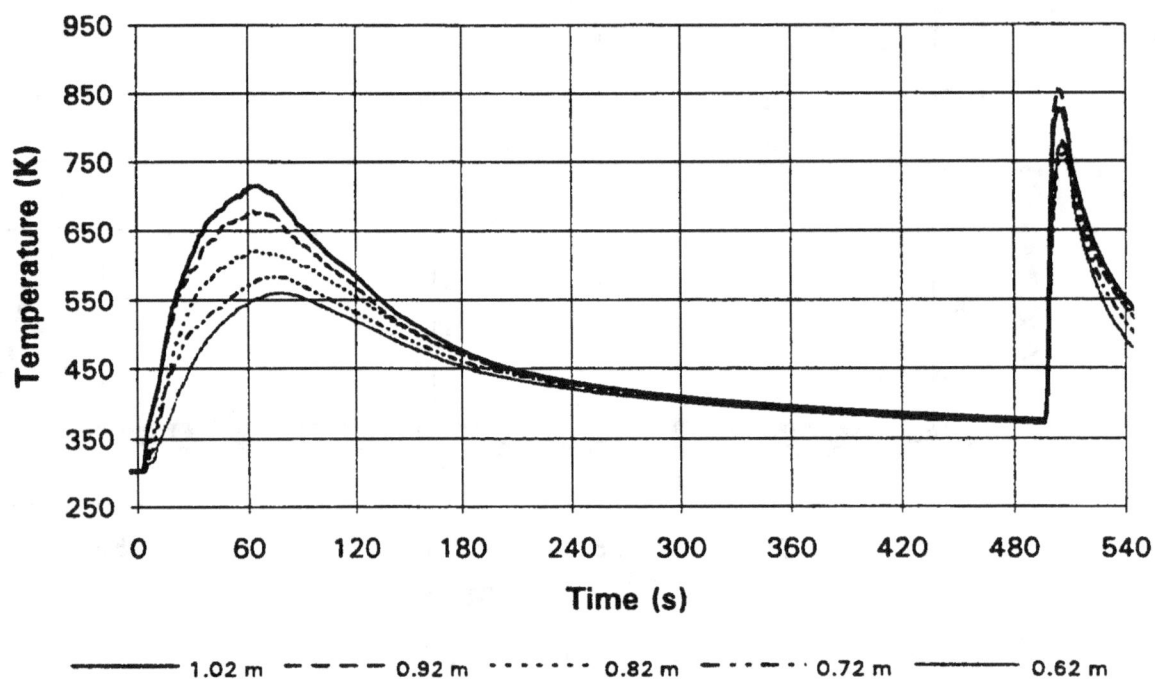

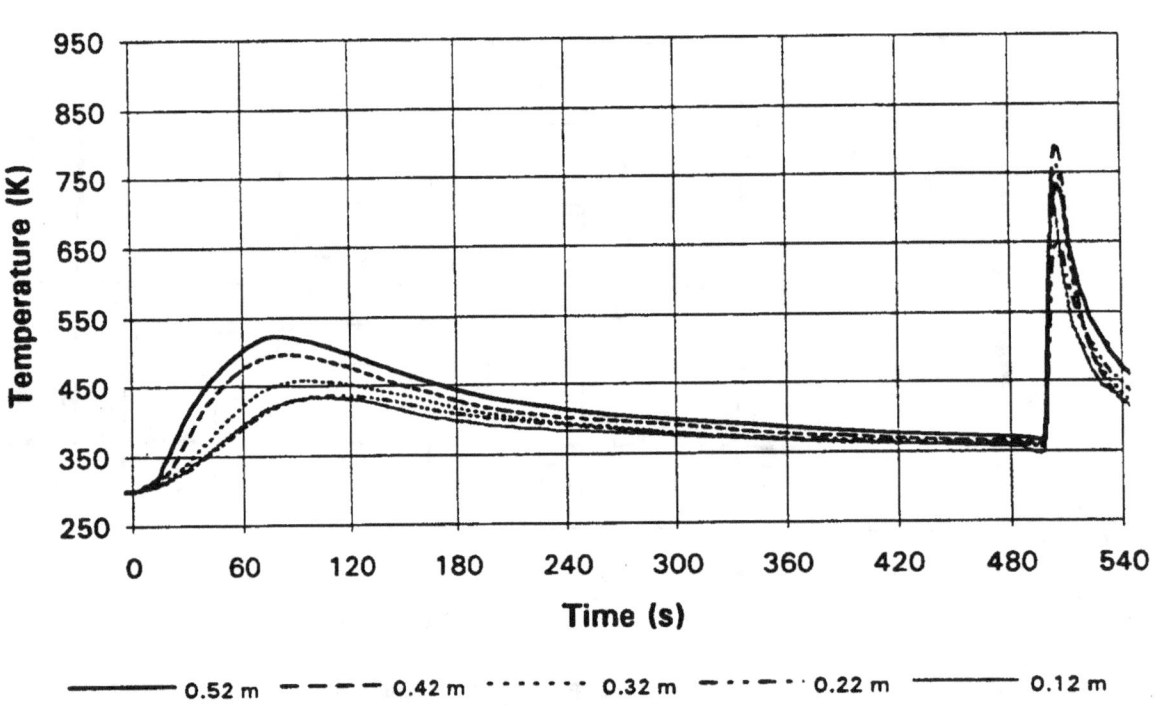

Run: P3EXP81

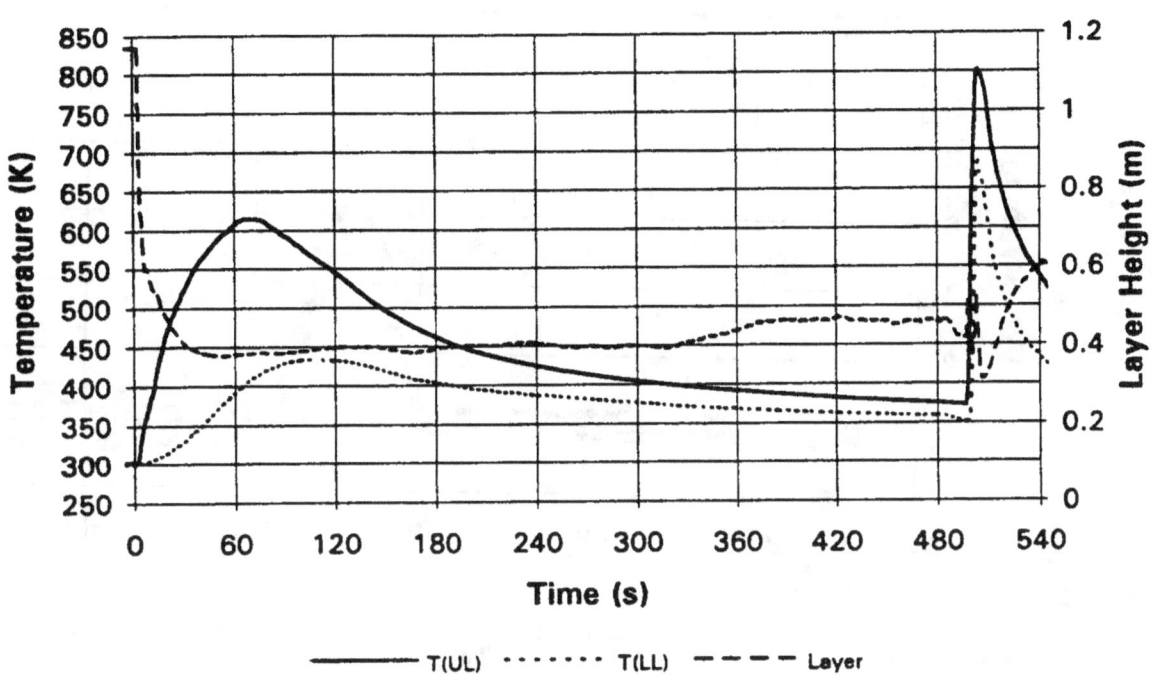

Two Zone Approximation - Temperature & Interface Histories

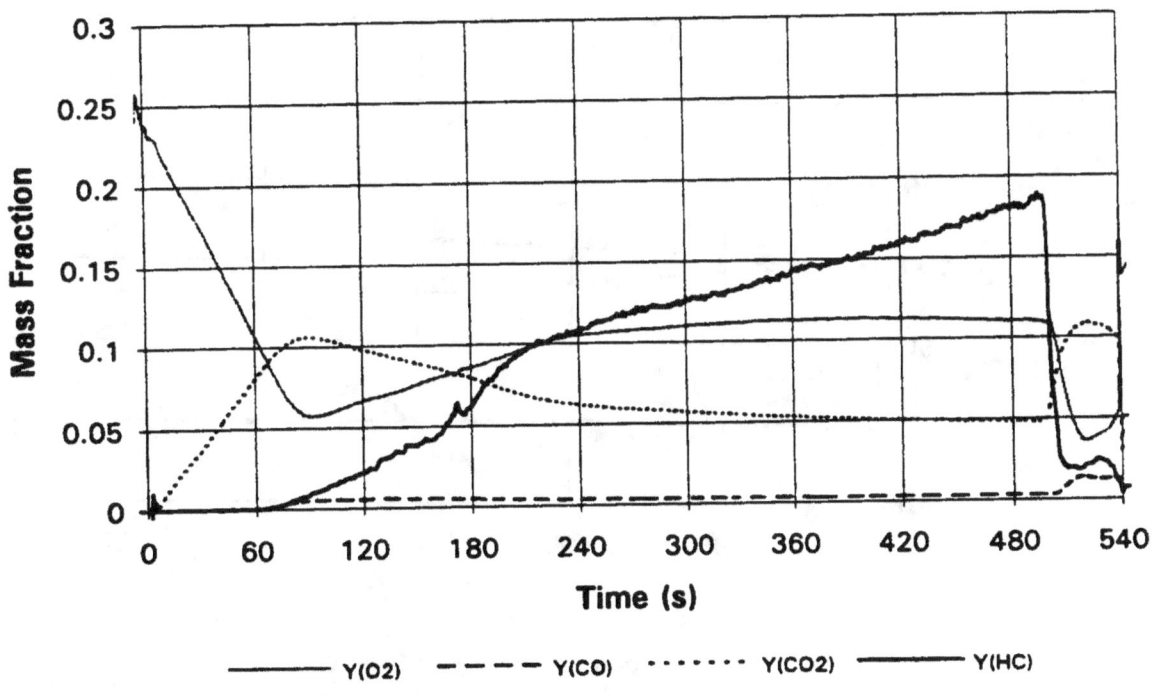

Mass Fraction History - Compartment Gases

Run: P3EXP81

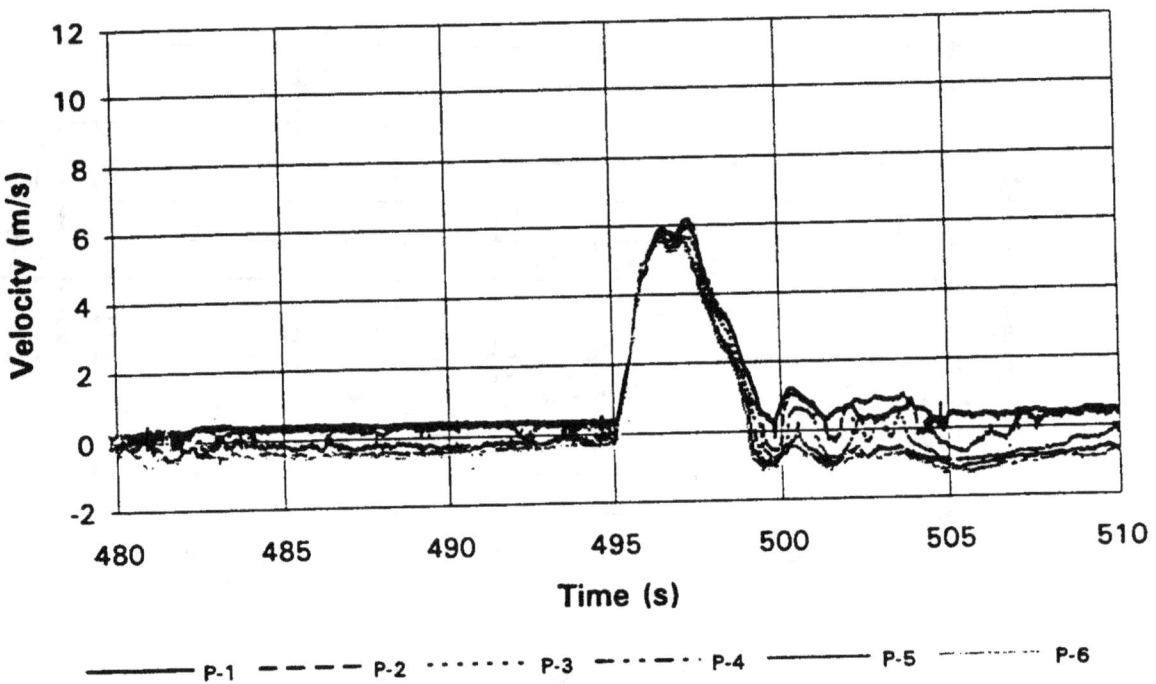

Bidirectional Probe Velocity History

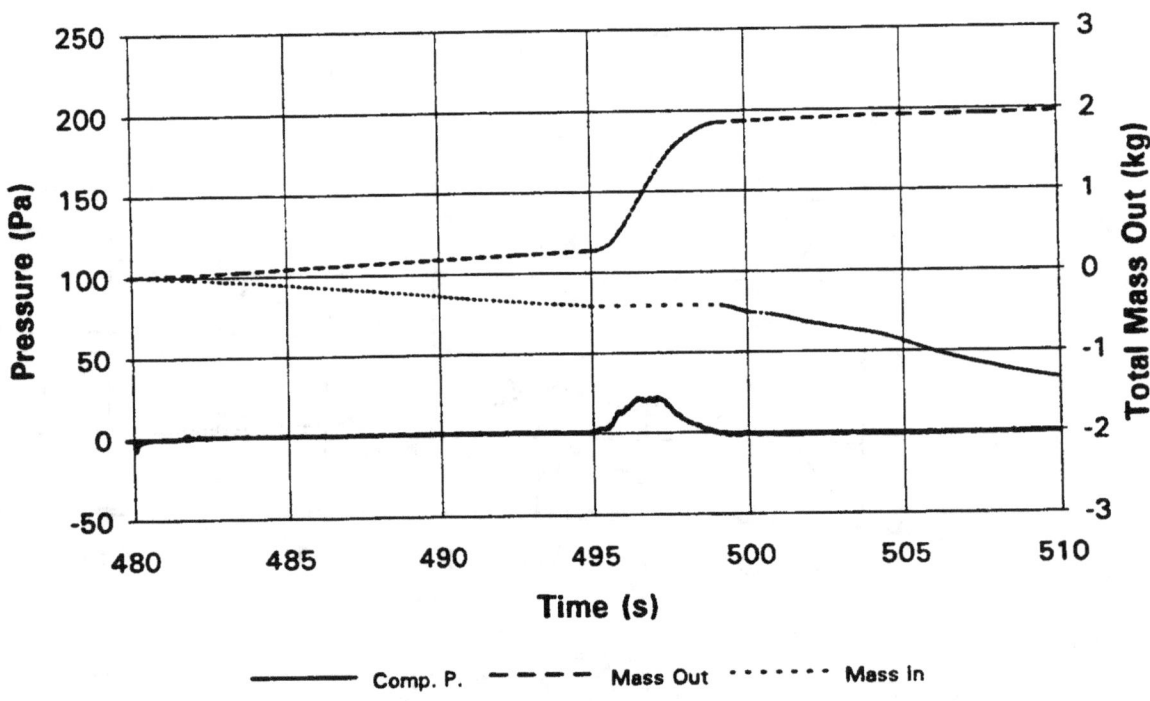

Pressure & Total Mass Out Histories

Run: P3EXP81

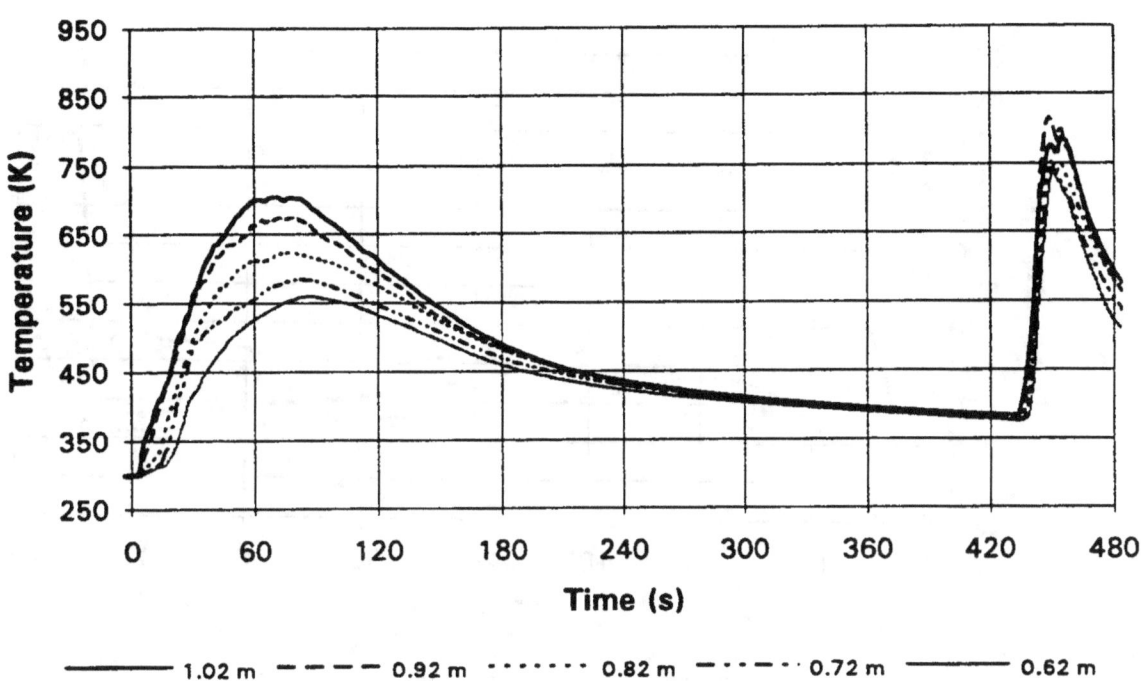
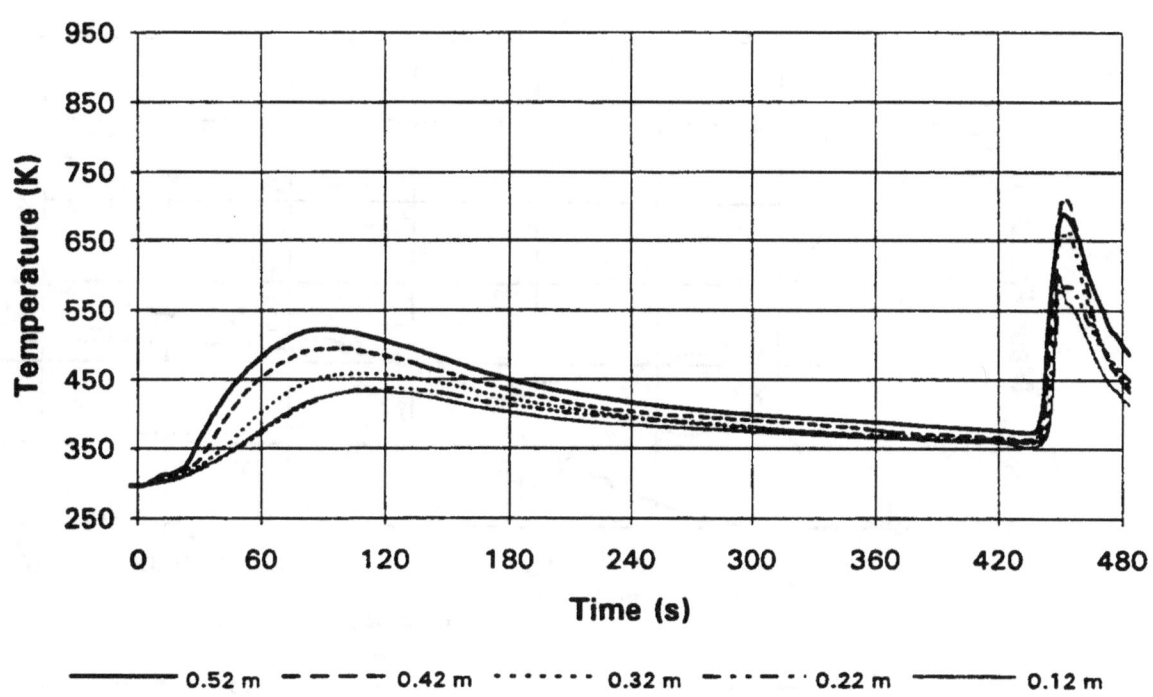

Run: P3EXP82

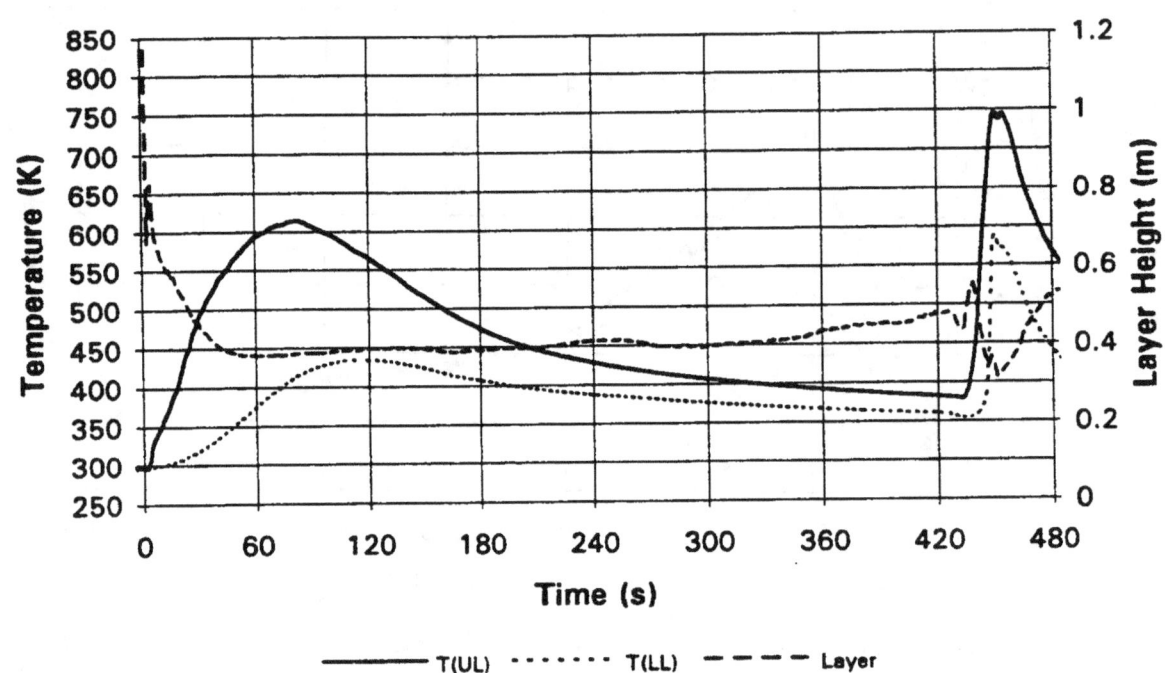

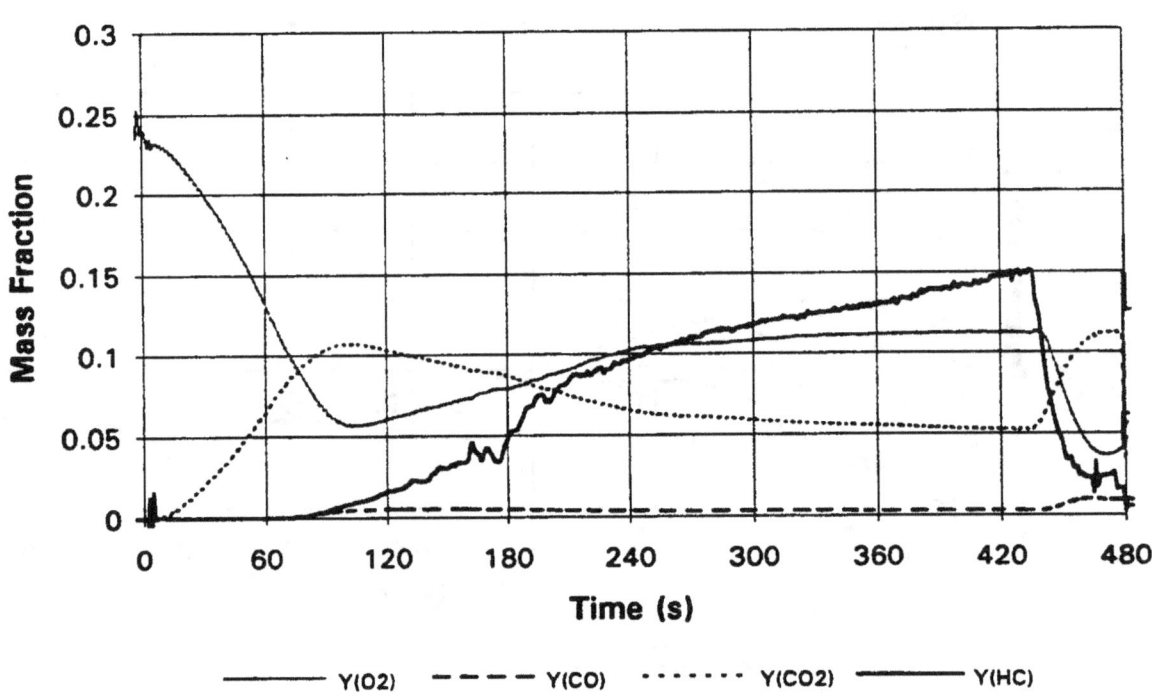

Run: P3EXP82

Bidirectional Probe Velocity History

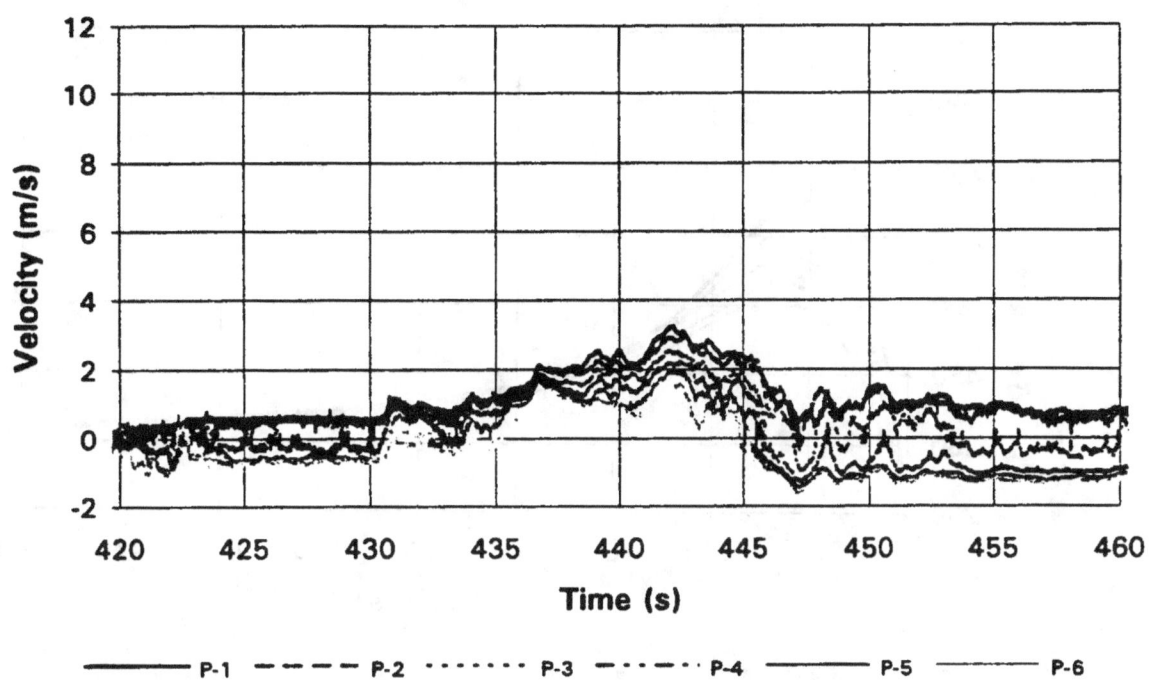

Pressure & Total Mass Out Histories

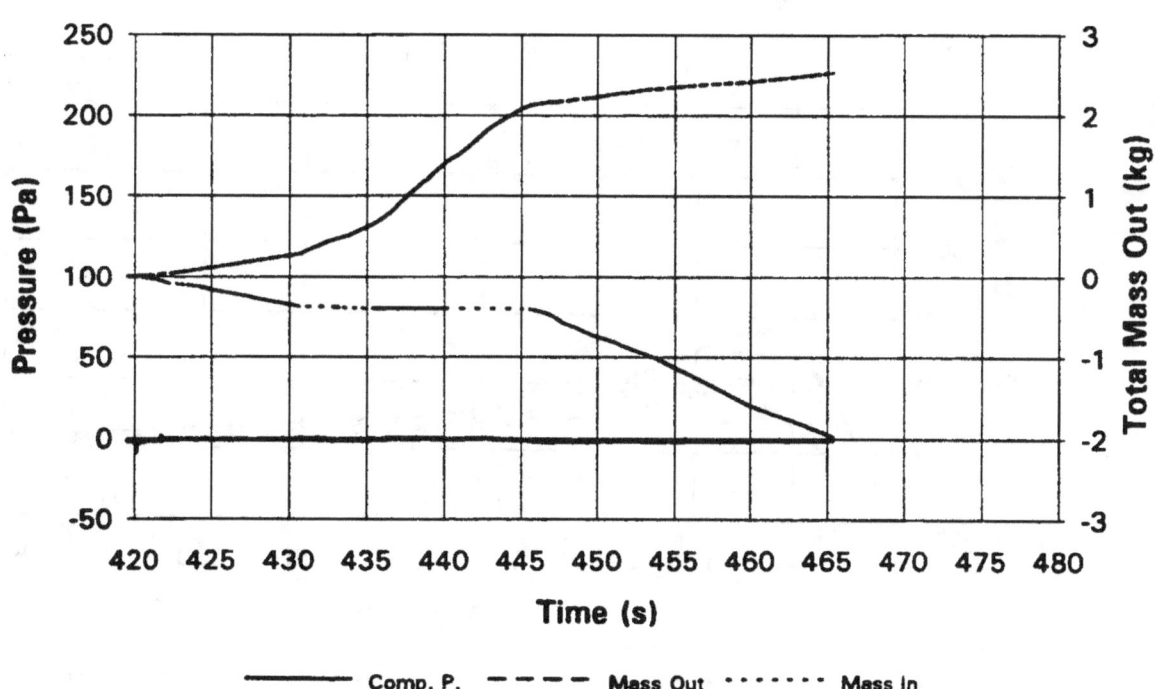

Run: P3EXP82

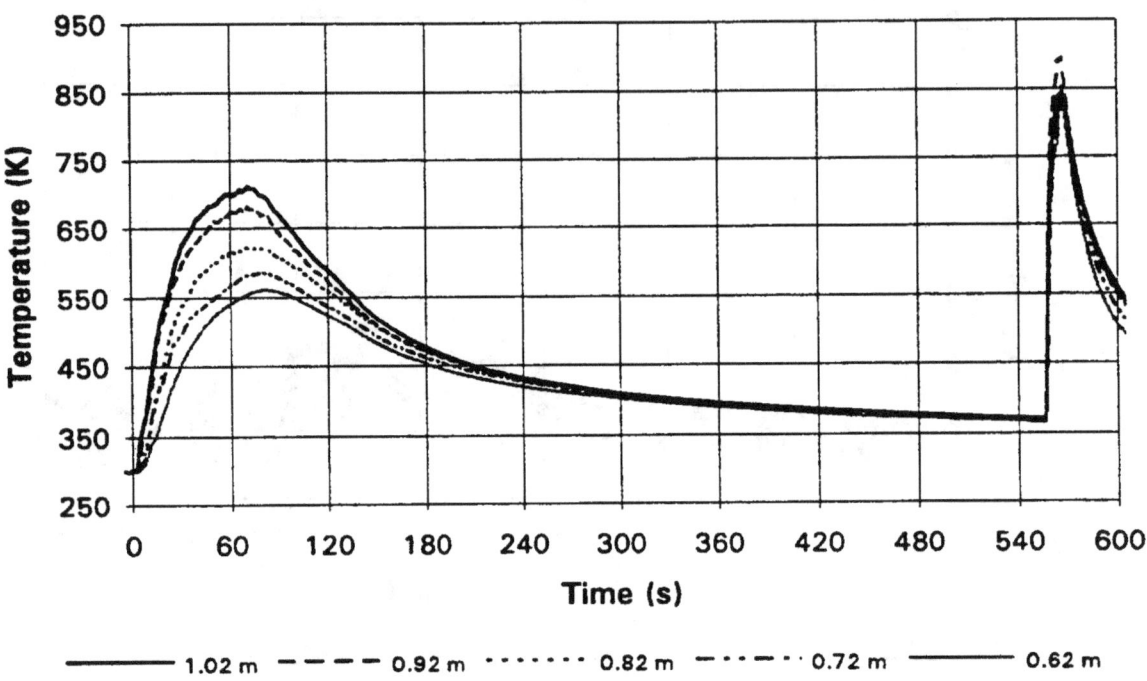

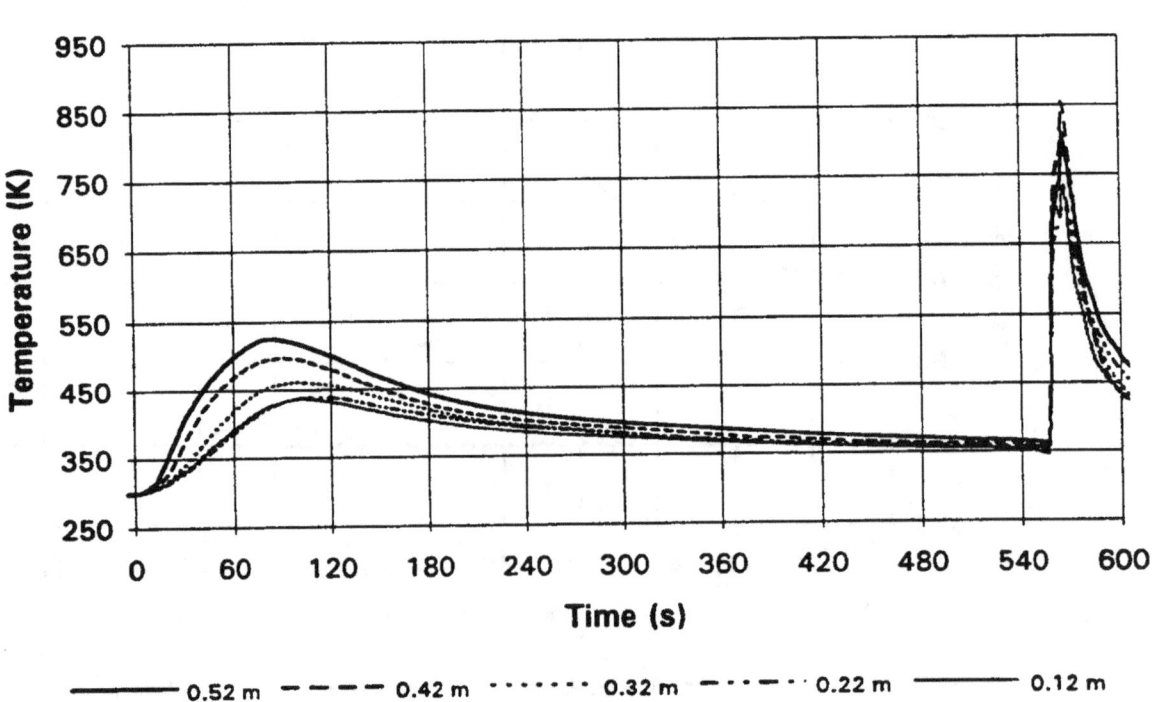

Run: P3EXP83

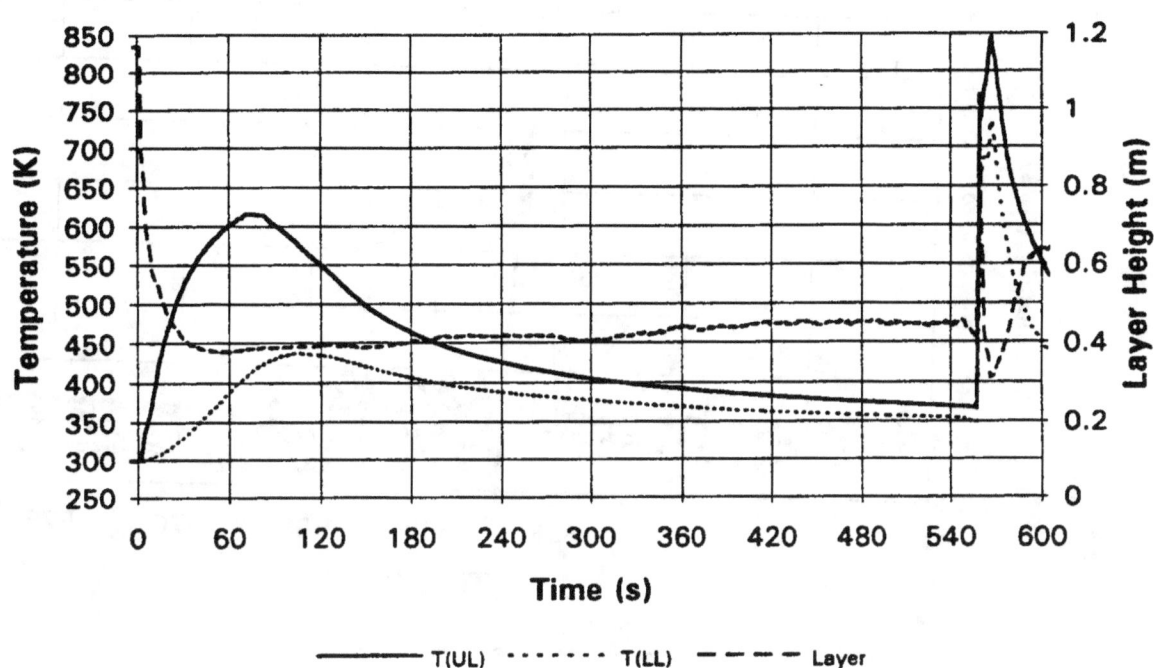

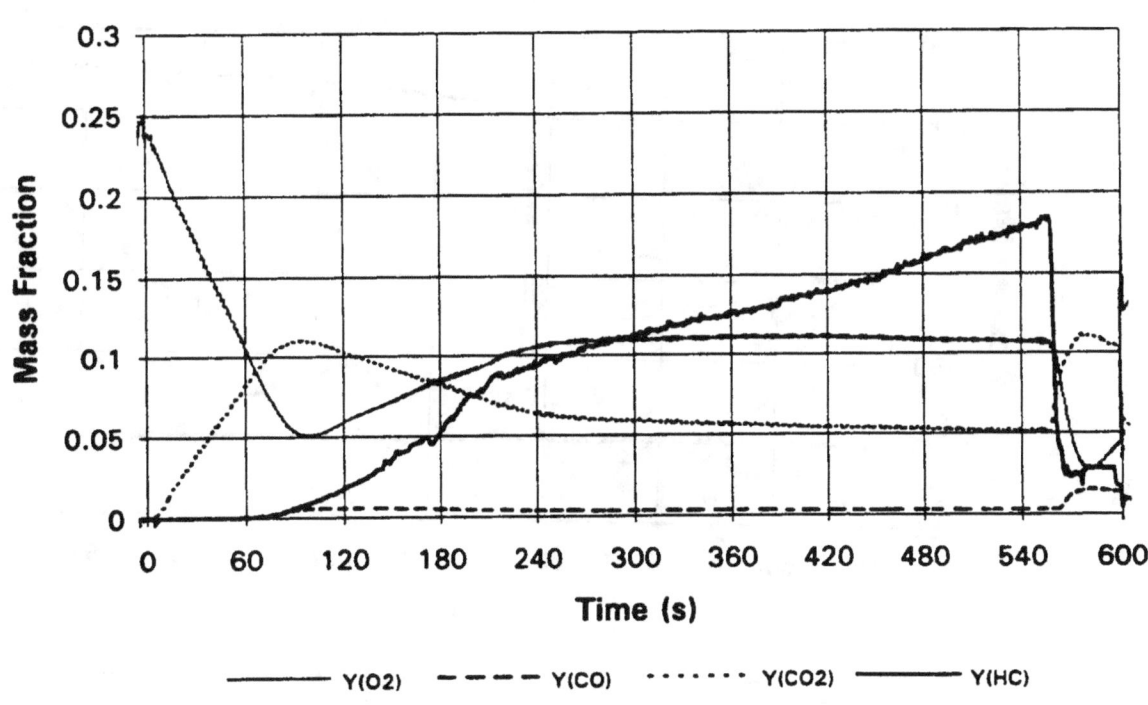

Run: P3EXP83

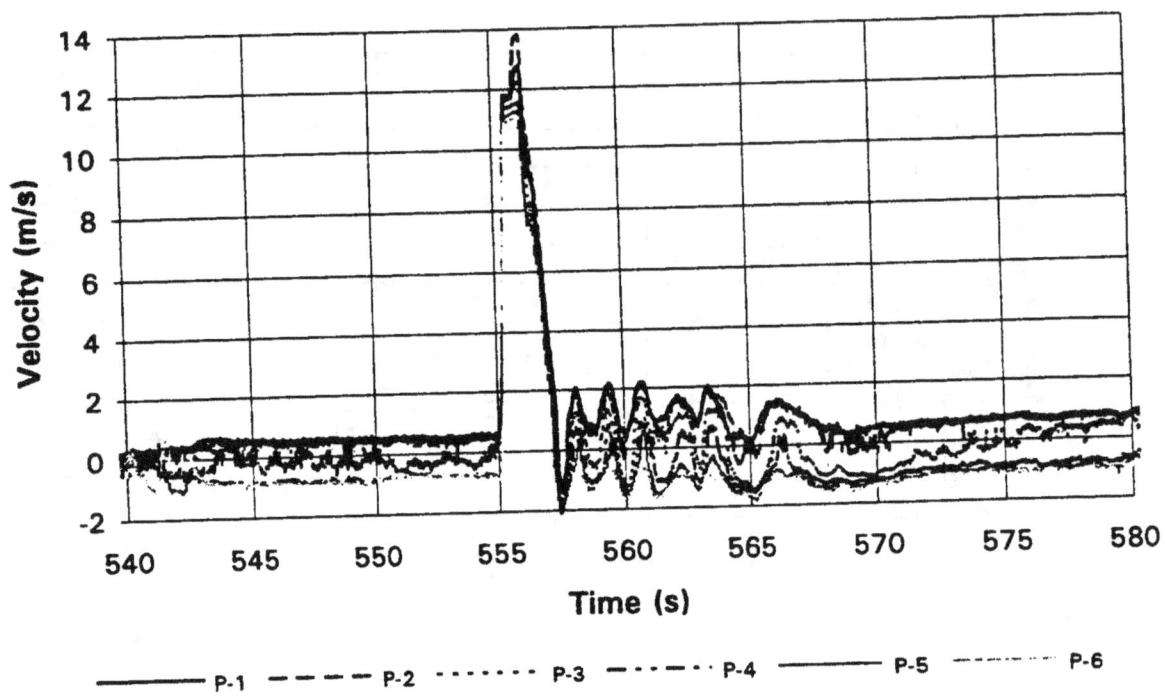

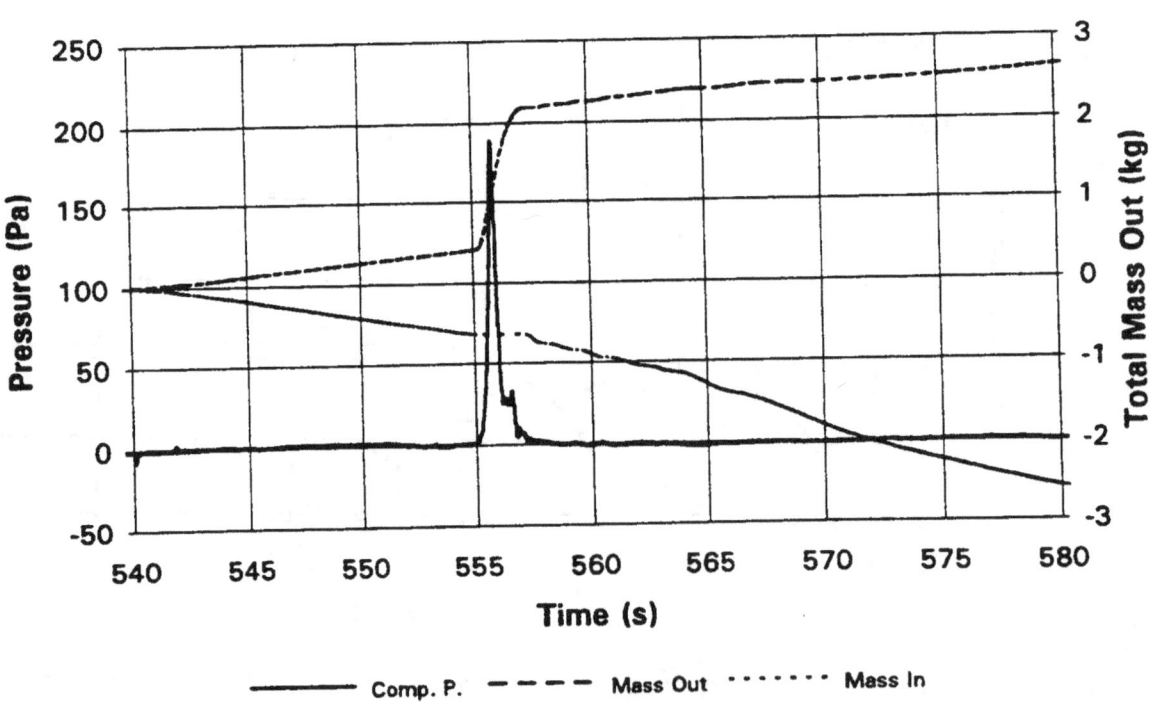

Run: P3EXP83

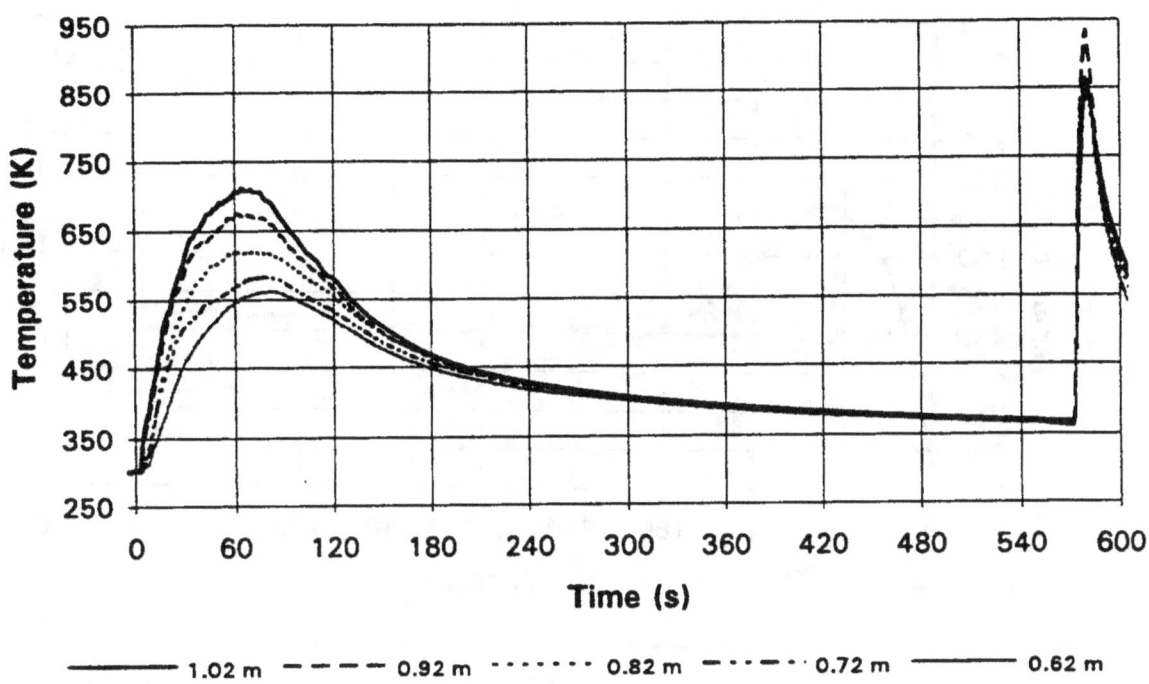

Layer Temperature History - TC Tree Data

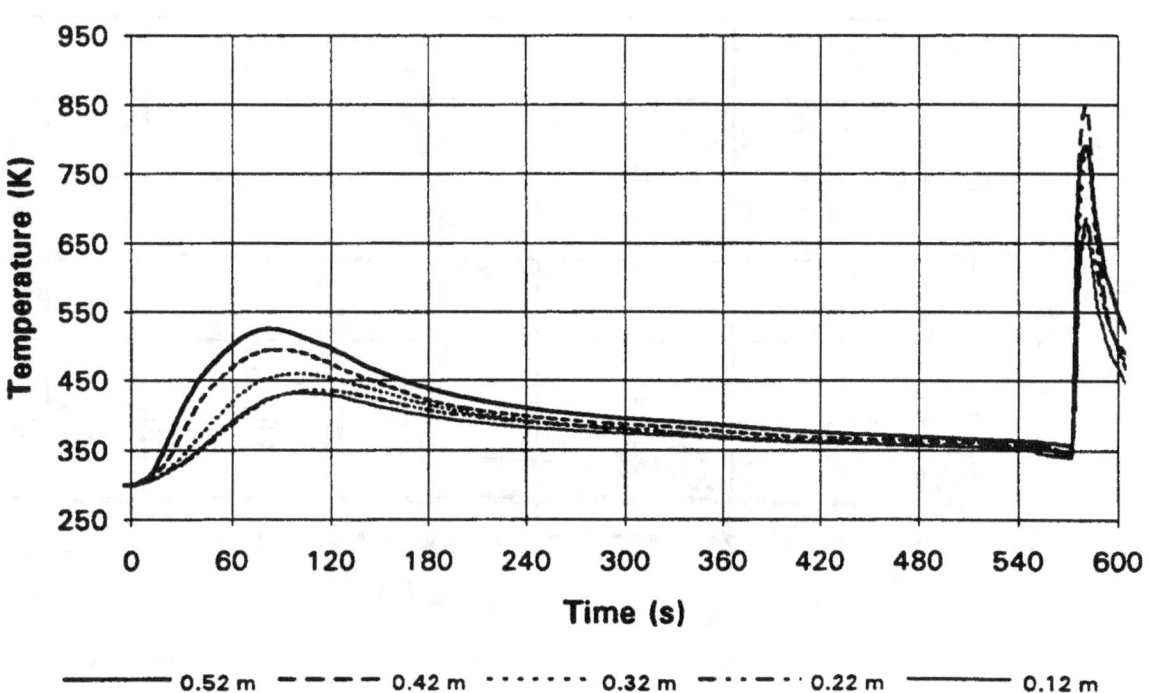

Layer Temperature History - TC Tree Data

Run: P3EXP84

Two Zone Approximation - Temperature & Interface Histories

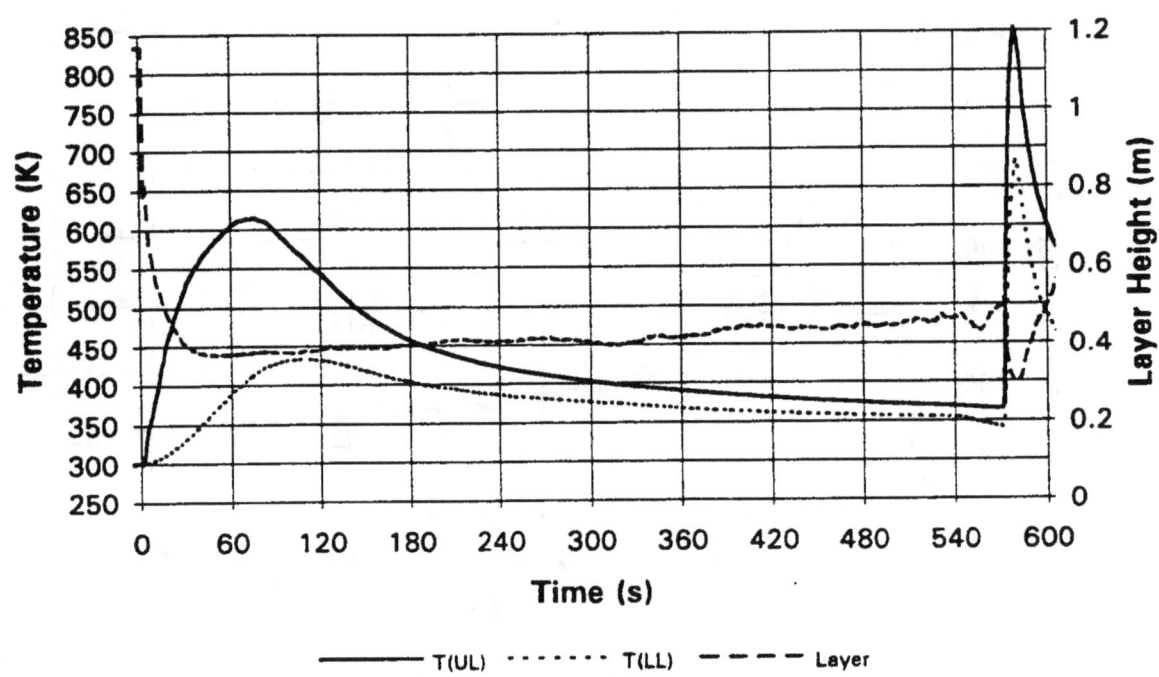

Mass Fraction History - Compartment Gases

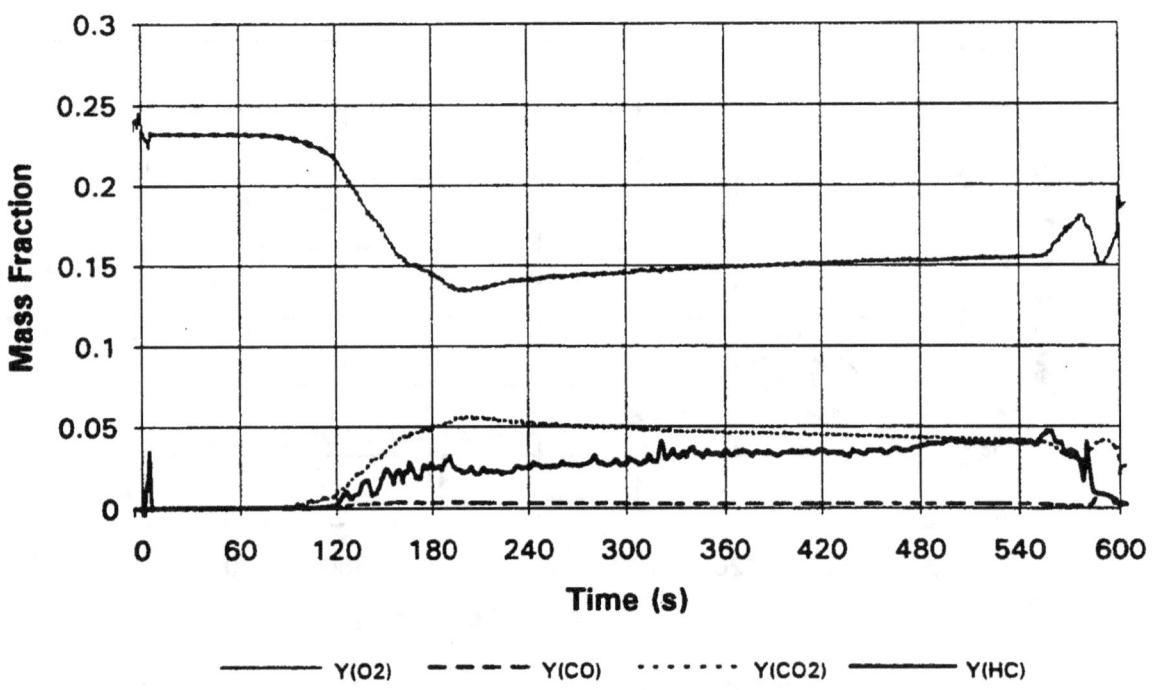

Run: P3EXP84

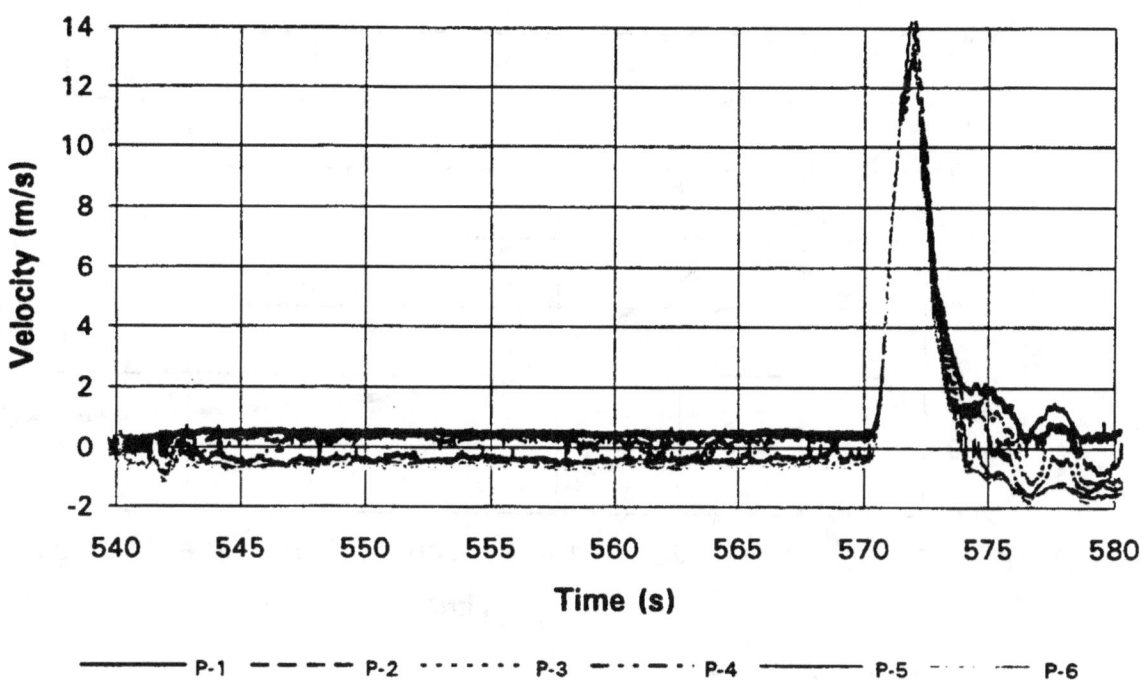

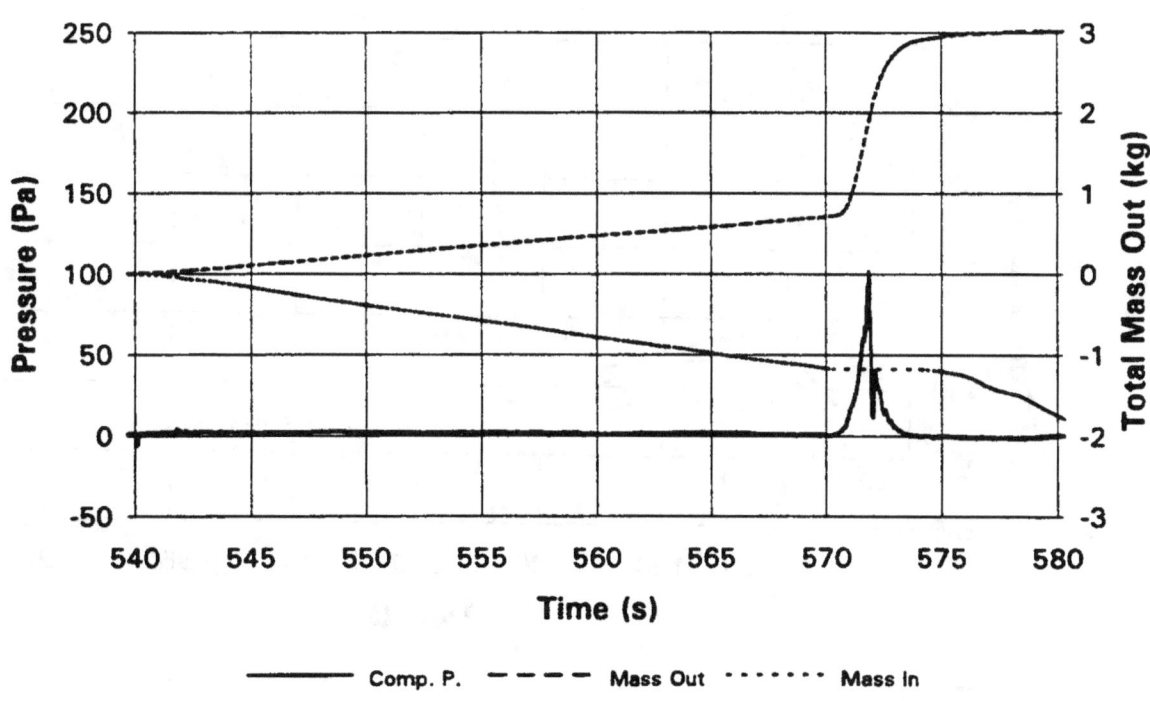

Run: P3EXP84

Layer Temperature History - TC Tree Data

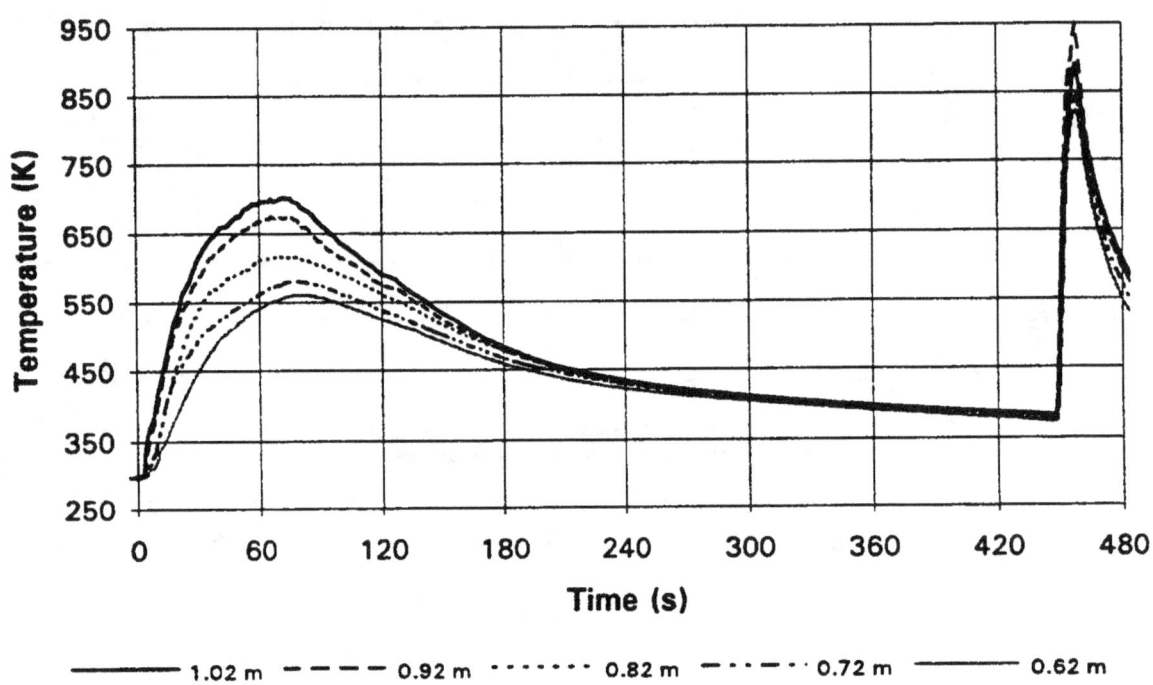

— 1.02 m ---- 0.92 m ····· 0.82 m —·· — 0.72 m —— 0.62 m

Layer Temperature History - TC Tree Data

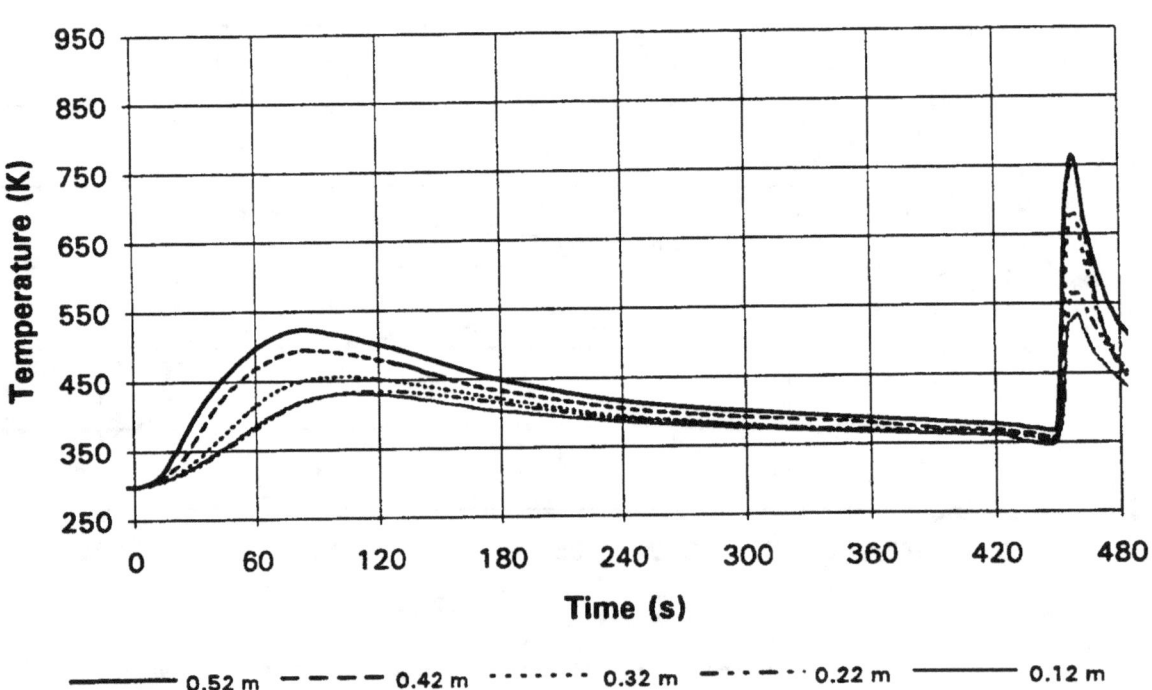

— 0.52 m ---- 0.42 m ····· 0.32 m —·· — 0.22 m —— 0.12 m

Run: P3EXP85

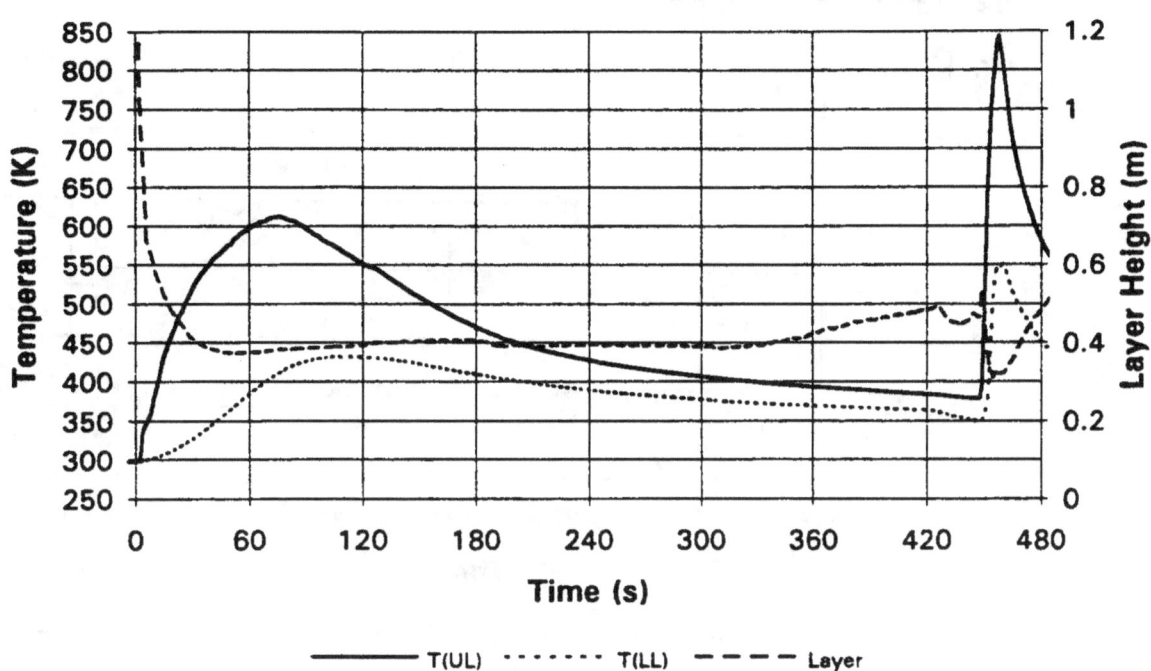

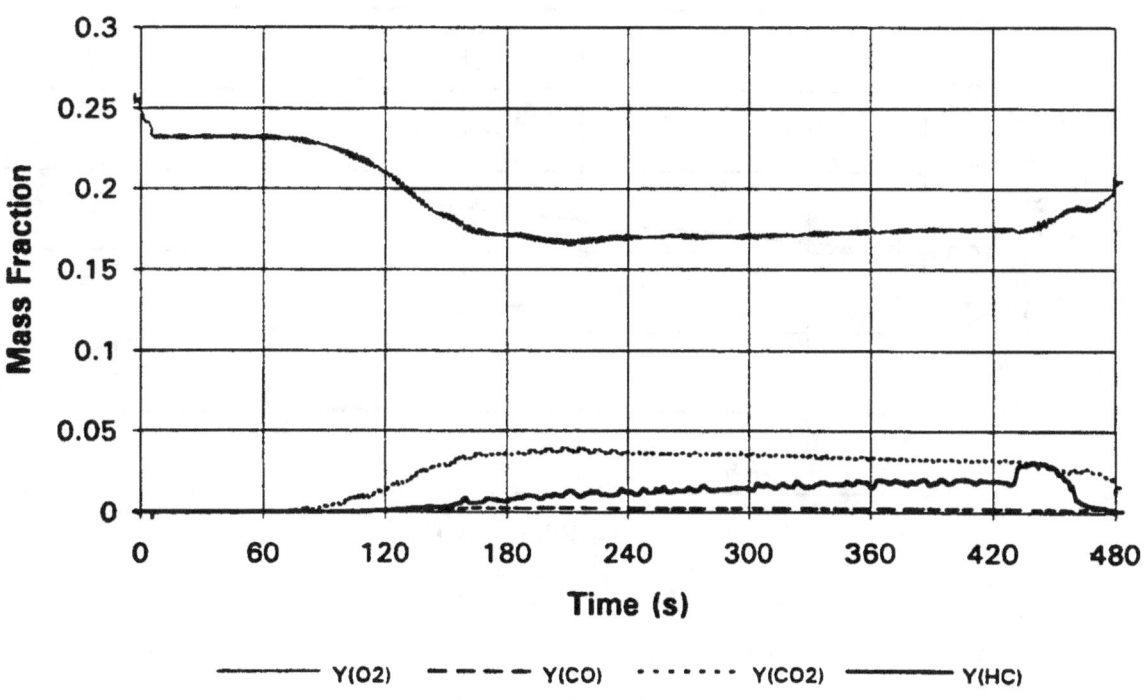

Run: P3EXP85

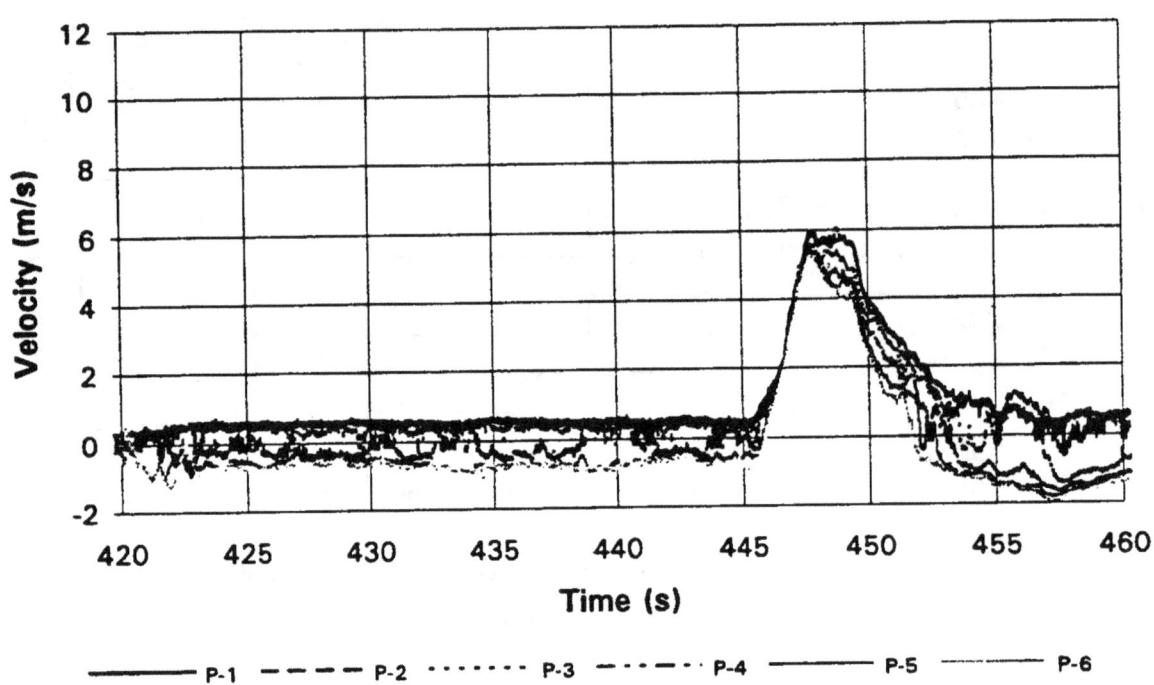

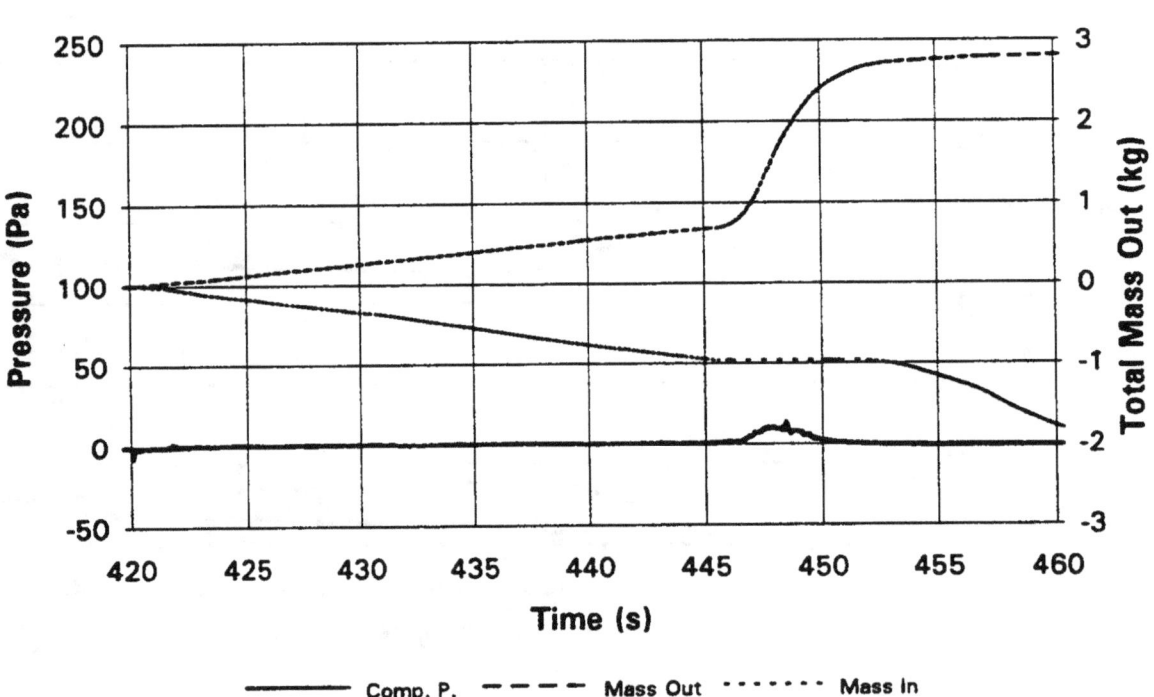

Run: P3EXP85

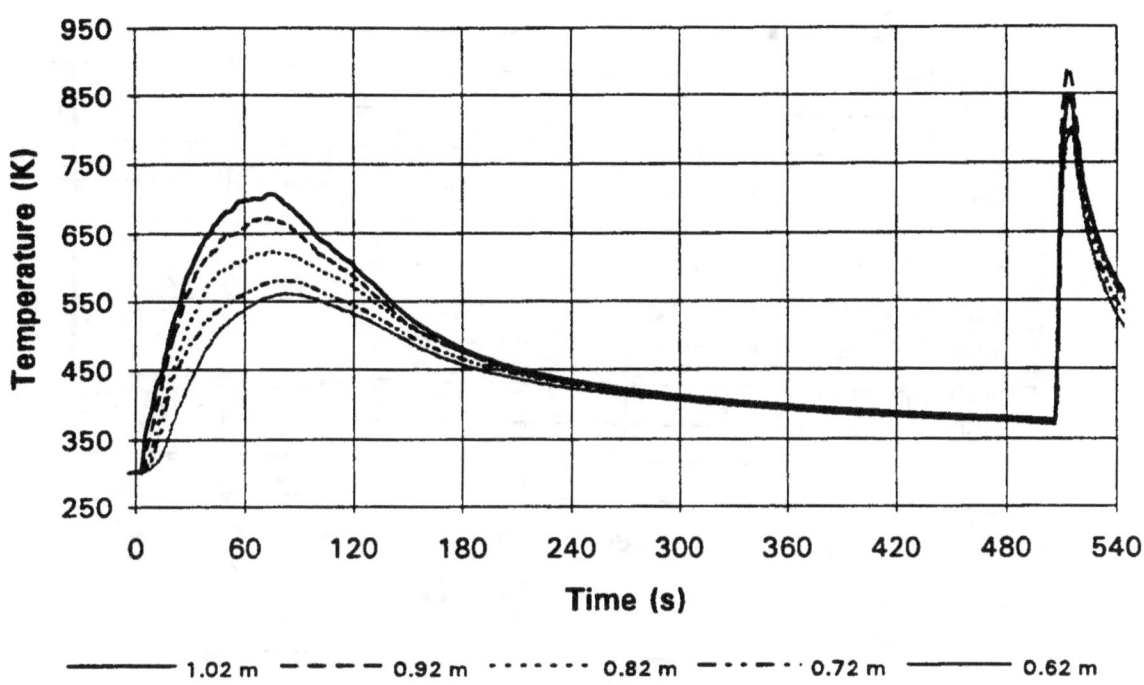

Layer Temperature History - TC Tree Data

(1.02 m, 0.92 m, 0.82 m, 0.72 m, 0.62 m)

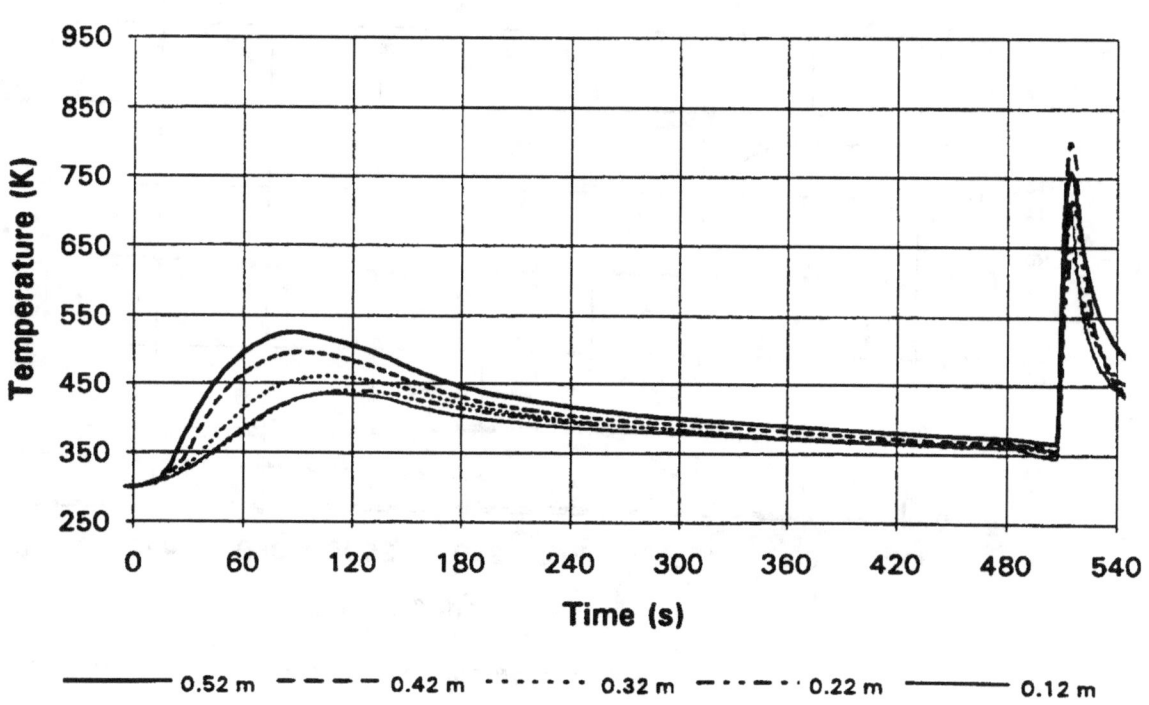

Layer Temperature History - TC Tree Data

(0.52 m, 0.42 m, 0.32 m, 0.22 m, 0.12 m)

Run: P3EXP86

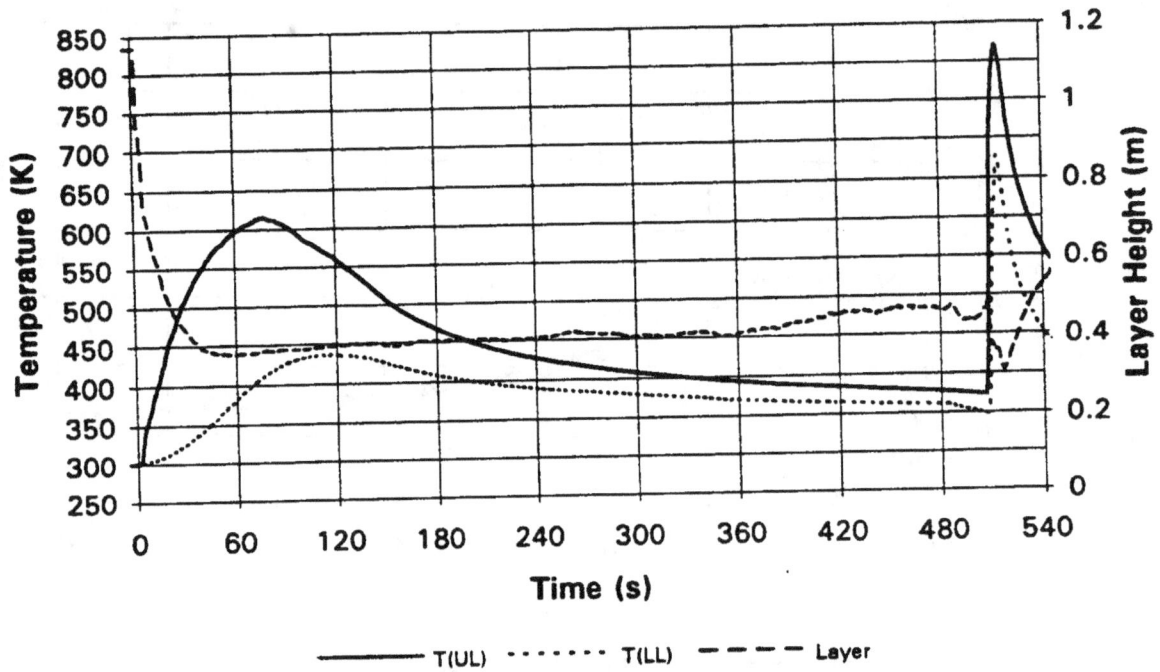

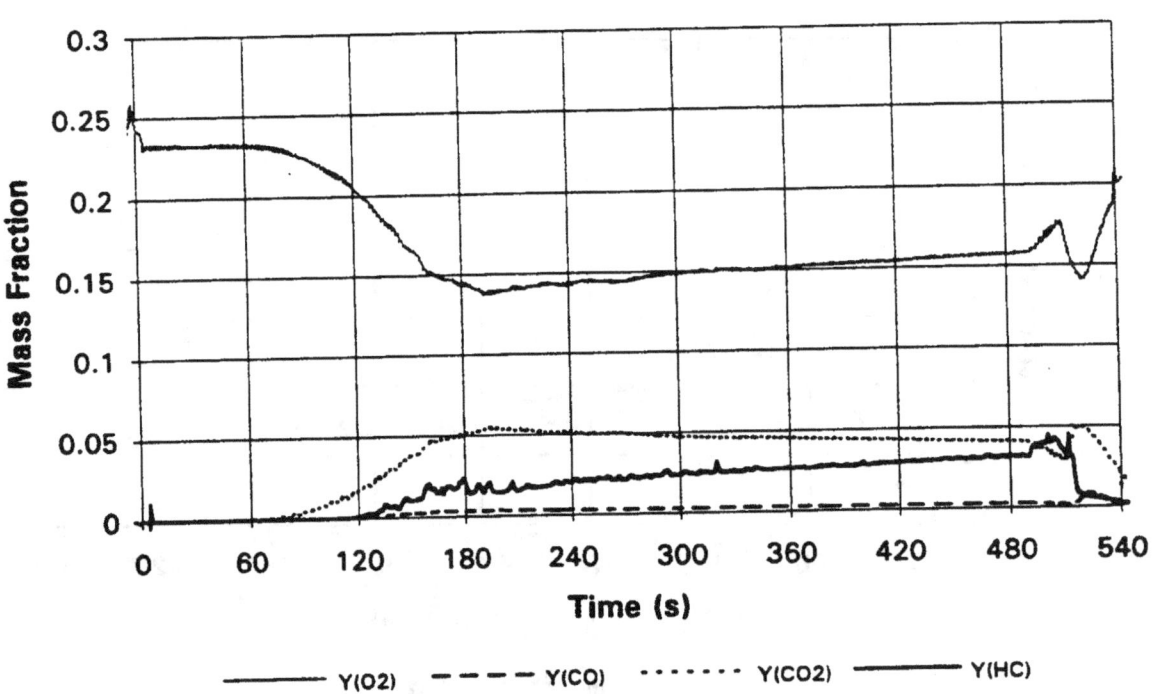

Run: P3EXP86

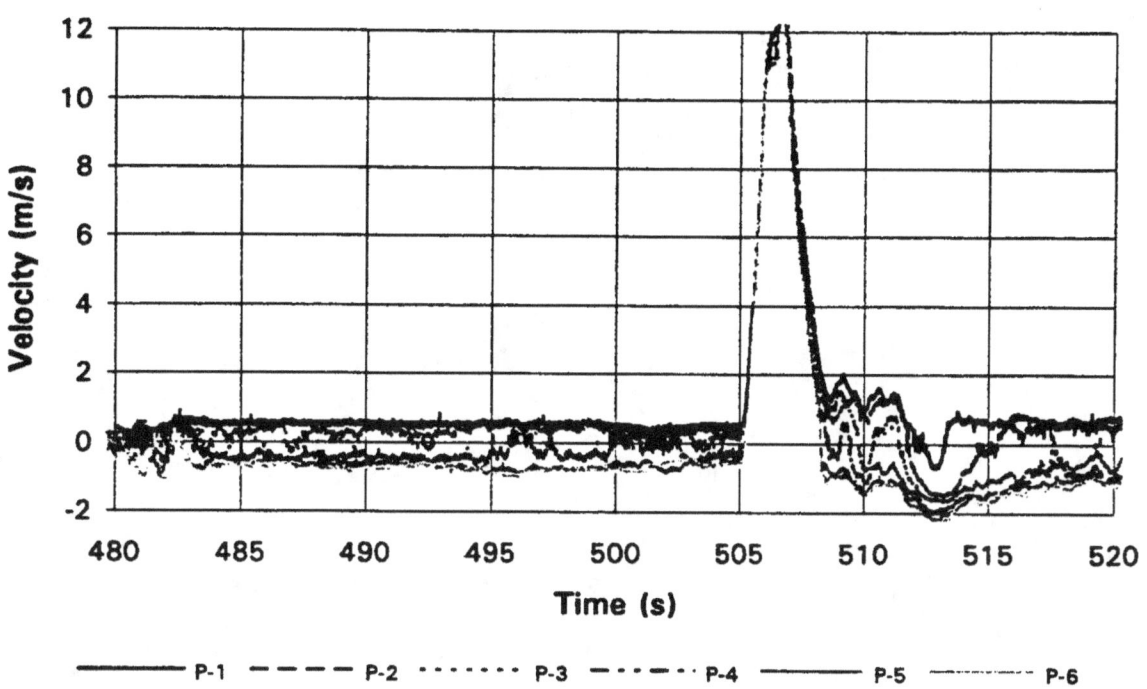

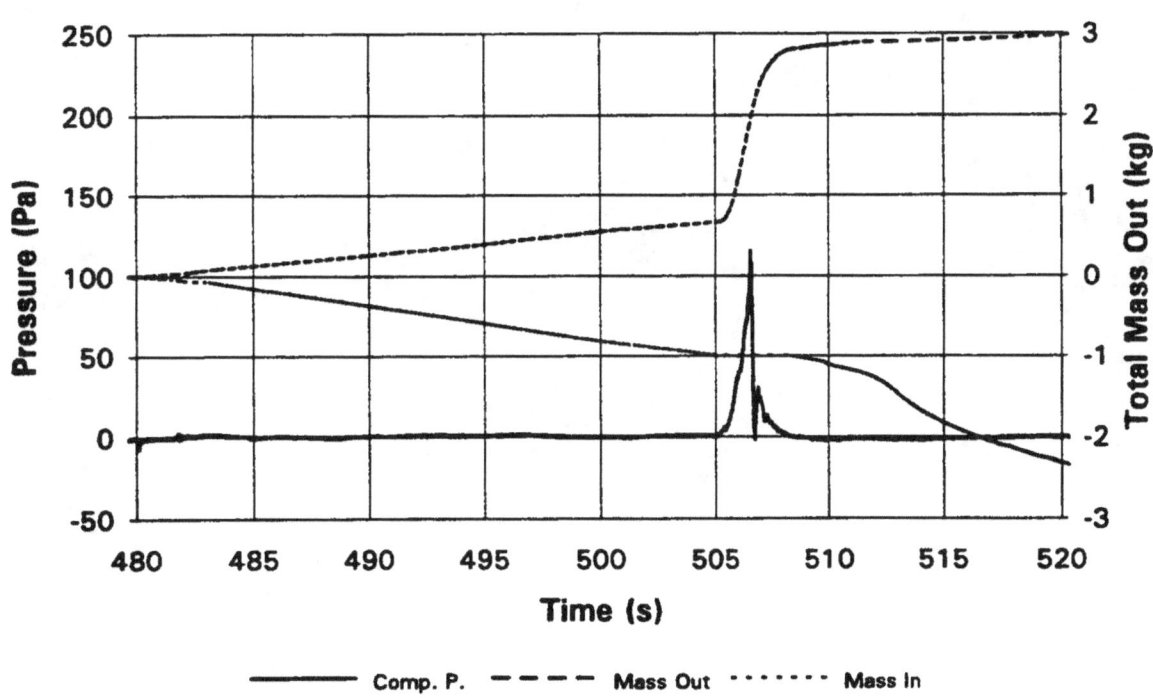

Run: P3EXP86

NIST-114	U.S. DEPARTMENT OF COMMERCE	(ERB USE ONLY)	
(REV. 9-92)	NATIONAL INSTITUTE OF STANDARDS AND TECHNOLOGY	ERB CONTROL NUMBER	DIVISION
ADMAN 4.09	**MANUSCRIPT REVIEW AND APPROVAL**	PUBLICATION REPORT NUMBER NIST-GCR-94-646	CATEGORY CODE
INSTRUCTIONS: ATTACH ORIGINAL OF THIS FORM TO ONE (1) COPY OF MANUSCRIPT AND SEND TO: THE SECRETARY, APPROPRIATE EDITORIAL REVIEW BOARD.		PUBLICATION DATE June 1994	NUMBER PRINTED PAGES

TITLE AND SUBTITLE (CITE IN FULL)

Backdraft Phenomena

CONTRACT OR GRANT NUMBER
60NANB0D1042

TYPE OF REPORT AND/OR PERIOD COVERED
Final Report, 1990 - 1992

AUTHOR(S) (LAST NAME, FIRST INITIAL, SECOND INITIAL)

Charles M. Fleischmann
University of California, Berkeley
Berkeley, CA 94720

PERFORMING ORGANIZATION (CHECK (X) ONE BOX)
- [ ] NIST/GAITHERSBURG
- [ ] NIST/BOULDER
- [ ] JILA/BOULDER

LABORATORY AND DIVISION NAMES (FIRST NIST AUTHOR ONLY)

SPONSORING ORGANIZATION NAME AND COMPLETE ADDRESS (STREET, CITY, STATE, ZIP)

U.S. Department of Commerce
National Institute of Standards and Technology
Gaithersburg, MD 20899

RECOMMENDED FOR NIST PUBLICATION

- [ ] JOURNAL OF RESEARCH (NIST JRES)
- [ ] J. PHYS. & CHEM. REF. DATA (JPCRD)
- [ ] HANDBOOK (NIST HB)
- [ ] SPECIAL PUBLICATION (NIST SP)
- [ ] TECHNICAL NOTE (NIST TN)
- [ ] MONOGRAPH (NIST MN)
- [ ] NATL. STD. REF. DATA SERIES (NIST NSRDS)
- [ ] FEDERAL INF. PROCESS. STDS. (NIST FIPS)
- [ ] LIST OF PUBLICATIONS (NIST LP)
- [ ] NIST INTERAGENCY/INTERNAL REPORT (NISTIR)
- [ ] LETTER CIRCULAR
- [ ] BUILDING SCIENCE SERIES
- [ ] PRODUCT STANDARDS
- [ ] OTHER NIST-GCR

RECOMMENDED FOR NON-NIST PUBLICATION (CITE FULLY) [ ] U.S. [ ] FOREIGN

PUBLISHING MEDIUM
- [ ] PAPER
- [ ] CD-ROM
- [ ] DISKETTE (SPECIFY)
- [ ] OTHER (SPECIFY)

SUPPLEMENTARY NOTES

ABSTRACT (A 1500-CHARACTER OR LESS FACTUAL SUMMARY OF MOST SIGNIFICANT INFORMATION. IF DOCUMENT INCLUDES A SIGNIFICANT BIBLIOGRAPHY OR LITERATURE SURVEY, CITE IT HERE. SPELL OUT ACRONYMS ON FIRST REFERENCE.) (CONTINUE ON SEPARATE PAGE, IF NECESSARY.)

The purpose of this project was to develop a fundamental physical understanding of backdraft phenomena. The research was divided into three phases: exploratory simulations, gravity current modeling, and quantitative backdraft experiments. The primary goal of the first phase was to safely simulate a backdraft in the laboratory. A half-residential-scale compartment was built to conduct exploratory experiments. The initial experiments concluded with a scenario describing the fundamental physics of backdrafts. The importance of the gravity current which enters the compartment after opening was identified. In the second phase, the gravity current speed and the extent of its mixed region was investigated in a series of scaled salt water experiments. The scaled compartment (0.3m x 0.15m x 0.15m) was fitted with a variety of end openings: full, slot, door, and window. Video and photo data indicate that the mixing layer which rides on the gravity current in the full opening case, expands to occupy nearly the entire current in the partial opening cases. The Froude number and nondimensional head height are independent of β and are in good agreement with numerical simulations and special limits from the literature.

KEY WORDS (MAXIMUM 9 KEY WORDS; 28 CHARACTERS AND SPACES EACH; ALPHABETICAL ORDER; CAPITALIZE ONLY PROPER NAMES)

backdraft; building fires; deflagration; fire research; fireballs; glass; gravity; room fires; ventilation; windows

AVAILABILITY

- [X] UNLIMITED
- [ ] FOR OFFICIAL DISTRIBUTION. DO NOT RELEASE TO NTIS.
- [ ] ORDER FROM SUPERINTENDENT OF DOCUMENTS, U.S. GPO, WASHINGTON, D.C. 20402
- [X] ORDER FROM NTIS, SPRINGFIELD, VA 22161

NOTE TO AUTHOR(S) IF YOU DO NOT WISH THIS MANUSCRIPT ANNOUNCED BEFORE PUBLICATION, PLEASE CHECK HERE. [ ]

ELECTRONIC FORM